Principles and Practices of In Situ Chemical Oxidation Using Permanganate

ROBERT L. SIEGRIST

MICHAEL A. URYNOWICZ

OLIVIA R. WEST

MICHELLE L. CRIMI

KATHRYN S. LOWE

Battelle Press

Columbus • Richland

Library of Congress Cataloging-in-Publication Data

Principles and practices of In Situ chemical oxidation using permanganate /
prepared by Robert L. Siegrist . . . [et al.].
   p. cm.
  Includes bibliographical references and index.
  ISBN 1-57477-102-7 (hard cover : alk. paper))
    1. Organochlorine compounds–Environmental aspects. 2. Oxidation. 3.
  Permanganates. 4. In situ remediation. 5. Hazardous waste site remediation.
  I. Siegrist, Robert L.

TD1066.O73 G85 2001
628.5—dc21   00-064180

Printed in the United States of America

Battelle Press
505 King Avenue
Columbus, Ohio 43201-2693
614-424-6393; 1-800-451-3543
Fax: 614-424-3819
Website: www.battelle.org/bookstore
E-mail: press@battelle.org

Contents

Preface

During the past decade, chemical oxidation has been developed and demonstrated as an in situ remediation approach for cleanup of contaminated sites. The underpinnings for this evolution are the extensive knowledge and experience gained in the municipal and industrial water and waste treatment industry. This work showed that many toxic organics were amenable to rapid and complete chemical destruction or to partial degradation as an aid to subsequent biological treatment. The literature now contains reports on the development and application of Fenton's reagent, ozone and permanganate oxidation systems for in situ treatment of organically contaminated soil and ground water.

This book provides guidance on the evaluation and design of in situ chemical oxidation systems with a focus on the use of potassium and sodium permanganate for remediation of organically contaminated sites. As an introduction to the topic, Chapter 1 presents an overview of chemical oxidation including its development and application for in situ treatment of contaminated sites. The oxidation chemistry of Fenton's reagent, ozone, and permanganate are highlighted along with optional methods of oxidant delivery for in situ application. The results of lab- and field-scale applications are also summarized. In Chapter 2, the principles and processes of chemical oxidation using potassium or sodium permanganate for organic chemical degradation are described, including reaction stoichiometry, equilibria, and kinetics as well as the effects of environmental factors. In Chapter 3, information is provided on the effects of permanganate on the behavior of metals. Chapter 4 provides a discussion of the potential for permeability loss and other secondary effects during in situ oxidation using permanganate. Chapter 5 describes optional methods of oxidant delivery for in situ remediation. A process for evaluation, design and implementation of permanganate systems is covered in Chapter 6. Chapter 7 presents a detailed description of five different applications of in situ chemical oxidation using potassium or sodium permanganate. Chapter 8 provides highlights of the current status and future directions of this remediation technology.

Acknowledgments

Preparation of this book was made possible and supported by several organizations and numerous individuals. Sponsorship of the work was provided in part through the Subsurface Contaminants Focus Area of the DOE Office of Science & Technology. The efforts and guidance of the following DOE program representatives are gratefully acknowledged:

Skip Chamberlain, DOE Headquarters, Germantown, MD

Jim Wright, DOE, Savannah River Operations Office, Aiken, SC

Elizabeth Phillips, DOE, Oak Ridge Operations Office, Oak Ridge, TN

Tom Early, Oak Ridge National Laboratory, Oak Ridge, TN

Also recognized are the staffs at the Portsmouth Gaseous Diffusion Plant in Piketon, Ohio and the Kansas Plant in Kansas City, MO where field applications of in situ chemical oxidation have occurred during the past decade.

The following individuals are recognized for their contributions through completion of laboratory experiments, modeling, or field applications, the results of which are included in this book:

Dianne Gates, Lawrence Livermore National Laboratory
(formerly with ORNL)

Steve Cline, Oak Ridge National Laboratory, Oak Ridge, TN

Frank Gardner, AimTech, Grand Junction, CO
(formerly with ORNL)

Mark Mumby, AimTech, Grand Junction, CO
(formerly with ORNL)

Doug Pickering, AimTech, Grand Junction, CO
(formerly with ORNL)

Bob Schlosser, AimTech, Grand Junction, CO
(formerly with ORNL)

John Zutman, AimTech, Grand Junction, CO
(formerly with ORNL)

Traci Case, Am. Water Works Assoc. Res. Fnd., Denver, CO
(formerly with CSM)

Helen Dawson, U.S. Environ. Protection Agency, Denver, CO
(formerly with CSM)

Rich Harnish, Colorado School of Mines, Golden, CO

Terry Jennings, Concurrent Technologies, Denver, CO
(formerly with CSM)

Amanda Struse, The IT Group, Denver, CO (formerly with CSM)

Sheila Van Cuyk, Colorado School of Mines, Golden, CO

Also acknowledged are private industries and design firms that have supported the research and collaborated on the field applications including Carus Chemical Company, Hayward Baker Environmental, Millgard Environmental Incorporated, GeoCon Inc., Schumacher Filters America, and FRx Inc.

The following individuals provided critical review of the draft manuscript and their comments were valuable in preparing the final version of the book:

Joe Baker, Honeywell, Kansas City, MO

Dr. Wilson Clayton, Aquifer Solutions, Denver, CO
(formerly with the IT Group)

Dr. Paul Kuhlmeier, Camp Dresser and McKee, Denver, CO

Ken Pisarczyk, Carus Chemical, Inc., Peru, IL

Dr. Tom Simpkin, CH$_2$M-Hill, Denver, CO

Dr. Robert Starr, Idaho National Engineering and Environmental Laboratory, Idaho Falls, ID

Dr. Neil Thomson, University of Waterloo, Waterloo, Ontario

Acronyms and Abbreviations

A_f	Arrhenius frequency factor
AOP	advanced oxidation processes
APHA	American Public Health Association
ASTM	American Society for Testing and Materials
AT	averaging time
ATSDR	Agency for Toxic Substances and Disease Registry
ben	benzene
bgs	below ground surface
Br^-	bromide ion
BTEX	benzene, toluene, ethylbenzene, and xylene
BW	body weight
^oC	degrees celsius
CA	concentration in air
CDI	chronic daily intake
CDIig	chronic daily intake through ingestion
CDIih	chronic daily intake through inhalation
CERCLA	Comprehensive Environmental Response Compensation and Liability Act
cf	cubic feet
cfm	cubic feet per minute
CFR	Code of Federal Regulations
cfs	cubic feet per second
CFU	colony forming units
ChemOx	abbreviation for the general technology of chemical oxidation
Cl^-	chloride ion
COC	constituent of concern, contaminant of concern

COD	chemical oxidant demand
CPF	cancer potency factor
CSM	Colorado School of Mines
CSTR	continuously stirred tank reactor
CVOC	chlorinated volatile organic compound
CW	concentration in water
D	diffusion coefficient
DCE	dichloroethene
1,1-DCE	1,1-dichloroethene
cis-1,2-DCE	cis-1,2-dichloroethene
DCG	derived concentration guideline
DI	deionized water
DNAPL	dense nonaqueous phase liquid
DNT	dinitrotoluene
DNX	dinitrosoamine
1-D	one dimensional
2-D	two dimensional
DO	dissolved oxygen
DOD	Department of Defense
DOE	Department of Energy
DQO	data quality objective
DSM	deep soil mixing
ECD	electron capture detector
Ea	activation energy
ED	exposure duration
EF	exposure frequency
Eh	redox potential
$E_H{}^o$	standard oxidation potential
eqn.	equation
EPA	Environmental Protection Agency
ESR	exchangeable sodium ratio
ESTCP	Environmental Security Technology Certification Program
ET	exposure time
Fe	iron

FID	flame ionization detector
f_{oc}	fractional organic carbon content
gpd	gallons per day
gpm	gallons per minute
GC	gas chromatography
GW	groundwater
h	hour
HA	humic acid
HBE	Hayward Baker Environmental
HE	high explosives
HMX	high melting explosives
HPLC	high pressure liquid chromatography
hr	hour
H_2O_2	hydrogen peroxide
I	ionic strength
IC_{50}	inhibitory concentration for 50%
ICP	inductively-coupled plasma emissions spectroscopy
IGR	ingestion rate
IHR	inhalation rate
INEEL	Idaho National Engineering and Environmental Laboratory
ISCO	in situ chemical oxidation
ISCOR	in situ chemical oxidation with recirculation
ITRD	innovative treatment remediation demonstration
J	mass flux
K	mass transfer coefficient
^oK	degrees Kelvin
K-40 (^{40}K)	radioisotope of potassium
KCP	DOE Kansas City Plant
K_G	cation exchange coefficient
Ksat	hydraulic conductivity
k	specific reaction rate constant
k_1	1st-order reaction rate constant
k_1'	pseudo-1st-order reaction rate constant
k_2	2nd-order reaction rate constant

$KMnO_4$	potassium permanganate
lb	pound
LMES	Lockheed Martin Energy Systems
LNAPL	light nonaqueous phase liquid
LPM	low permeability media
MCL	maximum contaminant level
MDL	method detection limit
min	minutes
mon	months
mM	millimolar
MNA	monitored natural attenuation
MnO_2	manganese dioxide
MnOx	manganese oxide
MnO_4^-	permanganate anion
Mn^{+2}	manganese ion
MNX	mononitrosoamines
MPIS	multipoint injection system
MRCO	mixed region chemical oxidation
mv	millivolt
MW	monitoring well or molecular weight
Na^+	sodium ion
$NaMnO_4$	sodium permanganate
NAPL	nonaqueous phase liquid
NATO	North Atlantic Treaty Organization
nd or ND	nondetectable
NEPA	National Environmental Policy Act
nm	nanometer
NOD	natural oxidant demand
NOM	natural organic matter
NPL	national priorities list
NRC	National Research Council
O_3	ozone
OH*	hydroxyl radical
OPM	oxidative particle mixture

ORNL	Oak Ridge National Laboratory
OSHA	Occupational Safety and Health Administration
PAH	polyaromatic hydrocarbon
PCB	polychorinated biphenyl
PCE	tetrachloroethene
pCi	picocurie
PCP	pentachlorophenol
PID	photoionization detector
POM	particulate organic matter
PORTS	DOE Portsmouth Gaseous Diffusion Plant
ppb	parts per billion
ppm	part per million
PRB	permeable reactive barrier
psi	pounds per square inch
PTI	permit to install
PTO	permit to operate
PTT	partitioning tracer test
PV	pore volume
PVC	polyvinyl-chloride
PZC	point of zero charge
RCRA	Resource Conservation and Recovery Act
RDX	cyclotrimethylenetrinatramine
REV	representative elemental volume
RfD	reference dose
RFI	RCRA facility investigation
RI	remedial investigation
ROI	radius of influence
rpm	revolutions per minute
SAR	sodium adsorption ratio
SEM	scanning electron microscope
SERDP	Strategic Environmental Research and Development Program
SSL	soil screening level
SVE	soil vapor extraction
SVOC	semivolatile organic compound

$t_{1/2}$	reaction half life
T	temperature
TCA	1,1,1-trichloroethane
TCB	trichlorobiphenyl
TCE	trichloroethene
TCLP	toxicity characteristic leaching procedure
TDS	total dissolved solids
TNT	trinitrotoluene
TNX	trinitrosoamines
TOC	total organic carbon
TPH	total petroleum hydrocarbons
trans-1,2-DCE	trans-1,2-dichloroethene
TSS	total suspended solids
TU	toxicity unit
UIC	underground injection control
UV	ultraviolet light
v	Darcy groundwater velocity
V	volt
VC	vinyl chloride
VOC	volatile organic compound
wt.	weight
ZHR	zero headspace reactor

Principles and Practices of In Situ Chemical Oxidation Using Permanganate

CHAPTER 1

In Situ Chemical Oxidation for Site Remediation

1.1. CONTAMINATED SITES AND IN SITU REMEDIATION

Subsurface contamination by toxic chemicals is a widespread problem in soil and groundwater at industrial and military sites in the U.S. and abroad (Riley et al. 1992, NRC 1994, NRC 1997, Siegrist and Van Ee 1994, ASTDR 1999, USEPA 1997). Awareness of the significance of contamination occurred in the 1970s when adverse health effects were observed at major sites across the U.S. (e.g., Times Beach, Love Canal, Valley of the Drums). Over the past 20 years in the U.S., about 500,000 sites with potential contamination have been reported. Of these about 300,000 have been found to be clean and others have been remediated. Over the next 50 years or more, major effort and funds will be spent cleaning up hazardous waste sites in the United States and around the world. Under current federal and state regulations in the U.S., the contaminated-site problem can be characterized as follows (USEPA 1997):

- Remediation is still required for over 217,000 sites.

- Time to complete ranges from 10 to 30 years or more.

- Cost of remediation is estimated at $187 billion (in 1996 dollars).

- Types of contamination to be remediated are similar in many respects:
  - Contaminated soil and/or groundwater with volatile organics (~2/3 sites) and
  - Metals and semivolatile organics are also prevalent.
- Driving force is often to protect drinking water derived from groundwater.

Organic chemicals are often the primary contaminants of concern (COCs). They commonly include volatile organic compounds (VOCs) such as tetrachloroethene (PCE), trichloroethene (TCE), and benzene (Ben), and semivolatile organic chemicals (SVOCs) such as polyaromatic hydrocarbons (PAHs) and pesticides (Riley et al. 1992, USEPA 1994, USEPA 1997). In addition, there may be metals (e.g., Pb, Cr) and radionuclides (U, Tc) as co-contaminants of concern (Evanko and Dzombak 1997). Figure 1-1 illustrates the prevalent categories of COCs at hazardous waste sites in the U.S.; Table 1-1 gives examples of specific COCs and their reference values for remediation.

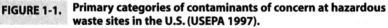

FIGURE 1-1.   **Primary categories of contaminants of concern at hazardous waste sites in the U.S. (USEPA 1997).**

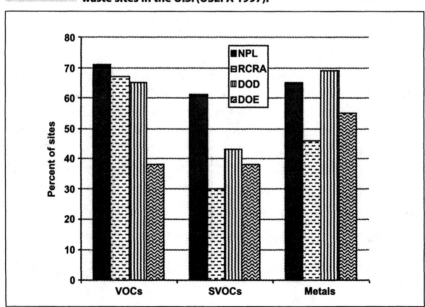

TABLE 1-1.	Examples of common COCs and the reference values for remediation.

Contaminant	ATSDR 1999 ranking[1]	Prevalance at NPL sites (no. of sites)[2]	U.S. Drinking Water MCLs (ug/L)	Contaminated site generic soil screening levels (mg/kg)[3]		
				Ingestion	Inhalation	Migration to GW
Arsenic	1	781 of 1300	50	0.4	750	1.0
Lead	2	922 of 1300	0[4]	400	—	—
Mercury	3	714 of 1467	2	23	10	0.1
Vinyl chloride	4	496 of 1430	2	0.3	0.03	0.0007
Benzene	5	813 of 1430	5	22	0.8	0.002
Polychlorinated byphenyls	6	383 of 1430	0.5	1	—	—
Cadmium	7	388 of 1300	5	78	1800	0.4
Polyaromatic hydrocarbons	9	600 of 1430	—	—	—	—
Trichloroethene	15	852 of 1430	5	58	5	0.003
Chromium	16	115 of 1300	100	390	270	2.0
Tetrachloroethene	31	771 of 1430	5	12	11	0.003
Pentachlorophenol	44	260 of 1416	1	3	—	0.001
Carbon tetrachloride	46	326 of 1416	5	5	0.3	0.003
Toluene	62	869 of 1416	1000	16000	650	0.6
1,1,1-Trichloroethane	86	696 of 1430	200	—	1200	0.1

[1] Agency for Toxic Substances and Disease Registry. 1999 Priority List of Hazardous Substances. Division of Toxicology, U.S. Dept. of Health and Human Services. Nov. 1999. http://www.atsdr.cdc.gov/cxcx3.html
[2] ToxFAQs. 1999. Agency for Toxic Substances and Disease Registry. April 1999. http://www.atsdr.cdc.gov/99list.html
[3] Reference values for cleanup are based on Soil Screening Guidance: Technical Background Document. U.S. EPA OSWER. EPA/540/R-95/128. May 1996.
[4] Maximum contaminant level goal

While varied exposure scenarios can present serious current or future health risks at Superfund and many other contaminated sites, baseline risk is often governed by inhalation exposures to toxic organics that volatilize from contaminated soils and by ingestion exposures from toxic organics that leach from wastes and soils into groundwater used for drinking water (Figure 1-2) (Labieniec et al. 1996, 1997, Sheldon et al. 1997, Sheldon 1999). During the early period of Superfund, appreciable levels of hazardous chemicals in uncontrolled environmental settings (e.g., soil or groundwater) were presumed to present a condition in need of action, and implicitly, an unacceptable risk. Risk reduction was achieved typically by excavation of soil and waste with subsequent treatment and disposal off-site, combined with pumping and treatment of contaminated groundwater (Mackay and Cherry 1989, USEPA 1997, USEPA 1999). Now, more explicit risk assessment and management underpins cleanup programs for Superfund and most other contaminated sites (USEPA 1989). Baseline risk assessments are first completed to define the need for and extent of site cleanup and to develop site-specific remediation alternatives to mitigate any unacceptable risks to an agreed upon goal.

FIGURE 1-2.   **Illustration of a site contaminated by volatile organic compounds where the transport pathways and exposure scenarios are associated with contaminated groundwater. (After Labieniec et al. 1996)**

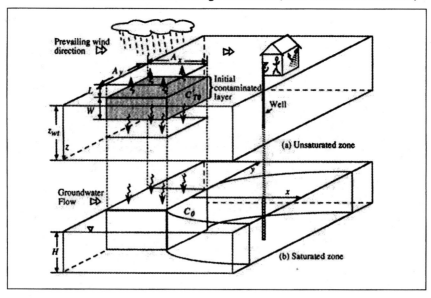

To mitigate current or future risks, remediation approaches increasingly employ in situ technologies comprised of engineered as well as natural attenuation systems (Figure 1-3) (NRC 1994, NRC 1997, USEPA 1997). In situ remediation in source areas of contaminated soil and groundwater is being accomplished using mass transfer and recovery methods (e.g., soil vapor extraction, air sparging, surfactant/cosolvent flushing) and in-place destruction methods (e.g., bioremediation, oxidation/reduction), sometimes aided by enabling techniques (e.g., soil mixing or fracturing, soil heating). For treatment of the distal regions of groundwater plumes, natural biogeochemical attenuation and permeable reactive barriers are two strategies that are evolving (Gavaskar et al. 1998, USEPA 1997, NRC 1997, NATO 1998, NATO 1999).

The application of in situ chemical oxidation (ISCO) for remediation of contaminated soil and groundwater has evolved over the past decade.

FIGURE 1-3. Remediation approaches for in situ application at a typical contaminated site.

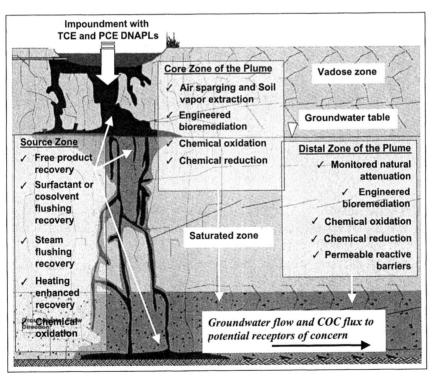

Research and development concerning ISCO has included laboratory bench- and pilot-scale studies followed by field demonstrations of varying scales. More recently, there have been an increasing number of full-scale applications. As a site remediation technology, the goal of ISCO has been to destroy target organic chemicals present in soil and groundwater systems and thereby reduce the mass, mobility, and/or toxicity of contamination. This chapter presents an overview of ISCO. Subsequent chapters focus on the use of permanganate oxidants.

1.2. EVOLUTION OF ISCO FOR SITE REMEDIATION

Chemical oxidation to destroy organic contaminants in water has been used for decades in the municipal and industrial water and waste treatment industry. Several common oxidants and their oxidation potentials are shown in Table 1-2. Due to the high oxidation potential of hydrogen peroxide (H_2O_2) and hydroxyl radicals ($OH^\bullet$), chemical oxidation was most often accomplished with the use of hydrogen peroxide or Fenton's reagent (H_2O_2 plus Fe^{+2}). Implemented in tank-based reactors, chemical oxidation was used for ex situ treatment of individual organics in water (Barbeni et al. 1987, Glaze and Kang 1988, Bowers et al. 1989, Watts and Smith 1991, Venkatadri and Peters 1993).

Subsequently, research began to explore peroxide and Fenton's reagent oxidation as applied in soil and groundwater environments (Watts et al. 1990, Watts and Smith 1991, Watts et al. 1991, Tyre et al. 1991, Ravikumar and Gurol 1994, Gates and Siegrist 1993, Gates and Siegrist 1995, Watts et al. 1997). Research also was initiated with alternative oxidants such as ozone (Bellamy et al. 1991, Nelson and Brown 1994, Marvin et al. 1998) and potassium permanganate ($KMnO_4$) (Vella et al. 1990, Vella and Veronda 1994, Gates et al. 1995, Schnarr et al. 1998, West et al. 1998a,b, Siegrist et al. 1998a,b, Siegrist et al. 1999, Yan and Schwartz 1996, 1999, Tratnyek et al. 1998, Urynowicz and Siegrist 2000).

As of this writing, a wide variety of organic contaminants in soil and groundwater have been successfully oxidized by hydrogen peroxide (including Fenton's reagent), ozone, and permanganate oxidants (Table 1-3). Alternative methods of oxidant delivery have included permeation by vertical lances (Jerome et al. 1997, Siegrist et al. 1998c, Moes et al. 2000), flushing by vertical and horizontal groundwater wells (Schnarr et al. 1998, West et al. 1998a,b, Lowe et al. 2000), and reactive zone emplacement by hydraulic fracturing (Murdoch et al. 1997a,b, Siegrist et al. 1999)

Oxidant	Oxidation potential (volts)	Oxidation power relative to chlorine	Equivalent weight[1] (lb)	Cost per equivalent[2] ($/lb)
Fluorine	3.03	2.23	—	—
Hydroxyl radical	2.80	2.06	—	—
Atomic oxygen	2.42	1.78	0.0176	0.0046
Ozone	2.07	1.52	0.0529	0.0532
Hydrogen peroxide	1.78	1.31	0.0375	0.0262
Perhydroxyl radical	1.70	1.25	—	—
Permanganate	1.68	1.24	0.1161	0.1509
Chlorine dioxide	1.57	1.15	0.1487	0.4587
Sodium chlorite	—	—	0.0498	—
Sodium or calcium hypochlorite	—	—	0.1641 or 0.1567	0.1615 or 0.1614
Persulfates	—	—	0.2509 – 0.2973	0.2007 –0.2943
Hypochlorous acid	1.49	1.10	—	—
Hypoiodous acid	1.45	1.07	—	—
Chlorine	1.36	1.00	0.0782	0.0099
Bromine	1.09	0.80	—	—
Iodine	0.54	0.39	—	—

TABLE 1-2. Oxidation potentials and costs of selected oxidants.

[1]Equivalent weights are given as the mass of oxidant for the same number of electrons transferred.
[2]Costs are based on March 1998 prices.

| TABLE 1-3. | Representative list of organics treated by chemical oxidants. |

Organic contaminant	Media treated	Oxidant[1]	References
Trichloroethene	Water (spiked)	H_2O_2	Bellamy et al. 1991
	Silica sand (spiked)	H_2O_2	Gurol and Ravikumar 1992
	Silty clay soil (spiked)	H_2O_2	Gates and Siegrist 1995
	Sand & clay soils (spiked)	H_2O_2 or $KMnO_4$	Gates et al.1995, Gates-Anderson et al. 2001
	Ground water (spiked)	$KMnO_4$	Case 1997, Yan and Schwartz 1998
	Ground water (field site)	$KMnO_4$	West et al. 1998, Schnarr et al. 1998
	Ground water (field site)	$NaMnO_4$	Lowe et al. 2000
	Silty clay soil (field site)	$KMnO_4$	Siegrist et al. 1999
Tetrachloroethene	Water (spiked)	H_2O_2	Bellamy et al. 1991
	Silica sand (spiked)	H_2O_2	Leung et al. 1992
	Sand, clay soils (spiked)	H_2O_2 or $KMnO_4$	Gates et al.1995, Gates-Anderson et al. 2001
	Ground water (field site)	$KMnO_4$	Schnarr et al. 1998
	Ground water (spiked)	$KMnO_4$	Yan and Schwartz 1996
	Ground water (field site)	Ozone	Dreiling et al. 1998
Carbon tetrachloride trans-1,2-DCE	Water (spiked)	H_2O_2	Bellamy et al. 1991
Pentachlorophenol	Silica sand (spiked)	H_2O_2	Gurol and Ravikumar 1992
	Natural soil (spiked)	H_2O_2	Watts et al. 1990
2,4-dichlorophenol, dinitro-o-cresol	Water (spiked)	H_2O_2	Bowers et al. 1989
Trifluralin, hexadecane, dieldrin	Soil (spiked)	H_2O_2	Tyre et al. 1991
Phenols	Water (spiked)	$KMnO_4$	Vella et al. 1990
Naphthalene, phenanthrene, pyrene	Clay, sandy soils (spiked)	H_2O_2 or $KMnO_4$	Gates et al.1995, Gates-Anderson et al. 2001
Octachlorodibenzo(p)-dioxin	Soil (spiked)	H_2O_2	Watts et al. 1991
Motor oil / diesel fuel	Soil (field site)	H_2O_2	Watts 1992
PAHs and pentachlorophenol	Soil and GW (field site)	Ozone	Marvin et al. 1998
BTEX and TPH	Soil and GW (field site)	Ozone H_2O_2	USEPA 1998
HMX, RDX, TNT, 2,6-DNT and 2,4-DNT	GW (field site)	$KMnO_4$	IT Corp and Stoller Corp 2000

[1] H_2O_2 implies H_2O_2 and Fenton's reagent.

(Figure 1-4). Key features of the oxidant systems are highlighted in Table 1-4 and details regarding principles and practices are described in the balance of this chapter and in the remainder of the book.

FIGURE 1-4.   **Examples of in situ chemical oxidation systems that have been deployed to remediate soil and groundwater.**

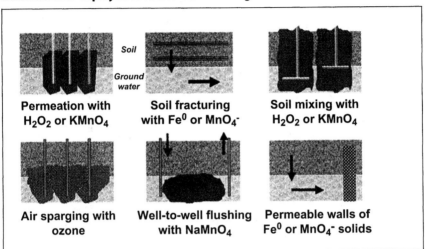

Soil

Ground water

Permeation with H_2O_2 or $KMnO_4$

Soil fracturing with Fe^0 or MnO_4^-

Soil mixing with H_2O_2 or $KMnO_4$

Air sparging with ozone

Well-to-well flushing with $NaMnO_4$

Permeable walls of Fe^0 or MnO_4^- solids

1.3. CHEMICAL OXIDANTS AND IN SITU APPLICATION

Hydrogen Peroxide and Fenton's Reagent

Studies of the peroxide degradation of toxic organics in soil and groundwater initially focused on petrochemicals (e.g., naphthalene, phenanthrene, pyrene, phenols) but later encompassed chlorinated solvents (e.g., TCE, PCE) (Murphy et al. 1989, Watts et al. 1990, Watts and Smith 1991, Watts et al. 1991, Tyre et al. 1991, Ravikumar and Gurol 1994, Gates and Siegrist 1993, Gates and Siegrist 1995, Watts et al. 1997, Gates-Anderson et al. 2001). Oxidation using H_2O_2 in the presence of native or supplemental Fe^{+2} produces Fenton's reagent, which yields free radicals $(OH^\bullet)$ that can rapidly degrade a variety of organic compounds (Table 1-3).

The application of hydrogen peroxide to soil and groundwater systems involves a variety of competing reactions as follows:

$$H_2O_2 + Fe^{+2} \;\rightarrow\; OH^- + Fe^{+3} + OH^\bullet \qquad (1.1)$$

$$H_2O_2 + Fe^{+3} \;\rightarrow\; HO_2^\bullet + H^+ + Fe^{+2} \qquad (1.2)$$

TABLE 1-4. Features of hydrogen peroxide, metal permanganates and ozone oxidants as used for in situ remediation.[1]

FEATURES	HYDROGEN PEROXIDE AND FENTON'S REAGENT	OZONE	POTASSIUM OR SODIUM PERMANGANATE
Reagent Characteristics:			
Form -	Liquid	Gas	Liquid or solid
Point of generation -	Offsite, shipped onsite	Onsite during use	Offsite, shipped onsite
Quantities available -	Small to large	Small to large	Small to large
Oxidation In Situ:			
Delivery Methods -	Wells, injection probes, mixing	Sparge wells	Wells, injection probes, mixing, fractures
Dose concentrations -	5 to 50 wt.% H_2O_2	Variable	0.02 to 4.0 wt.% MnO_4^-
Single / multiple dosing -	Multiple is common	Multiple	Single and multiple
Amendments -	Fe^{+2} and acid	Often ozone in air	None
Subsurface transport -	Advection	Advection	Advection and diffusion
Rate reaction / transport -	High or very high	Very high	Moderate to high
Companion technol.-	None required	Soil vapor extraction	None required
Oxidation Effectiveness:			
Susceptible organics -	BTEX, PAHs, Phenols, alkenes	Substituted aromatics, PAHs, phenols, alkenes	BTEX, PAHs, alkenes
Difficult to treat organics -	Some alkanes, PCBs	Alkanes, PCBs	Alkanes, PCBs
Oxidation of NAPLs -	Enhanced oxidation possible	Enhanced degradation possible	Enhanced degradation possible
Reaction products -	Organic acids, salts, O_2, CO_2	Organic acids, salts, O_2, CO_2	Organic acids, salts, MnO_2, CO_2
	Substantial gas evolution	Minimal gas evolution	Minimal gas evolution

continued

TABLE 1-4. (continued)

FEATURES	HYDROGEN PEROXIDE AND FENTON'S REAGENT	OZONE	POTASSIUM OR SODIUM PERMANGANATE
Subsurface Effects on Oxidation:			
Effect of NOM -	Demand for oxidant	Demand for oxidant	Demand for oxidant
Effect of pH -	Most effective in acidic pH	Most effective in acidic pH	Effective over pH 3.5 to 12
Effect of temperature -	Reduced rate at lower temp.	Reduced rate at lower temp.	Reduced rate at lower temp.
Effect of ionic strength -	Limited effects	Limited effects	Limited effects
Oxidation Effects on Subsurface:			
pH -	Lowered if inadeq. buffering	Lowered if inadeq. buffering	Lowered if inadeq. buffering
Temperature -	Minor to high increase	Minor to high increase	None to minor increase
Metal mobility -	Potential for redox metals	Potential for redox metals	Potential for redox/exch. metals
Permeability loss -	Potential for reduction due to gas evolution and colloids	Potential for reduction due to gas evolution and colloids	Potential for reduction due to MnO_2 colloid genesis
Microbiology -	Short-term reduction in bio-mass but post-treatment stimulation possible	Short-term reduction in bio-mass but post-treatment stimulation possible	Short-term reduction in bio-mass but post-treatment stimulation possible

[1]The information presented is for illustrative purposes only.

$$OH^{\bullet} + Fe^{+2} \rightarrow OH^{-} + Fe^{+3} \qquad (1.3)$$

$$HO_2^{\bullet} + Fe^{+3} \rightarrow O_2 + H^+ + Fe^{+2} \qquad (1.4)$$

$$H_2O_2 + OH^{\bullet} \rightarrow H_2O + HO_2^{\bullet} \qquad (1.5)$$

$$RH + OH^{\bullet} \rightarrow H_2O + R^{\bullet} \qquad (1.6)$$

$$R^{\bullet} + Fe^{+3} \rightarrow Fe^{+2} + products \qquad (1.7)$$

Hydrogen peroxide can also autodecompose in aqueous solutions with accelerated rates upon contact with mineral surfaces as well as carbonate and bicarbonate surfaces (Hoigne and Bador 1983) (eqn. 1.8).

$$H_2O_2 \rightarrow 2H_2O + O_2 \qquad (1.8)$$

The simplified stoichiometric reaction for hydrogen peroxide degradation of TCE is shown in eqn. 1.9.

$$3H_2O_2 + C_2HCl_3 \rightarrow 2CO_2 + 2H_2O + 3HCl \qquad (1.9)$$

Fenton's reagent oxidation is most effective under very acidic pH (e.g., pH 2 to 4) and becomes ineffective under moderate to strongly alkaline conditions and/or where free radical scavengers (e.g., CO_3^{-2}) are present in sufficient quantities. The reaction is strongly exothermic and can evolve substantial gas and heat. The oxidative reactions are extremely rapid.

The complexity of reactions that occur when H_2O_2 is added to a soil and groundwater environment make it difficult to describe the reaction kinetics explicitly. If it is assumed that the oxidation of target organics occurs by free radical processes, then the rate of reaction of Fenton's reagent with a target organic chemical can be described as second-order kinetics:

$$\frac{d[M]}{dt} = -k_2 [M][OH^{\bullet}] \qquad (1.10)$$

where k_2 is the rate constant for the reaction of the organic (M) and hydroxyl radicals ($OH^{\bullet}$), respectively. If hydroxyl radicals are maintained constant, the rate expression reduces to a pseudo first-order reaction:

$$\frac{d[M]}{dt} = -k_1' [M] \qquad (1.11)$$

where, k_1' is the overall apparent pseudo first-order reaction rate constant. The reaction rate is extremely fast with half-lives on the order of minutes.

For application in situ, there are several approaches for oxidant delivery, some of which have been patented based on reaction chemistry and/or mode of delivery (e.g., CleanOX, GeoCleanse, and ISOTEC). Several example applications are described in Table 1-5. While the features of an ISCO application with hydrogen peroxide or Fenton's reagent will be very site dependent, in situ chemical oxidation has typically included H_2O_2 concentrations in the range of 5 to 50 wt.% and where native iron has been lacking or unavailable, ferrous sulfate is often added at mM levels. In some cases, acetic or sulfuric acids also are added to reduce the pH to a more favorable acidic range. Delivery methods have included common groundwater wells or specialized injectors as well as deep soil mixing equipment (Figure 1-5). In many cases, multiple doses or application cycles are used to facilitate more uniform delivery of reagents and a high efficiency of treatment.

Ozone

Ozone (O_3) was first used in water treatment applications more than 100 years ago and is commonly used today (Langlais et al. 1991). More recently, ozone gas has been injected into the subsurface during air sparging to remediate organics in groundwater zones (Bellamy et al. 1991, Nelson and Brown 1994, Marvin et al. 1998). Ozone is slightly soluble in water and can oxidize contaminants directly or through formation of hydroxyl radicals ($OH^\bullet$), which are strong nonspecific oxidants. The decomposition of O_3 is affected greatly by pH, ozone concentration, and free radical scavengers (e.g., HCO_3^-). In the subsurface, inorganics (e.g., OH^-, Fe^{+2}) and organic compounds (e.g., humic substances) can initiate and promote the decomposition of ozone by a radical-type chain reaction. Ozone decomposition in water occurs by a chain process that is rate limited by the initial reaction of ozone with hydroxyl ions to yield hydroperoxide and superoxide radicals:

$$O_3 + OH^- \;\rightarrow\; HO_2^\bullet + O_2^- \qquad (1.12)$$

This chain process can be inhibited or disrupted by free radical scavengers such as bicarbonate and carbonate ions and some organics.

TABLE 1-5.	Example applications of in situ treatment using hydrogen peroxide and Fenton's reagent.	

Location (date) Delivery	Media and COCs	Application method and results
Ohio (1993) Deep soil mixing	Silty clay soil with TCE and VOCs.	H_2O_2 + compressed air injected during deep soil mixing to 15 ft. depth in 3 10-ft. diam. mixing zones. Up to 100 mg/kg mass reduced by 70% including 50% due to oxidation.
Colorado (1996) Injectors into GW	Ground water with BTEX.	H_2O_2 + chelated Fe injected via 8 to 14 lances and 7 trenches over 100 ft. x 100 ft. area. Four cycles at 4 to 6 days each. BTEX reduced from 25 mg/L to <0.09 mg/L. Property sold.
Massachusetts (1996) Injectors into GW	TCA and VC in ground water.	H_2O_2 + Fe + acid via 2 points over 3-days within 30 ft. D.W. TCA reduced from 40.6 to 0.4 mg/L, VC 0.40 to 0.08 or ND mg/L.
Alabama (1997) Injectors into GW	Soil with high levels of TCE, DCE, and BTEX.	H_2O_2 + $FeSO_4$ via 255 injectors into 8 to 26 ft. bgs zone of clay backfill in 2 acre waste lagoons. 120 days treatment time. 72000 lbs. of NAPLs treated down to soil screening levels.
South Carolina (1997) Injectors to GW	Deep GW zone with PCE and TCE DNAPLs in sandy clay aquifer.	H_2O_2 + $FeSO_4$ via 4 injectors into zone at 140 ft. bgs beneath old waste basin. 6-day treatment time. Treatment achieved 94% reduction in COCs with GW near MCLs. GW TCE reduced from 21 to 0.07 mg/L; PCE from 119 to 0.65 mg/L.

Contaminants oxidizable by ozone include aromatics, PAHs, pesticides, and aliphatic hydrocarbons (including saturated aliphatic molecules). The simplified decomposition reaction for stoichiometric reaction of ozone with TCE in water is given by eqn. 1.13.

$$O_3 + H_2O + C_2HCl_3 \rightarrow 2CO_2 + 3HCl \qquad (1.13)$$

FIGURE 1-5. **In situ chemical oxidation using Fenton's reagent at the Burlington Manufactured Gas Plant site in Burlington, Wisconsin (courtesy of Matt Dingens, Geo-Cleanse International, Inc.).**

(a)
View of the equipment at a typical hydrogen peroxide site. The 30-ft. long injection vehicle is in front of the 40-ft. long hydrogen peroxide trailer. The hoses run from the injection vehicle to the injector head.

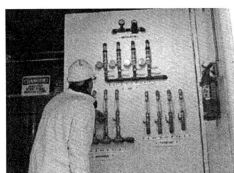

(b)
View of the injection control board showing the ability to control flow rates of air, hydrogen peroxide and catalyst at different flow rates to each of four injectors.

(c)
Injector head showing the hydrogen peroxide entering from the top, the catalyst entering from the left and the air entering from the right. This allows for a continuous and a simultaneous feed of reagents into the subsurface.

The rate of reaction of ozone with organics in a subsurface environment is dependent on the decomposition rate of ozone and the generation of free radicals. The O_3 decomposition rate in the presence of excess scavengers is given by eqn. 1.14:

$$\frac{d[O_3]}{dt} = -k_2\,[O_3][OH^-] \qquad (1.14)$$

where $[O_3]$ = ozone concentration (mol L^{-1}), OH^- = hydroxyl ion concentration (mol L^{-1}), and k_2 = 2nd-order rate constant (L mol^{-1} s^{-1}) (Langlais et al. 1991). The oxidation of a target organic can be described as a pseudo first-order reaction with respect to both the target organic and the oxidant, or second-order overall (eqn. 1.15) (Bellamy et al. 1991).

$$\frac{d[M]}{dt} = -k_1'\,[M][O_3] \qquad (1.15)$$

where $[M]$ = concentration of the organic being oxidized (mol L^{-1}), $[O_3]$ = the concentration of ozone in water (mol L^{-1}), and k_1' = O_3 second-order reaction rate constant (mol^{-1} L s^{-1}). Overall the rate of oxidation is also dependent on hydroxyl radical formation, so the rate expression is modified to include both direct oxidation and free-radical oxidation processes:

$$\frac{d[M]}{dt} = -k_1'\,[M][O_3] - k_{OH\bullet}\,[M][OH\bullet] \qquad (1.16)$$

where $k_{OH\bullet}$ is the second-order rate constant for the reaction of the organic (M) and hydroxyl radicals ($OH\bullet$). The $k_{OH\bullet}$ rate constants are very high (in the range of 10^8 to 10^{10} mol^{-1} L s^{-1}). If ozone and hydroxyl radicals remain relatively constant, the rate expression reduces to pseudo-1st order as shown in eqn. 1.17:

$$\frac{d[M]}{dt} = -k_1'\,[M] \qquad (1.17)$$

where, k_1' is the overall apparent pseudo first-order reaction rate constant. To ensure radical formation, hydrogen peroxide is often added to the ozone/water system. Bellamy et al. (1991) reported that at O_3 mass feed rates of 0.2 mg/L per min and with molar feed ratio of 0.5 H_2O_2/O_3, the pseudo first-order rate constants for VOC destruction in water (initial concentrations of 0.1 to 1.0 mg/L) were 0.0015 to 0.0025 sec^{-1} for PCE, 0.0030 to 0.0060 sec^{-1} for TCE and 0.0030 to 0.0090 sec^{-1} for DCE. These constants yield half-lives in the range of 1 to 10 min.

Due to ozone's high reactivity and instability, O_3 is produced onsite by electrical generators. In addition, the high reactivity of ozone and free radicals requires relatively closely spaced delivery points (e.g., air sparging wells). In some settings, scavengers for $OH^\bullet$ can reduce the reaction efficiency. Gas and heat can be evolved and particulates conceivably can be generated in fine-grained sediments. Ozone decomposition yields O_2 and provides beneficial oxygenation and aerobic biostimulation.

For application of ozone gas in situ, there are optional methods of delivery, most of which rely on ozone gas sparging approaches (Table 1-6). At least one process is patented based on mode of delivery, but

TABLE 1-6.	Example field applications of in situ treatment using ozone.

Location (date) Delivery	Media and COCs	Application method and results
Colorado (1997) *GW wells*	Soil and GW with BTEX and TPH	Former gas station site. Sand/gravel to 43 ft. bgs with GW at 28 ft. 3 wells to 50 ft. depth cycling air/ozone with water recirculation. 12 cycles per day. SVE also continued. TPH in soil from 90 to 2380 mg/kg and BTEX at 7.8 to 36.5 mg/kg. TPH in GW at 490 mg/L to NAPL. After 6 mon, GW below MCLs. No soils data. System shut down.
Kansas (1997) *Injectors into GW*	PCE in GW.	Old drycleaners site. GW at 14 to 16 ft. bgs in terrace deposits. One sparge point at 3 scfm at 35 ft. bgs. SVE wells in vadose zone. PCE in top 15 ft. of aquifer at 0.03 to 0.60 mg/L. Reduced 91% within 10 ft. of well. Comparisons with air only indicated 66 to 87% reductions.
California (1998) *Injectors into GW*	Soil and GW with PAHs and PCP	Wood treatment site 300 ft. by 300 ft. in area. Stratified sands and clays. 4 multilevel ozone injectors at up to 10 cfm. SVE wells in the vadose zone. After 1 mon, PAHs at 1800 mg/kg reduced by 67 to 99% and PCP at 3300 mg/kg reduced 39 to 98%.

engineered solutions using ozone are generally not proprietary. While the specifics of an application to a particular site will be dependent on site conditions, in situ chemical oxidation systems have been engineered to include ozone gas intermittently delivered along with compressed air into contaminated groundwater using conventional vertical and horizontal air sparging wells (Figure 1-6). Amendments to alter system pH or provide a catalyst generally are not utilized.

FIGURE 1-6. **In situ chemical oxidation using ozone sparging via vertical groundwater wells.**

(a) Ozone generation equipment.

(b) Ozone sparging well access port.

Potassium and Sodium Permanganate

Compared to hydrogen peroxide (and Fenton's reagent) and ozone, permanganate oxidation of soil and groundwater has been studied more recently for in situ treatment of chlorinated solvents (e.g., TCE, PCE) and petrochemicals (e.g., naphthalene, phenanthrene, pyrene and phenols) (Vella et al. 1990, Leung et al. 1992, Vella and Veronda 1994, Gates et al. 1995, Yan and Schwartz 1996, Schnarr et al. 1998, West et al. 1998a,b, Siegrist et al. 1998a,b, Lowe et al. 1999; Siegrist et al. 1999, Yan and Schwartz 1999, Struse and Siegrist 2000, Urynowicz and Siegrist 2000, Gates-Anderson et al. 2001). The stoichiometry and kinetics of permanganate oxidation at contaminated sites can be quite complex as there are numerous reactions in which Mn can participate due to its multiple valence states and mineral forms (more discussion in Chapter 2). For degradation of organic chemicals such as TCE, the oxidation involves direct electron transfer rather than free radical processes that characterize oxidation by Fenton's reagent or ozone. The stoichiometric reaction for the complete destruction of TCE is given in eqn. 1.18.

$$2KMnO_4 + C_2HCl_3 \rightarrow 2CO_2 + 2MnO_2 + 2K^+ + H^+ + 3Cl^- \qquad (1.18)$$

Thus for TCE oxidation, 1.81 g of MnO_4^- reacts with 1 g of C_2HCl_3 to produce 1.32 g of manganese dioxide (MnO_2), 0.67 g of carbon dioxide (CO_2) and 0.81 g of Cl^-.

The rate of organic chemical degradation by MnO_4^- in the absence of substantial natural organic matter (NOM) or other reductants, depends on the concentration of both TCE and also MnO_4^- and can be described by a second-order kinetic expression as given in eqn. 1.19:

$$\frac{d[M]}{dt} = - k_2 [M][MnO_4^-] \qquad (1.19)$$

where [M] = concentration of the organic being oxidized (mol L^{-1}), [MnO_4^-] = the concentration of permanganate in water (mol L^{-1}), and k_2 = second-order reaction rate constant (mol^{-1} L s^{-1}). The second order rate constants reported for TCE degradation are in the range of 0.6 to 0.9 mol^{-1} L s^{-1} and the reaction rate appears independent of pH (over a range of 4-8) and ionic strength (up to 1.57 M Cl-) (Yan and Schwartz 1999, Huang et al. 1999, Urynowicz 2000). The rate of reaction is temperature dependent as described by the Arrhenius eqn. with an activation energy on the order of 35 to 70 kJ/mol (Case 1997, Huang et al. 1999, Yan and Schwartz 1999).

For application of potassium or sodium permanganates in situ, a variety of delivery methods have been employed including those presented in Table 1-7. While there are some relevant patents in place or pending, the

TABLE 1-7.	Example applications of in situ treatment using permanganate.	
Location (date) *Delivery*	Media and COCs	Application method and results
Ohio (1997) *Horizontal well recirculation*	Ground water with TCE DNAPLs in a thin sandy aquifer.	KMnO$_4$ (2 to 4 wt.% feed) delivered by horizontal recirculation wells 200 ft. long and 100 ft. apart at 30 ft. bgs to treat 10^6 L zone of ground water over 30 days. TCE reduced from 820 mg/L to MCL in 13 of 17 wells. ~300 kg of TCE destroyed. Some MnO$_2$ particles generated. Aquifer heterogeneities noted.
Kansas (1996) *Deep soil mixing*	TCE and DCE in soil and ground water to 47 ft. depth.	KMnO$_4$ (3.1 to 4.9 wt.%) delivered by deep soil mixing (8 ft. augers) to 47 ft. bgs during 4 days. TCE reduced from 800 mg/kg by 82% in the vadose zone and 69% in the saturated zone (>8 ft. bgs). MnO$_4^-$ depleted. Microbes persisted. Comparison tests with mixed region vapor stripping yielded 69% reduction and bioaugmentation were 38% reduction.
Ohio (1998) *Vertical well recirculation*	TCE in silty sand and gravel ground water zone at 30 ft. bgs.	NaMnO$_4^-$ (250 mg/L) delivered by 5-spot vertical well recirculation system (ctr. well and 4 perimeter wells at 45 ft. spacing) for 3 pore volumes over 10 days. TCE reduced from 2.0 mg/L to MCL. Oxidant gradually depleted in 30 d and no Microtox toxicity. No permeability loss in formation.
Ohio (1996) *Hydraulic fracturing*	VOCs in silty clay soil from ground level to 18 ft. bgs.	KMnO$_4$ grout delivered by hydraulic fracturing to create multi-layered redox zones. Emplaced over 4 days but sustained oxidative zone for more than 15 mon. Dissolved TCE reduced from equiv. of 4000 mg/kg by 99% during 1 hr of contact.

use of permanganates for ISCO is not proprietary *per se* based on reaction chemistry and/or mode of delivery. While the specifics of an application to a particular site will be very site dependent, in situ chemical oxidation systems have been engineered to include potassium and sodium permanganate solutions at concentrations ranging from 100 to 40,000 mg/L delivered into contaminated groundwater using vertical and horizontal wells operated as injection wells or as injection/recovery/recirculation well networks (West et al. 1998, Schnarr et al. 1998, Lowe et al. 2000). For treatment of contaminated soil, potassium permanganate in concentrated solution (5000 to 40,000 mg/L) or solid form (~50% by wt.) has been delivered and dispersed by use of injection lances (Siegrist et al. 1998c) (Figure 1-7), deep soil mixing (Cline et al. 1997, Gardner et al. 1998), or hydraulic fracturing (Siegrist et al. 1999a) techniques. Amendments to alter system pH or provide a catalyst are not required.

FIGURE 1-7. Field application of in situ chemical oxidation using $KMnO_4$ delivered via vertical injection probes at Cape Canaveral, Florida.

(a) Overview of the LC-34 site at Cape Canaveral.

(b) $KMnO_4$ solution manifold and injection probes (IT Group).

(c) $KMnO_4$ feed equipment (Carus Chemical Co).

1.4. RISK REDUCTION ACHIEVED BY COC MASS DESTRUCTION

Identification of the benefits of in situ chemical oxidation or similar technologies is often based on a reduction in contaminant mass which yields a concomitant reduction in the contaminant flux that creates exposures to receptors (Freeze and McWhorter 1997). For most sites, the key exposures occur through inhalation of vapors that volatilize from soil or ingestion of drinking water that contains dissolved contaminants that have leached from soil or waste (e.g., Figure 1-2) (Labieniec et al. 1996, USEPA 1996, Sheldon et al. 1997). The incremental risk due to such a current or future exposure is often estimated by use of site characterization data and transport models. As shown in the following equations for carcinogenic risk, the estimated mass flux of contaminants to a human receptor leads to exposure concentrations (CA or CW) that are directly related to risk:

$$Risk = CDI*CPF \tag{1.20}$$

$$CDI_{ih} = CA * (IHR*ET*EF*ED) / (BW*AT) \tag{1.21}$$

$$CDI_{ig} = CW * (IGR*EF*ED) / (BW*AT) \tag{1.22}$$

where, CDI = chronic daily intake (mg/kg/d), CPF = cancer potency factor $(mg/kg/d)^{-1}$ (USEPA 1989). Chronic daily intakes from inhalation (CDI_{ih}) or drinking water ingestion (CDI_{ig}) are based on the average chemical concentration in air (CA, mg/m^3) and water (CW, mg/L) over the duration of exposure. The other terms in equations 1.21 and 1.22 (with the typical values prescribed for benzene shown in brackets) are: IHR = inhalation rate [15.2 m^3/d], IGR = ingestion rate [2 L/d], ET = exposure time [18.4 hr/d], EF = exposure frequency [365 d/yr], ED = exposure duration [30 yr], BW=body weight [71.8 kg], and AT = averaging time [27375 d]. Predicting the degree of treatment needed to achieve a desired reduction in risk is based on back-calculation (or iterative calculation) to set an acceptable residual contaminant level that will yield mass fluxes that, upon mixing into receiving mediums, do not exceed health-based exposure concentrations (e.g., CA or CW). While there are numerous intermedia and multimedia transport models being used for risk assessment and in some cases for prediction of the risk reduction benefits of in situ treatment, experimental validation of most models is only recently being completed (Sheldon 1999). Moreover, their use for assessing in situ treatment has yet to be evaluated.

1.5. ISCO SYSTEM FIELD APPLICATIONS AND EXPERIENCES

Field applications of ISCO are growing rapidly in the U.S. and abroad. Applications of ISCO using hydrogen peroxide (or Fenton's reagent), ozone, and permanganate have been highlighted in several recent survey documents (e.g., ESTCP 1999, USEPA 1998, Yin and Allen 1999, Siegrist et al. 2000a,b). General characteristics of 42 sites where ISCO has been deployed are summarized in Tables 1-8 and 1-9 while several representative applications are highlighted in Tables 1-5 to 1-7. In general, ISCO systems have been shown to be capable of achieving high treatment efficiencies (e.g., >90%) for common COCs such as chlorinated ethenes (e.g., TCE, PCE) and aromatic compounds (e.g., benzene, phenols, naphthalene), with very fast reaction rates (90% destruction in minutes). Field applications have demonstrated that ISCO can achieve destruction of COCs and achieve clean-up goals at some contaminated sites. However, some field-scale applications have had uncertain or poor in situ treatment performance. Poor performance is often attributed to poor uniformity of oxidant delivery caused by low permeability zones and site heterogeneity, excessive oxidant consumption by natural subsurface materials, presence of large masses of dense nonaqueous phase liquids (DNAPLs), and incomplete degradation. In some applications there continue to be concerns over secondary effects such as oxidation effects on metals, loss in permeability due to ISCO-produced particles, and gas evolution and fugitive emissions, as well as health and safety practices (Siegrist et al. 2000a,b).

1.6. ISCO SYSTEM SELECTION, DESIGN, AND IMPLEMENTATION

Selection of ISCO as the remediation technology of choice for a particular site is often based on the general benefits that ISCO can offer, including rapid and extensive reactions with various COCs, applicability to many subsurface environments, ability to tailor ISCO to a site from locally available components and resources, and compatibility with property transfers and Brownfields development projects. But also to be carefully considered are potential limitations for ISCO, including the potential need for large quantities of reactive chemicals to be introduced due to the oxidant demand of the target organics and the unproductive oxidant consumption of the formation, resistance of some COCs to chemical oxidation, and potential for ISCO-induced detrimental effects such as gas evolution, permeability loss, and secondary water quality effects.

TABLE 1-8.	Characteristics of 42 sites where in situ chemical oxidation has been deployed (ESTCP 1999).			

Features		DOD sites	DOE sites	Private sites	Total
Sites	**Total Number**	14	3	25	42
Contaminants	CVOC	6	3	10	19
	BTEX/TPH	5	–	16	21
	Both	1	–	–	1
	Unknown	1	–	–	1
Media treated	Soil only	0	0	0	0
	Ground water	2	0	17	19
	Both	10	3	7	20
	Unknown	2	–	1	3
Oxidant	H_2O_2	12	1	24	37
	$KMnO_4$	1	2	1	4
	Ozone	1	0	0	1
Vendor	GeoCleanse	8	1	4	13
	Clean-Ox	3	0	13	16
	ISOTEC	1	0	7	8
	Other	2	2	1	5
Scale	Pilot/demo	9	3	15	27
	Full only	1	0	4	5
	Both	4	0	6	10
Outcome[1]	Success	5	3	11	19
	Failure	6	0	0	6
	Uncertain	3	0	14	17

[1]Outcome determinations are relative and denote the technology's ability or lack thereof to satisfy facility-specific program performance objectives.

Once ISCO is selected as a viable alternative for a particular site, choices must be made between oxidant type and delivery method. Decisions in this regard are based on the oxidant systems and their viability for site-specific COCs (e.g., COC susceptibility to direct vs. free-radical oxidation) and subsurface conditions (e.g., soil pH) that impact oxidant reaction processes and transport behavior in the subsurface (Table 1-4). Remediation performance goals (e.g., COC mass

TABLE 1-9.	Summary observations from a survey of 42 sites where in situ chemical oxidation has been deployed (ESTCP 1999).

Characteristic	Comments
Total sites	42 sites were identified
Ownership	25 were private sites and 17 were government sites (3 DOE and 14 DOD)
Contaminants	19 sites were partially or primarily chlorinated VOCs with TCE being most prevalent
Oxidant system used	37 sites—Hydrogen peroxide 4 sites—Potassium permanganate 1 site—Ozone
Scope	14 DOD sites—pilot tests completed at all with 5 sites proceeding to full scale 3 DOE sites—demonstration projects 25 private sites—10 have proceeded to full scale
Status and Performance	Of the 14 DOD sites (all were hydrogen peroxide based) 6 were considered failures (including one explosion), while 5 are proceeding to full scale. Of the 25 private sites, several sites described as successes (e.g., based on no further action required by State) and several pilot tests are heading for full application. In many cases there is relatively little long-term data to assure that performance is sustained.
Cost of oxidants	$0.88/lb TCE (at 0.8 lb H_2O_2 to lb TCE; $0.88/lb of 50 wt% H_2O_2 (excl. other reagents) $4.32/lb TCE (at 2.4 lb of KMnO4 per lb TCE and $1.80/lb of $KMnO_4$)

reduction, endpoint concentrations) and regulatory constraints (e.g., concern over chemical injection and secondary drinking water standards) may influence the choice of oxidant and delivery method as well as the specifics of ISCO system design and implementation.

As noted through evaluation of field applications and experiences, a number of key issues may be relevant and need to be addressed during the evaluation, design, and implementation of ISCO, regardless of the oxidant and delivery system being employed (Table 1-10). Such issues

TABLE 1-10.	Factors to consider during the application of in situ chemical oxidation as identified during a survey of 42 sites where peroxide, permanganate, or ozone had been deployed (ESTCP 1999).

Category	Factors	Need
Site characterization	Total oxidant demand	Needed to support oxidant dose required for non-target materials (natural organic acids, reduce Fe and Mn, and sulfides)
	Contaminant delineation	Needed to target treatment zones and prevent untreated areas from impacting treated areas
	Mass estimates	Needed to permit evaluation of natural attenuation for post-chemical oxidation control of any diffusion-controlled rebound
	Vapor monitoring	Identify potentially explosive conditions in the subsurface
Design	Radius of influence	Better guidance on expected zone of influence beyond professional judgment
	Oxidant concentration	Rational basis for oxidant concentrations to degrade target organic without causing undesirable side reactions that cause excessive heat and foaming
	Enhanced mixing	Key design objective to increase contact of oxidant and contaminants
	Overall management strategy	Incorporate in situ chemical oxidation in source areas with other options for residuals after oxidative treatment
Operation	Multiple injections	Reinjection may be warranted in same or different locations even if initial results are promising
	Vapor monitoring	During drilling and injection monitoring of vapors should be required to prevent health and safety risks and ensure contaminants are not lost by volatilization
	Rebound monitoring	Monitoring of rebound and consideration of subsequent or coupled treatment

continued

TABLE 1-10.	(continued)	
Category	**Factors**	**Need**
Other	Health and safety	Needs careful attention to prevent serious health and safety incidents such as an explosion at a site with peroxide treatment
	Pilot testing	Pilot testing can be valuable but the direct link to full-scale application is not clear
	Fate and behavior of oxidant	Verification of effective treatment, determination of treatment mechanisms, and definition of subsurface conditions affecting technology application are still needed

include: (1) amenability of the target organic COCs to oxidative degradation, (2) destruction effectiveness for DNAPLs, (3) optimal oxidant loading (dose concentration and delivery) for a given organic in a given subsurface setting, (4) unproductive oxidant consumption, (5) potential adverse effects (e.g., mobilizing metals, forming toxic byproducts, reducing formation permeability, generating off-gases and heat), and (6) potential for ISCO to be coupled with other remediation technologies. For sites with common organics of concern that are present at lower levels (dissolved and sorbed phase only) in permeable groundwater formations with limited heterogeneity, current understanding of ISCO is adequate to support rational design and effective application. However, understanding is not complete for other more complex sites such as those with heterogeneous conditions (e.g., low permeability silt and clay zones), fractured bedrock, or source areas with large masses of DNAPLs.

ISCO system selection, design, and implementation practices should rely on an integrated effort based on typical site characterization data, ISCO-specific laboratory tests, laboratory and field pilot tests, and reaction and transport modeling. Figure 1-8 illustrates an approach to ISCO system implementation that the authors have developed and used. As illustrated, effective engineering of ISCO technologies must be done carefully with due attention to both reaction chemistry and in situ delivery to achieve a desired performance goal for a target COC without adverse oxidation-induced secondary effects.

FIGURE 1-8. Generalized approach for application of in situ chemical oxidation.

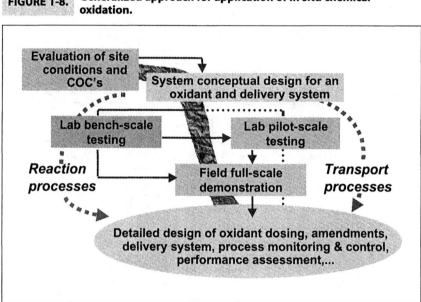

1.7. SUMMARY

In situ chemical oxidation is emerging rapidly as a viable remediation technology for mass reduction in source areas as well as for plume control and treatment. The oxidants most commonly employed to date include hydrogen peroxide, ozone, and permanganate systems, with subsurface delivery to groundwater by vertical or horizontal wells and sparge points and to soil by lance injectors and hydraulic fracturing. The potential benefits of in situ oxidation include the rapid and extensive reactions with various contaminants of concern, applicability to many biorecalcitrant organics and subsurface environments, ability to tailor treatment to a site from locally available components and resources, and facilitation of property transfers and Brownfields development projects. There are some potential limitations, including: potential need for large quantities of reactive chemicals to be introduced due to the oxidant demand of the target organics and the unproductive oxidant consumption of the formation, resistance of some COCs to chemical oxidation, and potential for process-induced detrimental effects including gas evolution, permeability loss, and mobilization of redox-sensitive and exchangeable sorbed metals. Full-scale deployment is accelerating, but care must be taken to achieve performance goals in a cost-effective manner while avoiding unforeseen adverse effects. Matching the *oxidant and delivery system* to the *COCs and site conditions* is critical to achieving performance goals.

1.8. REFERENCES

ATSDR (1999). Agency for Toxic Substances and Disease Registry (ATSDR). ToxFAQs. http://www.atsdr.cdc.gov. April 1999

Barbeni, M., C. Nfinero, E. Pelizzetti, E. Borgarello, and N. Serpon (1987). Chemical degradation of chlorophenols with Fenton's reagent. *Chemosphere*, 16, pp. 2225-37.

Bellamy, W.D., P.A. Hickman, and N. Ziemba (1991). Treatment of VOC-contaminated groundwater by hydrogen peroxide and ozone oxidation. *Res. Journal WPCF*, 63, pp. 120-28.

Bowers, A.R., P. Gaddipati, W.W. Eckenfelder, and R.M. Monsen (1989). Treatment of toxic or refractory wastewaters with hydrogen peroxide. *Wat. Sci. Tech.*, 21, pp. 477-86.

Case, T.L. (1997). Reactive permanganate grouts for horizontal permeable barriers and in situ treatment of groundwater. M.S. Thesis, Colorado School of Mines, Golden, CO.

Cline, S.R., O.R. West, N.E. Korte, F.G. Gardner, R.L. Siegrist, and J.L. Baker (1997). $KMnO_4$ chemical oxidation and deep soil mixing for soil treatment. *Geotechnical News*. December. pp. 25-28.

Environmental Security Technology Certification Program (1999). Technology Status Review: In Situ Oxidation. http://www.estcp.gov.

Evanko, C.R. and D.A. Dzombak (1997). Remediation of metals-contaminated soils and groundwater. Technology Evaluation Report, TE-97-01. Groundwater Remediation Technologies Analysis Center, Pittsburgh, PA.

Freeze, R.A. and D.B. McWhorter (1997). A framework for assessing risk reduction due to DNAPL mass removal from low-permeability soils. *Ground Water*, 35(1):111-123.

Gardner, F.G., N.E. Korte, J. Strong-Gunderson, R.L. Siegrist, O.R. West, S.R. Cline, and J. Baker (1998). Implementation of deep soil mixing at the Kansas City Plant. Final project report by Oak Ridge National Laboratory for the Environmental Restoration Program at the DOE Kansas City Plant. ORNL/TM-13532.

Gates, D.D., and R.L. Siegrist (1993). Laboratory evaluation of chemical oxidation using hydrogen peroxide. Report from The X-231B Project for In Situ Treatment by Physicochemical Processes Coupled with Soil Mixing. Oak Ridge National Laboratory. ORNL/TM-12259.

Gates, D.D., and R..L. Siegrist (1995). In situ chemical oxidation of trichloroethylene using hydrogen peroxide. *J. Environmental Engineering.* Vol. 121, pp. 639-44.

Gates, D.D., R.L. Siegrist and S.R. Cline (1995a). Chemical oxidation of contaminants in clay or sandy Soil. Proceedings of ASCE National Conference on Environmental Engineering. Am. Soc. of Civil Eng., Pittsburgh, PA.

Gates-Anderson, D.D., R.L. Siegrist and S.R. Cline (2001). Comparison of potassium permanganate and hydrogen peroxide as chemical oxidants for organically contaminated soils. *J. Environmental Engineering.* 127(4):337-347.

Gavaskar, A.R., N. Gupta, B.M. Sass, R.J. Janosy and D. O'Sullivan (1998). Permeable Barriers for Groundwater Remediation: Design, Construction, and Monitoring. Battelle Press, Columbus, OH.

Glaze, W.H. and J.W. Kang (1988). Advanced oxidation processes for treating groundwater contaminated with TCE and PCE: laboratory studies. *J. Amer. Water Works Assoc.* Vol. 5, pp. 57-63.

Hoigne, J. and H. Bador (1983). Rate constants of reaction of ozone with organic and inorganic compounds in water. *Water Research.* 17:173.

Huang, K., G.E., Hoag, P. Chheda, B.A. Woody, and G.M. Dobbs (1999). Kinetic study of oxidation of trichloroethylene by potassium permanganate. *Environmental Engineering Science.* 16(4): 265-274.

Jerome, K.M., B. Riha and B.B. Looney (1997). Demonstration of in situ oxidation of DNAPL using the Geo-Cleanse technology. WSRC-TR-97-00283. Westinghouse Savannah River Company, Aiken, SC.

Labieniec, P.A., D.A. Dzombak, and R.L. Siegrist (1996). SoilRisk: A risk assessment model for organic contaminants in soil. *J. Environmental Engineering.* 122(5): 388-398.

Labieniec, P.A., D.A. Dzombak, and R.L. Siegrist (1997). Evaluation of uncertainty in a site -specific risk assessment. *J. Environmental Engineering.* 123(3):234-243.

Langlais, B., D.A. Reckhow and D.R. Brink (1991). *Ozone in Water Treatment.* Lewis Publishers. CRC Press, Boca Raton, FL.

Leung, S.W., R.J. Watts and G.C. Miller (1992). Degradation of perchloroethylene by Fenton's reagent: speciation and pathway. *J. Environ. Qual.* 21, 377-81.

Lowe, K.S., F.G. Gardner, R.L. Siegrist, and T.C. Houk (2000). Field pilot test of in situ chemical oxidation through recirculation using vertical wells at the Portsmouth Gaseous Diffusion Plant. EPA/625/R-99/012. U.S. EPA ORD, Washington, DC. pp. 42-49.

MacKay, D.M. and J.A. Cherry (1989). Groundwater contamination: limits of pump-and-treat remediation. *Environ. Sci. & Technol.* 23: 630-636.

Marvin, B.K., C.H. Nelson, W. Clayton, K.M. Sullivan, and G. Skladany (1998). In situ chemical oxidation of pentachlorophenol and polycyclic aromatic hydrocarbons: from laboratory tests to field demonstration. Proc. 1st International Conf. Remediation of Chlorinated and Recalcitrant Compounds. May 1998, Monterey, CA. Battelle.

Moes, M. C. Peabody, R. Siegrist, and M. Urynowicz (2000). Permanganate injection for source zone treatment of TCE DNAPL. In: Wickramanayake, G.B., A.R. Gavaskar, and A.S.C. Chen (ed.). Chemical Oxidation and Reactive Barriers. Battelle Press, Columbus, OH. pp. 117-124.

Murdoch, L., W. Slack, R. Siegrist, S. Vesper, and T. Meiggs (1997a). Advanced hydraulic fracturing methods to create in situ reactive barriers. Proc. International Containment Technology Conference and Exhibition. February 9-12, 1997, St. Petersburg, FL.

Murdoch, L., B. Slack, B. Siegrist, S. Vesper, and T. Meiggs (1997b). Hydraulic fracturing advances. *Civil Engineering.* May 1997. pp. 10A-12A.

Murphy, A.P., W.J. Boegli, M.K. Price, and C.D. Moody (1989). A Fenton-like reaction to neutralize formaldehyde waste solutions. *Environ. Sci. Technol.* 23, pp. 166-69.

National Research Council (NRC) (1994). Alternatives for groundwater cleanup. National Academy Press, Washington, D.C.

National Research Council (NRC) (1997). Innovations in ground water and soil cleanup. National Academy Press, Washington, D.C.

NATO (1998). Treatment walls and permeable reactive barriers. NATO/CCMS Pilot Study Special Session, February 1998, Vienna. EPA/542/R-98/003. May 1998. www.nato.int.ccms

NATO (1999). Monitored natural attenuation. NATO/CCMS Pilot Study Special Session, May 1999, Angers, France. EPA/542/R-99/008. www.nato.int.ccms

Nelson, C.H. and R.A. Brown (1994). Adapting ozonation for soil and ground water cleanup. *Chemical Engineering.* McGraw-Hill, Inc.

Ravikmur, J.X., and M. Gurol (1994). Chemical oxidation of chlorinated organics by hydrogen peroxide in the presence of sand. *Environ. Sci. Technol.* 28, 394-400.

Riley, R. G. and J. M. Zachara (1992). Nature of chemical contaminants on DOE lands. DOE/ER-0547T. Office of Energy Research, U.S. Department of Energy, Washington, D.C.

Schnarr, M., C. Truax, G. Farquhar, E. Hood, T. Gonully, and B. Stickney (1998). Laboratory and controlled field experimentation using potassium permanganate to remediate trichloroethylene and perchloroethylene DNAPLs in porous media. *Journal of Contaminant Hydrology*. 29:205-224.

Sheldon, A.B., H.E. Dawson, and R.L. Siegrist (1997). Performance and reliability of intermedia transfer models used in human exposure assessment. In: Environmental Toxicology: Modeling and Risk Assessment. ASTM STP 1317, Philadelphia, PA.

Sheldon, A.B. (1999). Experimental validation and reliability evaluation of selected soil voc leaching and volatilization models used in environmental health risk assessment. Ph.D. Dissertation, Colorado School of Mines, Golden, CO.

Siegrist, R.L. and J.J. van Ee (1994). Measuring and interpreting vocs in soils: state of the art and research needs. EPA/540/R-94/506. U.S. EPA ORD, Washington, D.C. 20460.

Siegrist, R.L., K.S. Lowe, L.D. Murdoch, W.W. Slack, and T.C. Houk (1998a). X-231A demonstration of in situ remediation of DNAPL compounds in low permeability media by soil fracturing with thermally enhanced mass recovery or reactive barrier destruction. Oak Ridge National Laboratory Report. ORNL/TM-13534.

Siegrist, R.L. K.S. Lowe, L.C. Murdoch, T.L. Case, D.A. Pickering, and T.C. Houk (1998b). Horizontal treatment barriers of fracture-emplaced iron and permanganate particles. NATO/CCMS Pilot Study Special Session on Treatment Walls and Permeable Reactive Barriers. EPA 542-R-98-003. May 1998. pp. 77-82.

Siegrist, R.L., K.S. Lowe, D.R. Smuin, O.R. West, J.S. Gunderson, N.E. Korte, D.A. Pickering, and T.C. Houk (1998c). Permeation dispersal of reactive fluids for in situ remediation: field studies. Project Report prepared by Oak Ridge National Laboratory for the U.S. DOE Office of Science & Technology. ORNL/TM-13596.

Siegrist, R.L., K.S. Lowe, L.C. Murdoch, T.L. Case, and D.A. Pickering (1999). In situ oxidation by fracture emplaced reactive solids. *J. Environmental Engineering*. Vol.125, No.5, pp.429-440.

Siegrist, R.L., M.A. Urynowicz, and O.R. West (2000a). An Overview of In Situ Chemical Oxidation Technology Features and Applications. EPA/625/R-99/012. U.S. EPA ORD, Washington, DC. pp. 61-69.

Siegrist, R.L., M.A. Urynowicz, and O.R. West (2000b). In Situ Chemical Oxidation for Remediation of Contaminated Soil and Ground Water. *Ground Water Currents*. Issue No. 37, U.S. EPA Office of Solid Waste and Emergency Response, EPA 542-N-00-006, September 2000. http://www.epa.gov/tio.

Struse, A.M. and R.L. Siegrist (2000). Permanganate transport and matrix interactions in silty clay soils. In: Wickramanayake, G.B., A.R. Gavaskar, and A.S.C. Chen (ed.). Chemical Oxidation and Reactive Barriers. Battelle Press, Columbus, OH. pp. 67-74.

Tratnyek, P.G., T.L. Johnson, S.D. Warner, H.S. Clarke, and J.A. Baker (1998). In situ treatment of organics by sequential reduction and oxidation. Proc. First Intern. Conf. on Remediation of Chlorinated and Recalcitrant Compounds. Monterey, Calif. pp. 371–376.

Tyre, B.W., Watts, R.J., and Miller, G.C. (1991). Treatment of Four Biorefractory Contaminants in soils using catalyzed hydrogen peroxide. *J. Environ. Qual.* 20, pp. 832-38.

Urynowicz, M.A. 2000 (2000). Reaction kinetics and mass transfer during in situ oxidation of dissolved and dnapl trichloroethylene with permanganate. Ph.D. dissertation, Environmental Science & Engineering Division, Colorado School of Mines. May.

Urynowicz, M.A. and R.L. Siegrist (2000). Chemical degradation of TCE DNAPL by permanganate. In: Wickramanayake, G.B., A.R. Gavaskar, and A.S.C. Chen (ed.). Chemical Oxidation and Reactive Barriers. Battelle Press, Columbus, OH. pp. 75-82.

USEPA (1989). Risk assessment guidance for Superfund, Volume I: Human health evaluation manual (Part A). EPA/540/1-89/002. U.S. EPA Office of Emergency and Remedial Response, Washington, D.C.

USEPA (1996). Soil screening guidance: technical background document. EPA/540/R95/128. U.S. EPA Office of Solid Waste and Emergency Response. Washington, D.C. May 1996.

USEPA (1997). Cleaning up the Nation's waste sites. markets and technology trends. EPA 542-R-96-005. Office of Solid Waste and Emergency Response. Washington, D.C.

USEPA (1998). In situ remediation technology: in situ chemical oxidation. EPA 542-R-98-008. Office of Solid Waste and Emergency Response. Washington, D.C.

USEPA (1999). Groundwater cleanup: overview of operating experience at 28 sites. EPA 542-R-99-006. Office of Solid Waste and Emergency Response. Washington, D.C.

Vella, P.A., G. Deshinsky, J.E. Boll, J. Munder, and W.M. Joyce (1990). Treatment of low level phenols with potassium permanganate. *Res. Jour. WPCF.* 62 (7): 907-14.

Vella, P.A. and B. Veronda (1994). Oxidation of Trichloroethylene: A comparison of potassium permanganate and Fenton's reagent. 3rd Intern. Symposium on Chemical Oxidation. In: In Situ Chemical Oxidation for the Nineties. Vol. 3. Technomic Publishing Co., Inc. Lancaster, PA. pp. 62-73.

Venkatadri, R. and R.W. Peters (1993). Chemical oxidation technologies: ultraviolet light/hydrogen peroxide, Fenton's reagent, and titanium dioxide assisted photocatalysis. *J. Haz. Waste Haz. Materials.* 10(2):107-149.

Watts, R.J., R.A. Rausch, S.W. Leung, and M.D. Udell (1990). Treatment of pentachlorophenol contaminated soils using Fenton's reagent. *Haz. Waste and Haz. Mater.* 7, pp. 335-45.

Watts, R.J., S.W. Leung, and M.D. Udell (1991). Treatment of contaminated soils using catalyzed hydrogen peroxide. Proceeding to the First International Symposium on Chemical Oxidation. Technomic, Nashville, TN.

Watts, R.J. and B.R. Smith (1991). Catalyzed hydrogen peroxide treatment of octachlorobidenzo-pdioxin (occd) in surface soils. *Chemosphere,* 23, 949-55.

Watts, R.J., M.D. Udell, and R.M. Monsen (1993). Use of iron minerals in optimizing the peroxide treatment of contaminated soils. *Water Environ. Res.,* 65, 839-44.

Watts, R.J., A.P. Jones, P. Chen, and A. Kenny (1997). Mineral-catalyzed Fenton-like oxidation of sorbed chlorobenzenes. *Water Environ. Res.,* 69, 269-275.

West, O.R., S.R. Cline, W.L. Holden, F.G. Gardner, B.M. Schlosser, J.E. Thate, D.A. Pickering, and T.C. Houk (1998a). A full-scale field demonstration of in situ chemical oxidation through recirculation at the X-701B site. Oak Ridge National Laboratory Report, ORNL/TM-13556.

West, O.R., S.R. Cline, R.L. Siegrist, T.C. Houk, W.L. Holden, F.G. Gardner, and R.M. Schlosser (1998b). A field-scale test of in situ chemical oxidation through recirculation. Proc. Spectrum '98 International Conference on Nuclear and Hazardous Waste Management. Denver, Colorado, Sept. 13-18, pp. 1051-1057.

Yan, Y.E., and F.W. Schwartz (1996). Oxidation of chlorinated solvents by permanganate. Proc. Intern. Conf. on Remediation of Chlorinated and Recalcitrant Compounds, Ohio: Battelle Press. pp. 403-408.

Yan, Y.E. and F.W. Schwartz (1999). Oxidative degradation and kinetics of chlorinated ethylenes by potassium permanganate. *Journal of Contaminant Hydrology.* 37, pp. 343-365.

Yin, Y. and H.E. Allen (1999). In situ chemical treatment. Ground Water Remedation Technology Analysis Center, Technology Evaluation Report, TE-99-01. July, 1999.

CHAPTER 2

Permanganate Oxidation of Organic Chemicals

This chapter describes in detail the principles of permanganate oxidation of organic chemicals as well as experimental observations that have enhanced the understanding of in situ chemical oxidation using potassium or sodium permanganate. The results and findings from field applications also are highlighted herein with more detailed case studies provided in Chapter 7. As appropriate, comparisons are made with other oxidants. Much of the work described in this Chapter was carried out by the authors of this book, although comparisons with other relevant findings are included.

2.1. BACKGROUND ON PERMANGANATE AND ITS OXIDATION PROPERTIES

Properties of Potassium and Sodium Permanganate

In the United States, potassium and sodium permanganate are manufactured by Carus Chemical Company located in Peru, Illinois, and distributed to end users via chemical suppliers across the U.S. Potassium permanganate ($KMnO_4$) is a crystalline solid (Table 2-1) from which aqueous solutions of a desired concentration (mg/L up to wt.%) can be prepared on site using groundwater or tapwater, thereby avoiding the cost of transporting large quantities of diluted aqueous chemicals. Several grades of $KMnO_4$ (e.g., technical grade, pharmaceutical grade), which have different particle sizes and physical properties, are commercially

TABLE 2-1.	Properties and characteristics of potassium permanganate (CAIROX).[1]

Property	Value and/or comments
Chemical formula	$KMnO_4$
Purity (% by weight)	Technical Grade = 98%, Free Flowing Grade = 97%, USP Grade = 99%
Molecular weight	158.03 g/mol
Form and features	Dark purple solid with metallic luster, sweetest astringent taste, odorless, granular crystalline, oxidizer
Specific gravity—solid	2.703 g/cm³
Specific gravity—6% sol.	1.039 g/cm³
Bulk density	90 to 100 lb/ft³
Solubility in distilled water:	
0C	27.8 g/L
20C	65.0 g/L
40C	125.2 g/L
60C	230.0 g/L
Heat of solution	10.2 Kcal/mole
Packaging	25 kg pail, 50 kg drum, 150 kg drum, plus special packaging
Hazardous class (ID no.)	Oxidizer (UN1490)
Stability	Stable indefinitely if held in cool dry area in sealed containers
Incompatibilities	Avoid contact with acids, peroxides, and all combustible organics or readily oxidizable materials
Materials compatibility	In neutral or alkaline conditions $KMnO_4$ is not corrosive to iron, mild steel or stainless steel. However chloride corrosion may be accelerated. Plastics such as polypropylene, PVC, epoxy resins, Lucite, Viton A, and Hypalon are suitable but Teflon FEP and TFE, and Telzel ETFE are best. Natural rubbers and fibers are often incompatible

[1]Refer to Appendix A for additional manufacturer's information.

available. Sodium permanganate ($NaMnO_4$) is supplied by Carus as a concentrated liquid (min. 40 wt.% as $NaMnO_4$, Table 2-2). In this form, MnO_4^- ion is provided without the potassium (for sites where ^{40}K is a concern) and without dusting hazards associated with dry $KMnO_4$ solids.

The composition of potassium permanganate has two facets that are relevant to water quality effects but that are unrelated to in situ oxidation

TABLE 2-2.	Properties and characteristics of sodium permanganate (LIQUOX).[1]
Property	**Value and/or comments**
Chemical formula	$NaMnO_4$
Purity	40.0 % by weight minimum as $NaMnO_4$
Molecular weight	141.93 g/mol
Form and features	Dark purple liquid with metallic luster, sweetest astringent taste, odorless, granular crystalline, oxidizer
Specific gravity	1.36 to 1.39 g/cm^3 for a 40% solution
Solubility in water	Miscible with water in all proportions
Insolubles	100 to 1900 ppm
Potassium	1000 to 2200 ppm
pH	6.0 to 7.0
Packaging	18.9L Jerrican, 18.9L steel drum, 208L steel drum
Hazard class (ID no.)	Oxidizer (UN3214)
Stability	Stable for >18 mon
Incompatibilities	Avoid contact with acids, peroxides, and all combustible organics or readily oxidizable materials
Materials compatibility	In neutral or alkaline conditions $NaMnO_4$ is not corrosive to carbon and 316 stainless steel. However chloride corrosion may be accelerated. Plastics such as Teflon, polypropylene, HDPE, and EDPM are compatible, but Teflon FEP and TFE, and Telzel ETFE are best

[1]Refer to Appendix A for additional manufacturer's information.

processes. Since $KMnO_4$ solids are derived from mined potassium ores, there are impurities in commercially available $KMnO_4$ crystals (Table 2-3). These impurities include salts and metals for which there may be potentially relevant criteria and standards (e.g., ambient water

TABLE 2-3.	Inorganic composition of permanganate products.[1]			
	CAIROX Potassium Permanganate			LIQUOX 40% Sodium Permanganate Solution
	Technical	Free-Flowing	USP	
Aluminum	65	85		<1
Antimony	<1	<1	<1	<1
Arsenic	4	4	<1	<1
Barium	10	15	2	<0.1
Boron	6	6	3	2
Cadmium	<1	<1	<1	<0.1
Chloride	100	100	40	
Calcium	85	95	<10	3
Chromium	13	10	<0.5	<0.5
Cobalt	2	2	<0.5	<0.1
Copper	15	15	2	<0.1
Iron	30	30	3	<0.1
Lead	2	2	1	0.3
Magnesium	5	5	1	1
Mercury	0.06	0.1	N.D.	<0.02
Molybdenum	10	10	<1	<0.1
Nickel	3	3	<1	<0.2
Selenium	0.3	0.3	0.1	1
Silicon	1,000	3,000	<200	<75
Silver	1	1	<1	<1
Sodium	270	300	<50	major
Sulfate	175	250	30	<400
Strontium	<1	<1	<1	<0.1
Zinc	6	6	<1	0.1

[1]Concentrations are listed as milligrams per kilogram of product.

quality criteria, drinking water standards). Depending on the concentration of $KMnO_4$ used during in situ remediation compared to the water quality standards and criteria that might be applicable to the site as well as their point of application, these impurities may or may not be of concern. Another facet of $KMnO_4$ composition that may be of interest, particularly at nuclear facilities, involves the presence of ^{40}K, a radioactive isotope of potassium. The natural abundance of ^{40}K is 0.0117% or 0.029 g of ^{40}K per kg of $KMnO_4$. The potential for elevated concentrations of ^{40}K in groundwater that is used for drinking water consumption is remote but conceivably a potential concern. To exceed a DOE derived exposure guideline for ^{40}K of 7000 pCi/L would require consumption of water containing a $KMnO_4$ concentration of nearly 3.5 wt.%.

Permanganate Reactions

The primary redox reactions for permanganate are given in equations 2.1 to 2.3. The half-cell reaction for permanganate under acidic conditions involves a 5-electron transfer as shown in eqn. 2.1. In the pH range of 3.5 to 12, which is typical of environmental settings, the half-cell reaction involves a 3-electron transfer as shown in eqn. 2.2. At high pH (pH > 12), a single electron transfer occurs as given in eqn. 2.3. In these three reactions, manganese is reduced from Mn^{+7} to either Mn^{+2} (eqn. 2.1), Mn^{+4} (eqn. 2.2), or Mn^{+6} (eqn. 2.3). Eqn. 2.2, which represents the typical half-cell reaction under common environmental conditions, leads to the formation of a manganese oxide solid.

$$MnO_4^- + 8H^+ + 5e^- \rightarrow Mn^{+2} + 4H_2O \qquad \text{(for pH < 3.5)} \qquad (2.1)$$
$$MnO_4^- + 2H_2O + 3e^- \rightarrow MnO_2(s) + 4OH^- \qquad \text{(for pH 3.5 to 12)} \qquad (2.2)$$
$$MnO_4^- + e^- \leftrightarrow MnO_4^{-2} \qquad \text{(for pH > 12)} \qquad (2.3)$$

The Mn^{+2} cations formed under low pH (eqn. 2.1) can be oxidized subsequently by excess permanganate (eqn. 2.4).

$$3Mn^{+2} + 2MnO_4^- + 2H_2O \rightarrow 5MnO_2(s) + 4H^+ \qquad (2.4)$$

In acidic solutions, the Mn^{+4} in MnO_2 can be reduced slowly to yield Mn^{+2} as shown in eqn. 2.5.

$$MnO_2(s) + 4H^+ + 2e^- \rightarrow Mn^{+2} + 2H_2O \qquad (2.5)$$

Permanganate can also react with water but at very slow rates (see eqn. 3.1). MnO_4^- decomposition and disproportionation reactions can also occur, but only at appreciable rates under extremely high pH (e.g., as the concentration of NaOH increases above 4 to 5 moles/L).

The half-cell reactions of permanganate as a function of pH (eqn. 2.1–2.4) yield the potential for oxidation of reductants in the system. These reductants can include reduced inorganics (e.g., Fe^{+2}, Mn^{+2}), natural organic matter (NOM), or target organic chemicals of concern (e.g., TCE, PCE).

Permanganate in Chemical Manufacturing and The Study of NOM

Insight into the processes and interactions occurring during ISCO with permanganate has been gained through research and practice in organic chemistry including chemical manufacturing and the characterization of natural organic matter (NOM).

Permanganate as an Oxidant in Chemical Manufacturing. Manganese is somewhat unique in that it has a large number of oxidation states (e.g., +2,3,4,5,6,7), each of which has a different oxidation potential. In the context of this document, the permanganate ion (MnO_4^-) has an oxidation potential of 1.68 volts. This is lower than that of many other common oxidants including hydrogen peroxide, ozone, and hydroxyl radicals (Table 1-2). However, permanganate has been used for decades as an oxidizer in organic chemical manufacturing. Arndt (1981) describes the use and reactions of manganese compounds as oxidizing agents in organic chemistry and includes extensive discussion of the permanganate ion, highlights of which are included herein.

Permanganate oxidation of many organic compounds can be achieved without cleavage of the carbon framework. For example, the oxidation of toluene and substituted toluenes involves the oxidation of the methyl groups that leads to the formation of corresponding benzoic acids, but only a small amount of ring cleavage. Electron-donating substituents facilitate the reaction while electron-withdrawing substituents diminish it. Similarly, other methyl-substituted aromatic compounds (e.g., 1,2,4-trimethylbenzene, 2-chloro-6-nitrotoluene) and methyl-substituted pryidines (e.g., 2,3,6-trimethyl pyridine) are oxidized to the corresponding carboxylic acids. However, methyl and acetyl groups that are not located on the aromatic ring, or nitrogen within a 5-member heteroaromatic ring, are resistant to oxidation by permanganate. An acetyl group (-CO-CH3) located on the aromatic or heteroaromatic ring is oxidized by permanganate to the oxalyl group (-CO-COOH). Carbon to carbon double bonds within a ring system can be oxidized to give cis-1,2-glycols. Similarly, oxidation of olefins (e.g., TCE) can result in the production of epoxide intermediates and diols such as 1,2-glycols. Oxidation of cycloalkenes under basic conditions can produce diols. Oxidation of PAHs such as naphthalene under acidic conditions can lead to formation of phthalic acid. Oxidation of phenanthrene can produce carboxylic acids and quinones. Compounds of the general formula RCH_2X, where X = halogen or NO_2, are also oxidized by $KMnO_4$ to carboxylic acids. Secondary carbon atoms (which are attacked before primary carbon atoms) can be oxidized to ketones. Tertiary carbon-hydrogen bonds (R_3CH) can be oxidized (R_3COH). In some cases, new carbon-to-carbon bonds can be created during oxidation by dehydrogenative dimerization.

Permanganate as an Oxidant for Study of Natural Organic Matter. Chemical oxidants, including $KMnO_4$, have been used to study the composition of NOM in soil and water environments for over 40 years. Valuable insight has been gained on the interaction of oxidants with NOM. Nriagu (1976) and Khan and Schnitzer (1978) present summaries of early studies. Khan and Schnitzer (1972), Ortiz De Serra and Schnitzer (1973), Almendros et al. (1989) and Hanninen (1992) provide additional research findings that demonstrate that the same oxidation procedures applied to different types of soil and NOM can result in differing break-down products. Griffith and Schnitzer (1975) demonstrated that for some soil types, differences may be negligible. There is agreement that the major compounds produced by the oxidation of methylated and unmethy-lated humic and fulvic acids include aliphatic carboxylic, benzenecar-boxylic, and phenolic acids. In addition, smaller amounts of n-alkanes, substituted furans, and dialkyl phthalates are also generated. Similar results have been obtained for other types of NOM. Some findings specif-ically related to $KMnO_4$, which has most commonly been utilized under alkaline oxidation conditions, are highlighted below.

Ortiz De Serra and Schnitzer (1972) and Khan and Schnitzer (1971) reported that oxidation of fulvic acid with $KMnO_4$ resulted in the pro-duction of 3-11 times the amount of methylated phenolic, benzenecar-boxylic acid, and other breakdown products compared to that derived from non-oxidative methods. Khan and Schnitzer (1978) demonstrated that alkaline CuO oxidation of unmethylated humic materials is adept at producing phenolic structures but is inefficient at degrading aromatic, C-C bonded structures containing little oxygen. Oxidation with alkaline

$KMnO_4$ degraded the latter structures much more dramatically, but
$-OCH_3$ groups present on oxidation products did resist further $KMnO_4$
oxidation. The presence of $KMnO_4$ in the system resulted in the greatest
number of break-down products.

Manganese oxides in soils can also interact with NOM and, potentially,
other chemicals. Pohlman and McColl (1989) studied the redox processes
between polyhydroxyphenolic acid (a component of polymeric humic
substances) and soil manganese oxide suspensions. They found that the
acid containing hydroquinone and catechol-type structures with phenolic
$-OH$ groups para- or ortho- to each other can readily be oxidized by manganese surfaces. Hydroquinones and catechols first are oxidized to p- and
o-benzoquinones then to polymeric oxidation products, following second-order kinetics. Polyhydroxyphenolic acid with para-and ortho- phenolic-OH groups also was readily oxidized, while those with
phenolic-OH groups in the meta position were not. The lack of reactivity
of the meta- group was attributed to the inability to be oxidized to benzo-
quinone derivative intermediates and to subsequent polymeric oxidation
products. The conclusion was supported by Mn^{2+} mobility data. It was
also demonstrated that some of the oxidation products had metal-
chelating properties, and that demethylation of the acids resulted in
increased metal bonding. Makino et al. (1998) investigated manganese
oxide oxidation of Cr(III) in soils and determined it was impacted by
coexistent anions (e.g., acetate, sulfate, phosphate), reaction pH, Cr(III)
concentration, and reaction time. A coexistent anion was found to sup-
press both oxidation and adsorption in soil. The researchers also found
that the degree of oxidation depended more on soil NOM content than on
Mn content in some samples, but not all.

Permanganate as an Oxidant for Environmental Applications.
Permanganate has been shown to oxidize a wide variety of organic and
inorganic compounds in water under varied conditions (Cherry 1962,
Spicher and Skrinde 1963, Stewart 1965, Lee 1980, Colthurst and Singer
1982, Fatiadi 1987). Permanganate ion has a long history of successful
application in the drinking water and wastewater treatment industries for
the oxidation and removal of organic contaminants (Weber 1972, Steel
and McGhee 1979) including halogenated compounds (Singer et al. 1980).
Early studies of MnO_4^- for remediation of contaminated sites demonstrated the viability of $KMnO_4$ for organic contaminant destruction. For
example, Vella and Veronda (1994) compared potassium permanganate

(KMnO$_4$) and Fenton's reagent for the chemical oxidation of TCE in deionized water and tap water. Potassium permanganate concentrations ranged from 10 to 50 mg/L and both oxidants were found to be capable of near complete mineralization of TCE, but KMnO$_4$ was effective over a much wider pH range. During the past ten years, a growing number of investigators have conducted studies exploring different facets of the permanganate oxidation chemistry focused on different chlorinated ethenes under different conditions (e.g., Schnarr and Farquhar 1992, Gates et al. 1995, Yan and Schwartz 1996, Case 1997, Schnarr et al. 1998, West et al. 1998a,b, Yan and Schwartz 1998, Tratnyek et al. 1998, Huang et al. 1999, Siegrist et al. 1999, Struse 1999, Yan and Schwartz 1999, Urynowicz 2000).

2.2. OXIDATION OF CHLORINATED ETHENES DISSOLVED IN WATER

Studies of the permanganate degradation of toxic chemicals have focused on chlorinated ethenes (e.g., TCE, tetrachloroethene (PCE)) dissolved in water. Initial efforts have been focused this way because chlorinated ethenes are very common COCs in groundwater at contaminated sites and they are also susceptible to oxidative degradation (Vella and Veronda 1994, Yan and Schwartz 1996, Case 1997, Schnarr et al. 1998, Huang et al. 1999, Yan and Schwartz 1999, Urynowicz and Siegrist 2000). As discussed in this section, the reaction stoichiometry and kinetics for common chlorinated solvents (e.g., TCE, PCE) in aqueous systems are relatively well understood as are the effects of environmental conditions (e.g., pH, temperature). However, reaction processes occurring in situ during soil and groundwater remediation at contaminated sites can be quite complex, and questions remain.

Reaction Stoichiometry

Combining the permanganate half-cell reaction shown in eqn. 2.2 with that for TCE oxidation given in eqn. 2.6 yields the complete redox reaction shown in eqn. 2.7:

$$2MnO_4^- + 4H_2O + 6e^- = 2MnO_2(s) + 8OH^- \quad \text{(for pH 3.5 to 12)} \quad (2.2)$$

$$C_2HCl_3 + 8OH^- = 2CO_2 + 4H_2O + H^+ + 3Cl^- + 6e^- \quad (2.6)$$

$$2MnO_4^- + C_2HCl_3 = 2MnO_2(s) + 2CO_2 + 3Cl^- + H^+ \quad (2.7)$$

The stoichiometric reactions for the complete destruction of several common chlorinated organic solvents in an aqueous system are given in eqn. 2.8 to 2.11.

Tetrachloroethene (PCE)
$$4KMnO_4 + 3C_2Cl_4 + 4H_2O \rightarrow 6CO_2 + 4MnO_2 + 4K^+ + 8H^+ + 12Cl^- \qquad (2.8)$$

Trichloroethene (TCE)
$$2KMnO_4 + C_2HCl_3 \rightarrow 2CO_2 + 2MnO_2 + 2K^+ + H^+ + 3Cl^- \qquad (2.9)$$

Dichloroethene (DCE)
$$8KMnO_4 + 3C_2H_2Cl_2 \rightarrow 6CO_2 + 8MnO_2 + 8K^+ + 6Cl^- + 2OH^- + 2H_2O \qquad 2.10)$$

Vinyl chloride (VC)
$$10KMnO_4 + 3C_2H_3Cl \rightarrow 6CO_2 + 10MnO_2 + 10K^+ + 3Cl^- + 7OH^- + H_2O \qquad (2.11)$$

Based on the above stoichiometry, the oxidant demand and product formation for chemical oxidation of the four chlorinated ethenes are given in Table 2-4. Note that the reactions are comparable in permanganate demand whether it is supplied in the potassium form or sodium form. As shown in Table 2-4, on a unit mass basis, halocarbons with higher chlorine substitution (e.g., PCE vs. DCE) consume less oxidant (per the stoichiometric requirement) and produce less MnO_2 solids.

Reaction Pathways

The hydroxylation of olefins by MnO_4^- has been known for many years (Wagner 1895). It is generally accepted that under neutral to acidic pHs, MnO_4^- initially reacts with the carbon-carbon double bond to form a five-member cyclic hypomanganate ester (Figure 2-1). That reaction consumes 1 mole of permanganate for each mole of olefin, and that cycloaddition has been reported to be the rate-determining step in degradation (Wiberg and Saegebarth 1957, Stewart 1965, Freeman 1976). Yan and Schwartz (1998, 1999) recently proposed a similar reaction scheme for the chemical oxidation of TCE in which the cyclic ester can then undergo decomposition to carbon dioxide along several oxidative or hydrolysis pathways depending on pH. Several carboxylic acid intermediates including formic, oxalic, glyoxylic and glycolic acids were also identified. In highly alkaline solutions, hydroxyl radicals, which can contribute to oxidative destruction, also may be formed. The degradation

TABLE 2-4.	Stoichiometric requirements for mineralization of organic compounds by permanganate.[1]		
Target compound	Compound molecular weight (g/mol)	Oxidant demand (g MnO_4^- / g of target)	MnO_2 produced (g MnO_2 / g target oxidized)
Tetrachloroethene	165.6	0.96	0.70
Trichloroethene	131.2	1.81	1.32
Dichloroethene	96.8	3.28	2.39
Vinyl chloride	62.4	6.35	4.64
Phenol	94.1	11.8	8.62
Naphthalene	128.2	14.8	10.8
Phenanthrene	178.2	14.7	10.7
Pyrene	202.3	14.5	10.6

[1]Molecular weight of MnO_4^- = 118.9 g, MnO_2 = 86.9 g, $KMnO_4$ = 158 g, and $NaMnO_4$ = 141.9 g.

FIGURE 2-1. Pathway for oxidation of olefins such as TCE.

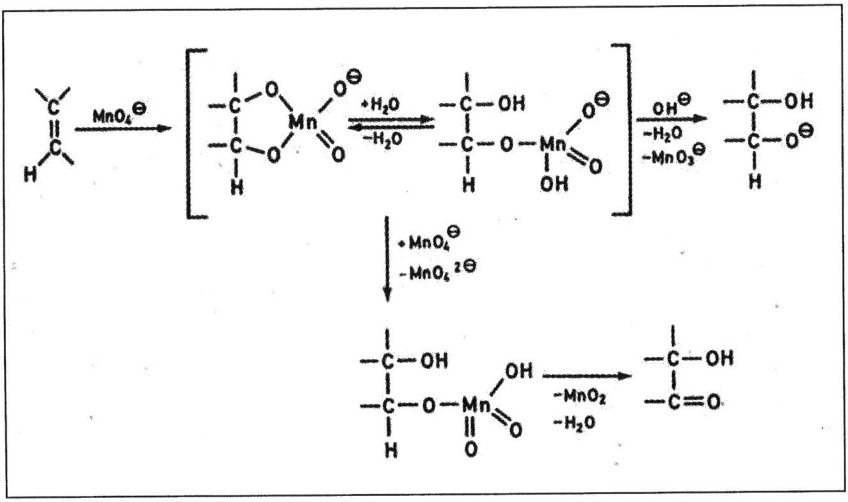

then can include destruction by direct electron transfer or free radical advanced oxidation.

Reaction Kinetics

The rate of reaction of permanganate with ethenes such as TCE can impact the design and performance of a remediation system involving chemical oxidation with permanganate. In some remediation schemes, a fast rate of reaction is desired. One example is where contaminated groundwater is extracted and addition of permanganate is done in-line before delivery to the subsurface via a groundwater injection well. The kinetics of reaction will determine what degree of treatment can be accomplished above-ground and whether a target efficiency can be achieved during a given contact time (e.g., 90% destruction in 5 min). The reaction may also generate MnO_2 precipitates that will require effective management. In contrast, there are situations where a slow rate of reaction is advantageous. A common situation is where permanganate is delivered in relatively clean water (i.e., with little to no oxidant demand) into the subsurface via injection probes or wells. The kinetics of reaction with target organics and also NOM and other reductants can affect the rate and extent of oxidant transport away from the injection location. If the goal is to disperse the oxidant throughout a subsurface region away from the point of delivery, it is advantageous to have a fast transport rate compared to the rate of reaction.

For chemical oxidation to occur, MnO_4^- must collide with the target (e.g., TCE) and, as shown in Figure 2-1, this involves 1 mole of MnO_4^- reacting with each mole of target to produce an intermediate that further combines with another mol of MnO_4^- to produce products such as CO_2, MnO_2, and salts. This is represented in eqn. 2.12:

$$A + B \rightarrow C \rightarrow D + E + B \rightarrow F + E + P \qquad (2.12)$$

where A = target organic (e.g., TCE), B = MnO_4^-, C = cyclic ester, D = organic intermediate, E = MnO_2, F = CO_2, and P = other products (e.g., Cl^-). Since the first step is the rate limiting step, the rate of reaction is controlled by the concentration of both reactants (A + B), and the degradation of [A] can be best described as a second-order irreversible bimolecular reaction with the kinetics given by following rate law,

$$r_A = \frac{d[A]}{dt} = -k[A]^\alpha [B]^\beta \qquad (2.13)$$

where, k = specific reaction rate constant (units determined by α and β), [A] = concentration of the target (mol L^{-1}), [B] = concentration of the oxidant (mol L^{-1}), α = reaction order with respect to A, B = reaction order with respect to β, and t = time. For the second-order degradation of TCE by MnO_4^-, the general eqn. 2.14 can be written as:

$$\frac{d[TCE]}{dt} = k_2 [TCE] [MnO_4^-] \qquad (2.14)$$

where, k_2 = second-order reaction rate constant (L mol^{-1} s^{-1}), $[TCE]$ = concentration of TCE (mol L^{-1}), $[MnO_4^-]$ = concentration of the MnO_4^- (mol L^{-1}), $\alpha = 1$ for first-order reaction with respect to TCE, and $\beta = 1$ for first-order reaction with respect to MnO_4^-.

Findings of several investigators have confirmed that, in the absence of substantial NOM or other reductants that participate in parallel competing reactions, the reaction is first-order with respect to TCE and also to MnO_4^-, or second-order overall (eqn. 2.14).

If MnO_4^- is maintained in excess (e.g., greater than 5x the stoichiometric requirement based on the TCE present), the reaction can be adequately described by pseudo first-order kinetics (eqn. 2.15). This is accomplished by setting the pseudo first-order rate constant, k_1' (units of T^{-1}), equal to the product of the second-order reaction rate constant, k_2, and the initial concentration of MnO_4^- (eqn. 2.16). It is noted that eqn. 2.16 is applicable to relatively clean systems where parallel competing reactions for permanganate are not substantial.

$$\frac{d[TCE]}{dt} = k_1' [TCE] \qquad (2.15)$$

$$k_1' = k_2 [MnO_4^-] \qquad (2.16)$$

Case (1997) and Siegrist et al. (1999) completed laboratory and field studies to examine the oxidation of TCE in water by $KMnO_4$ dissolved in water and also in the form of a new oxidative particle mixture (OPM). The OPM is comprised of a mixture of $KMnO_4$ solids suspended in a mineral gel such that it is pumpable (Siegrist et al. 1999; U.S. patent no. 6,102,621). During initial experiments with dissolved phase TCE in groundwater, the starting TCE concentrations were 10, 100 and 800 mg/L. A $KMnO_4$ dose of 25,000 mg/L was delivered by adding either permanganate crystals to the groundwater or an equivalent mass of OPM. In these reactions, the TCE was so rapidly degraded that concentrations were below detection limits (~1 mg/L) within the first 1 to 2 min. Additional experiments were performed with ~5:1 molar concentrations of KMnO$_4$ to TCE and nearly complete degradation (>99%) still occurred in less than 30 min (Figure 2-2).

Yan and Schwartz (1998, 1999) conducted a study investigating the oxidative treatment of chlorinated ethenes (tetrachloroethene, tri-chloroethene, cis-dichloroethene, trans-dichloroethene and 1,1-dichloro-ethene) with MnO$_4^-$. Using a sealed and water-jacketed spherical glass reaction vessel in a series of systematic batch experiments, the authors reported that chemical oxidation was rapid with pseudo first-order rate constants greater than 6.5 x 10^{-4} s^{-1} at 1 mmol L^{-1} of [MnO$_4^{-1}$]. Tetra-chloroethene was the exception with a relatively low rate constant of 4.5 x 10^{-5} s^{-1}. With respect to TCE, the reaction was reported to be second-order (k_2 = 0.67 ± 0.03 mol^{-1} L s^{-1}) and independent of pH over the range of pH 4 to 8.

Huang et al. (1999) also studied the kinetics of chemical oxidation of TCE by MnO$_4^-$ using zero head-space syringe reactors. The authors reported a second-order rate constant of 0.89 ± 0.03 mol^{-1} L s^{-1} and an activation energy of 35 ± 2.9 kJ mol^{-1}.

Urynowicz (2000) performed oxidation studies using static micro-reaction vials and mixed syringe reactors. With TCE in a phosphate buffered de-ionized water (20C, pH = 6.9; I = 0.05 mol L^{-1}), the reaction

---

FIGURE 2-2.   **Experimentally observed oxidation of TCE in simulated ground-water (temperature = 20C; final TCE concentration = 0.6 mg/L).**

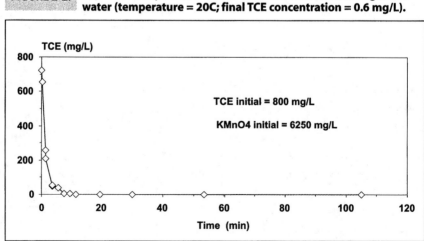

was shown to be first-order with respect to MnO_4^- and TCE, and second-order overall (k_2 = 0.89±0.07 mol^{-1} L s^{-1}). No significant difference between the batch or CSTR experiments was observed.

An irreversible second-order kinetic model can be used to explore the degradation of an ethene like TCE over time at different initial concentrations of MnO_4^- (it is noted that this analysis applies to the initial reaction of TCE and MnO_4^- and not for complete mineralization; see eqn. 2.14). The solution to eqn. 2.14 combined with the stoichiometry of the reaction (eqn. 2.9) is given in eqn. 2.17 and this can be used for determining reductant (TCE) and oxidant (MnO_4^-) concentrations with time.

$$\frac{1}{([TCE]_o - [MnO_4^-]_o)} \ln \left\{ \frac{[MnO_4^-]_o [TCE]_t}{[TCE]_o [MnO_4^-]_t} \right\} = k_2 t \qquad (2.17)$$

Equation 2.17 can be rearranged and made nondimensional with respect to the degradation of TCE as shown in Figure 2-3. This illustrates the degradation of an initial concentration of TCE at 1000 mg/L (7.41 mmol L^{-1}) at different ratios of the initial molar concentrations of MnO_4^- to TCE (i.e., $[MnO_4^-]_o/[TCE]_o$) plotted against the absolute value of ($[TCE]_o-[MnO_4^-]_o)k_2 t$). Figure 2-3 illustrates that as the $[MnO_4^-]_o$ /$[TCE]_o$ ratio becomes larger (e.g., >3) the kinetics of TCE degradation do not change as the permanganate concentration is increased further, and the TCE degradation can be described as a pseudo first-order reaction. Figure 2-4 illustrates the time-dependent degradation of 1000 mg/L of initial TCE at different initial concentrations of MnO_4^-. Consistent with the information shown in Figure 2-3, as the initial concentration of MnO_4^- exceeds that of the TCE, the half-life of the degradation is on the order of 2 minutes or less (Figure 2-4) and the differences in the kinetics become increasingly smaller at higher $[MnO_4^-]_o$ concentrations. Figures 2-5 and 2-6 present comparisons of the model predictions to laboratory and field experimental data. These examples illustrate that in a relatively "clean" aqueous system (i.e., one without substantial NOM or other reductants present), the rate of degradation of a chlorinated ethene such as TCE can be predicted by using second-order kinetic expressions (eqn. 2-17).

Effects of Oxidant Type and Dose Concentration

There is evidence that oxidant form has no effect on organic degradation rate or extent but oxidant dose concentration does. That is, as the dose concentration of permanganate is increased, under otherwise

FIGURE 2-3. Calculated change of concentration of TCE in a second-order reaction with MnO_4^- (concentrations are in mol/L).

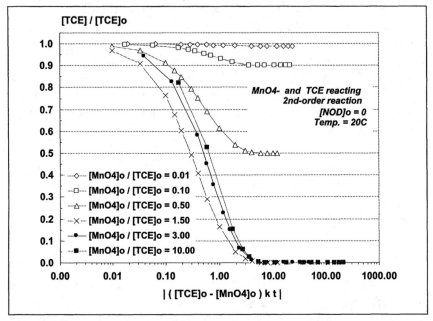

FIGURE 2-4. Calculated change of concentration of TCE versus time during a second-order reaction with MnO_4^-.

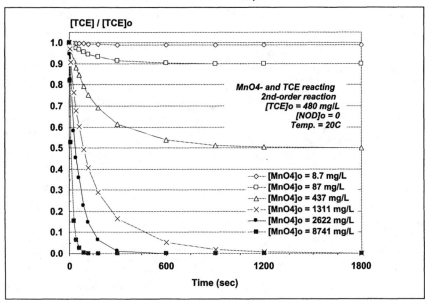

FIGURE 2-5. Predicted and observed change in concentration of TCE versus time during a second-order reaction with MnO$_4^-$ in contaminated groundwater from a field site.

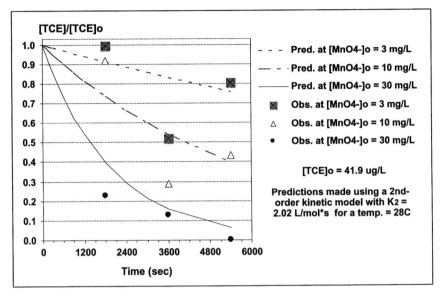

FIGURE 2-6. Predicted and observed change in concentration of TCE versus time during a second-order reaction with MnO$_4^-$ in simulated groundwater.

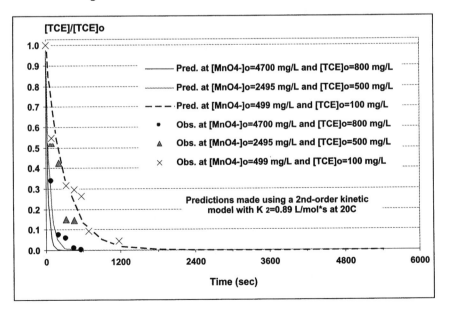

comparable conditions (i.e., same reductants and matrix conditions), the rate of organic degradation increases. In addition, the magnitude of the MnO_4^- consumed increases. Siegrist et al. (2000) at CSM investigated this phenomenon using batch experiments and a duplicated 2^5 factorial design to examine oxidant consumed as affected by oxidant type ($NaMnO_4$ vs. $KMnO_4$), oxidant concentration (300 vs. 3000 mg-MnO_4/L, TCE concentration (7 vs. 54 mg/L), reaction time (15 vs. 300 min) and presence of silt/clay particles (7.5 vs. 750 mg/L). Regression analysis of the experimental data (Figure 2-7) revealed that the MnO_4^- consumed was dependent on the TCE concentration and the oxidant dose (for both oxidant forms plus both reaction times combined) following the relationship shown in eqn. 2.18 ($R^2 = 0.96$).

$$MnO_4^- \text{ consumed} = 0.196 \, (MnO_4^- \text{ dose}) + 2.09 \, (\text{TCE conc.}) \quad (2.18)$$

FIGURE 2-7. Oxidant consumption (mg-MnO_4^-/L) as affected by oxidant dose concentration during chemical oxidation of TCE.

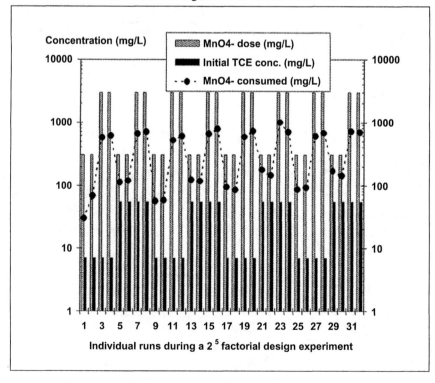

where all three concentrations are in mg/L. The constant, 2.09, is in the range of the stoichiometric requirement for mineralizing TCE (1.81 g MnO_4^- per g of TCE). This relationship indicates that increasing the oxidant dose can substantially increase the oxidant consumed. Results of natural oxidant-demand tests with soil and aquifer sediments have revealed similar consumption vs. dose behavior (see discussion below and data in Table 2-5). These results suggest that increasing the oxidant dose far in excess of that needed for oxidation of the target organics (and also the NOD as noted below) may be counterproductive. The mechanism underlying this consumption behavior could be associated with decomposition of MnO_4^- by the MnO_2 solids produced by the oxidation reaction.

Effects of Environmental Factors

There are several key environmental factors that can influence the rate and extent of degradation of chlorinated ethenes dissolved in water, including: (1) the natural oxidant demand of the groundwater matrix, (2) the environmental temperature, and (3) the pH and ionic composition of the groundwater. These factors are discussed below.

Natural Oxidant Demand. Natural oxidant demand (NOD) represents the consumption of permanganate in reactions that are unrelated to degradation of the target organic COCs. When NOM or other reductants are present in soil and groundwater along with TCE (or other target chemicals), the NOD can exert a competing parallel demand for permanganate. This natural oxidant demand can be important because it increases the oxidant dose required and can alter the kinetics and extent of reaction with the target organics if excess oxidant is not provided. Similar to the reaction kinetics illustrated in Figures 2-3 and 2-4, when $[MnO_4^-]_o \gg ([TCE]_o + [NOD]_o)$, it can satisfy both demands, and the kinetics of degradation of TCE are not markedly altered by the NOD. However, if the $[MnO_4^-]_o$ dose is not sufficient ($[MnO_4^-]_o < \{[TCE]_o + [NOD]_o\}$), the oxidant may be depleted by the combined demands such that the kinetics of the reaction with the target organic are reduced. More importantly, if the oxidant is depleted entirely, the target chemicals may not be degraded to the level desired.

The reaction of permanganate with NOM (and some soil and geologic media) appears rapid and on a mass/mass basis, in the range of that occurring with target compounds such as TCE. For example, during batch experiments with a model humic acid (HA) (Fluka peat extract) in a

TABLE 2-5.	Summary of natural oxidant demand data.			
Media	**Conditions**	**MnO$_4^-$ demand[1]**	**Comments**	**Reference**
Groundwater (simulated)	MnO$_4^-$ = 450 mg/L TOC = 50mg/L Temp. = 20C Time = 1hr	6.0 mg/mg TOC	Residual MnO$_4^-$ to TOC was 12:1 mg/L to mg/L	Lab experiments at CSM during 1998-99.
	MnO$_4^-$ = 900 mg/L TOC = 550 mg/L Temp. = 20C Time = 1 hr	2.3 mg/mg TOC	MnO$_4^-$ was exhausted so residual MnO$_4$- to TOC was <1:1	Lab experiments at CSM during 1998-99.
Groundwater (site in CA)	MnO$_4^-$ = 500 mg/L TOC = 20 mg/L Temp. = 20C Time = 15.7 days	9.8 mg/mg TOC	MnO$_4^-$ was not exhausted. Cl$^-$ conc. in water was ~6 g/L	Lab experiments at CSM during 1998-99. Unpublished project results.
	MnO$_4^-$ = 5000 mg/L TOC = 20 mg/L Temp. = 20 C Time = 15.7 days	146 mg/mg TOC	MnO$_4^-$ was not exhausted. Cl$^-$ conc. in water was ~6 g/L	Lab experiments at CSM during 1998-99. Unpublished project results.
Soil (site in OH)	MnO$_4^-$ = 500 mg/L TOC = 0.9 mg/g soil Temp. = 20C Time = 14 days	3.3 mg/mg TOC		Struse 1999
	MnO$_4^-$ = 5000 mg/L TOC = 0.9 mg/g soil Temp. = 20C Time = 14 days	12.4 mg/mg TOC		Struse 1999
Soil (site in CA)	MnO$_4^-$ = 5.0 mg/g TOC = 3.0 mg/g soil Temp. = 20C Time = 15.7 days	≥0.25 mg/mg TOC	MnO$_4^-$ depleted between 0.2 to 1 days	Lab experiments at CSM during 1998-99. Unpublished project results.
	MnO$_4^-$ = 50.0 mg/g TOC = 3.0 mg/g soil Temp. = 20C Time = 15.7 days	≥2.5 mg/mg	MnO$_4^-$ depleted between 2 to 7 days	Lab experiments at CSM during 1998-99. Unpublished project results.
Basalt Crushed basalt Sediment (INEEL, ID)	MnO$_4^-$ = 0.01%, 0.10% Time = 1 day for all	50 mg/mg basalt 100 mg/mg cr. basalt 200 mg/mg sed.		Cline et al. 1999 *continued*

TABLE 2-5.	(continued)			
Media	**Conditions**	**MnO$_4^-$ demand[1]**	**Comments**	**Reference**
Soil (various field aquifer media)	MnO$_4^-$ = 500 mg/L TOC = not reported Temp. = 20C Time = up to 10 days	≥ 0.5 mg/g sample for all samples	MnO$_4^-$ exhausted by 10 days	Li and Schwartz 2000
Soil (silt, clay, sand from site in CA)	MnO$_4^-$ = 10 mg/L TOC = not reported Temp. = not reported Time = 1 day, 14 days	Silt 4.27 mg/g (1d) 7.61 mg/g (14d) Clay: 4.34 mg/g (1d) 7.16 mg/g (14d) Sand: 2.19 mg/g (1d) 4.49 mg/g (14d)	Fast initial rate (0 to 1 d) followed by slower rate (1-14d) Natural material demand much greater (10x) than COCs would require	Chambers et al. 2000
Soil (site in OH)	Not reported	60 mg/kg soil	Demand exerted by sandy soil. Humic acids and minerals exerted negligible demand	Oberle and Schroder 2000
Soil (site in MI)	Not reported	15,000 to 200,000 mg/kg soil	Demand attributed to elevated TOC and Fe levels	Oberle and Schroder 2000

[1]Oxidant demand is expressed as the mg oxidant per mg of TOC in the system at time zero and not the demand per mg of TOC consumed.

simulated groundwater, this organic matter was observed to exert a rapid and substantial demand for MnO$_4^-$ as illustrated in Figure 2-8. The extent of permanganate NOD has been measured directly or indirectly in a number of experiments and field tests as summarized in Table 2-5. Results are presented in mg of oxidant per mg of TOC wherever possible. Inspection of Table 2-5 suggests that the magnitude of the permanganate demand by natural media varies from 2 to over 100 mg MnO$_4^-$ per mg of TOC and is highly variable and not easily predicted. This demand is equal to or greater than that of ethenes such as TCE or PCE. Also, the data shown in Table 2.5 indicate that, under otherwise comparable conditions, the MnO$_4$- consumption increases as the oxidant dose concentration increases.

The consumption of permanganate by NOD can reduce the degradation efficiency (rate and extent of destruction) for a target chemical if the permanganate dose is not sufficient to satisfy all of the demand in the system. This is illustrated in Figure 2-9 where the time to 99% degradation of an

FIGURE 2-8.   **Permanganate oxidant demand and destruction of humic acid in groundwater. (a) MnO$_4^-$ initial = 450 mg/L and HA initial = 100 mg/L and (b) MnO$_4^-$ initial = 900 mg/L and HA initial = 1000 mg/L.**

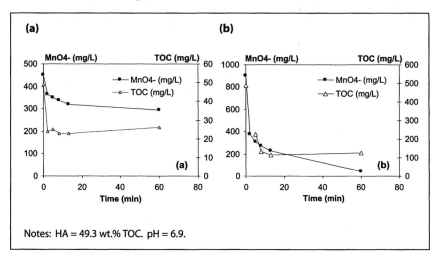

Notes: HA = 49.3 wt.% TOC. pH = 6.9.

FIGURE 2-9.   **Time to 99% destruction of initial TCE in groundwater as affected by the concentrations of TCE, NOM and permanganate dose.**

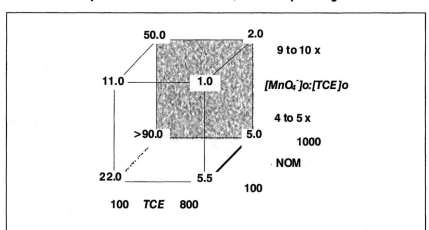

Note: each corner of the cube represents the time in min. to 99% destruction of TCE measured during a single batch experiment under the initial conditions shown on the axes of the cube: TCE = 100 or 800 mg/L, NOM = 100 or 1000 mg-HA/L, [MnO$_4^-$]$_o$/[TCE]$_o$ = 4 to 5x, or 9 to 10x.

initial TCE concentration of 100 or 800 mg/L in groundwater with 100 or 1000 mg/L of humic acid added is given for a permanganate dose of 5x or 10x that of the stoichiometric requirement based on the TCE only. If the TCE concentration is high there is adequate oxidant added at the 5x or 10x dose such that the kinetics are slowed somewhat (longer time to 99% destruction) but the reaction still proceeds to destroy 99% of the initial TCE. However, in the case of the low TCE concentration (100 mg/L) and a higher concentration of NOM in the system (1000 vs. 100 mg/L), the achievement of 99% degradation takes 50 minutes at a $[MnO_4^-]_0/[TCE]_0$ dose of 10:1 and is not achieved in 90 min at a dose of 5:1.

While NOM can react with permanganate, some fraction of the original NOM and/or products of its degradation have been shown to be resistant to complete mineralization, even in the presence of high residual MnO_4^- levels over long time periods (see Figure 2-8). In lab and field experimentation where soil and aquifer media have been exposed to 5000 mg/L or higher MnO_4^- concentrations over two to 15 months, an appreciable fraction of the initial soil TOC was observed to persist (Struse 1999, Siegrist et al. 1999). The absence of complete mineralization is not surprising as organic acids can be produced during permanganate oxidation of NOM (see Section 2.1) and some of these organic compounds could be resistant to permanganate degradation under typical environmental conditions and experimental time frames. Alternatively, some organics could be coated with MnO_2 solids (as described in Chapter 3) and thereby protected from further oxidation by MnO_4^- in the surrounding aqueous phase.

Effects of Temperature. The relationship of reaction rate with temperature often can be described by the Arrhenius equation (eqn. 2.19) (Logan 1996, Case 1997):

$$k = A_f e^{\{-E_a / RT\}} \tag{2.19}$$

where A_f is the Arrhenius frequency factor, E_a is the activation energy of the reaction (kJ mol^{-1}), T is the temperature (K), and R is the gas constant (8.314 J K^{-1} mol^{-1}). Activation energy, E_a, can be determined from the kinetic rate constants, k, of an elementary reaction at two or more different temperatures. The rate constants determined for $KMnO_4$ oxidation of TCE by Case (1997) at 20°C and 10°C revealed $E_a = 78$ kJ mol^{-1}, which is relatively high for most oxidation reactions (Logan, 1996). Yan and Schwartz (1996) determined the reaction rate constants of the $KMnO_4$ oxidation of 10 mg/L TCE at two different temperatures (10°C, 25°C) and

E_a was calculated to be 73 kJ mol^{-1}. Huang et al. (1999) also studied the kinetics of chemical oxidation of TCE by MnO_4^- and reported an activation energy of 35 ± 2.9 kJ mol^{-1}.

The temperature dependency of the second-order reaction of TCE with MnO_4^- is given by eqn. 2.20 which is a rearrangement of eqn. 2.19.

$$\ln k = \ln A_f - \frac{E_a}{R} \left\{ \frac{1}{T} \right\} \qquad (2.20)$$

The degradation of TCE at several temperatures and the second-order rate constants is given in Figure 2-10. At 20°C as compared to 10°C, the second-order rate constant increases by three-fold.

Effects of pH and Ionic Composition. The effects of system pH on permanganate oxidation reactions are of interest and, conversely, so are the effects of the oxidation reactions on the system pH. Permanganate degradation reactions appear to be relatively independent of pH in the range typical of environmental settings. Yan and Schwartz (1998, 1999) reported a 2nd-order reaction for TCE degradation by MnO_4^- that was independent of pH in the pH 4 to 8 range. Urynowicz (2000) evaluated the kinetic effects of pH buffering during a series of batch experiments

FIGURE 2-10. Temperature effects on the degradation of TCE during a second-order reaction with MnO_4^- at different temperatures.

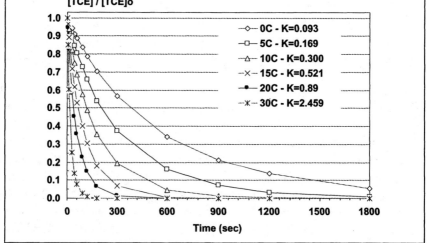

with $[MnO_4^-]_o = 0.00628$ mol L^{-1} and $[TCE]_o = 0.00076$ mol L^{-1}. Deionized water with phosphate buffer (pH ~ 6.9) was compared to deionized water with no buffer. Experimental conditions were such that the MnO_4^- concentration remained relatively constant over the course of the reaction (i.e., $[MnO_4^-]_t >> [TCE]_t$) and pseudo first-order rate constants were determined. The rate constant for the unbuffered system where the pH declined to less than pH 4 was somewhat higher at $k_1 = 0.0072$ s^{-1} compared to that of the buffered system where $k_1 = 0.0053$ s^{-1}.

Oxidation can yield acidity and result in a depression in system pH, the extent of which is dependent on the reaction intensity and the buffering capacity of the system. In relatively unbuffered systems, the pH can rapidly drop to pH 2 to 3 (Case 1997). This dramatic pH shift could effect the oxidation reaction and also the behavior of other constituents in the system (more discussion in Chapter 3).

Regarding ionic composition, the presence of excess chloride ions in the reacting solution (e.g., from DNAPL degradation) might be expected to push the equilibrium toward reactants, which may also tend to slow the reaction kinetics (Yan and Schwartz 1996). Additionally, the complexation of contaminant ions and compounds with reactants and reaction intermediates may raise the activation energy of the reaction and slow the reaction kinetics. While limited, research involving different soils and groundwaters does not suggest a strong influence on rate or extent of reaction of permanganate with target organic chemicals such as TCE due to ionic composition in groundwater with high chloride (up to 6,000 mg/L) or otherwise high ionic strength ($I = 1.57$ M) (Case 1997, Urynowicz 2000).

2.3. OXIDATION OF CHLORINATED ETHENES THAT ARE SORBED OR PRESENT AS NAPLS

Most of the initial studies of MnO_4^- oxidation of chlorinated ethenes focused on the chemical oxidation of organics dissolved in water. More recently, studies have begun to explore the oxidation of chlorinated ethenes when they are present in a sorbed or NAPL phase. Information has been gained through batch and flow-through experiments using aqueous and soil slurry systems (Gates et al. 1995, Urynowicz 2000, Gates-Anderson et al. 2001) as well as packed columns and tanks containing porous media (McKay et al. 1998, Schnarr et al. 1998, Drescher et al.

1999, Struse and Siegrist 2000) and field experimentation (Schnarr et al. 1998, West et al. 1998a,b, Lowe et al. 2000).

Sorbed Phase Organics

Permanganate appears capable of oxidizing organic chemicals that are sorbed to soils and aquifer sediments though the reaction mechanisms are not fully understood. It is speculated that the oxidation occurs in the aqueous phase after organics desorb from the media, with the rate of desorption increased due to the shift in equilibrium partitioning that results as the aqueous phase concentration of the target organic is depleted. In addition, since organic chemicals like TCE can sorb to NOM coatings on the soil and aquifer mineral matter, as the NOM is oxidized by the permanganate the associated chemicals could be oxidized as well. Evidence related to oxidation of sorbed organics is revealed by the fact that permanganate oxidation of organics like TCE does occur in groundwater even when there are high levels of dissolved and particulate organic matter present and presumably having sorbed a substantial fraction of the TCE mass in the system (e.g., see Figure 2-9). There also have been laboratory experiments, as noted below.

Gates et al. (Gates et al. 1995, Gates-Anderson et al. 2001) completed laboratory tests to compare the treatment efficiency of $KMnO_4$ with H_2O_2 (alone or with amendments) for sand and silty clay soil contaminated with either a mixture of VOCs (TCE, PCE, and 1,1,1-TCA) or SVOCs (naphthalene, phenanthrene, and pyrene). The relative treatment effects of soil type, oxidant loading rate and dosing, reaction period, as well as the use of surfactant or iron amendments and pH adjustment were examined in batch experiments with contaminated soil slurries (Figure 2-11 and 2-12). When $KMnO_4$ was applied to low organic carbon, acidic or alkaline soils, at loading rates of 15 to 20 g/kg, it was found to consistently degrade 90% or more of the chlorinated ethenes (TCE and PCE) and 99% of the polyaromatic SVOCs (naphthalene, pyrene and phenanthrene). H_2O_2 was more sensitive to contaminant and soil type and the VOC treatment efficiencies were somewhat lower as compared to $KMnO_4$ under comparable conditions, particularly with the sandy soil and even when supplemental iron was added. In clay soil, H_2O_2 with iron addition degraded over 90% of the SVOCs present compared to near zero in sandy soil, unless the pH was depressed to pH 3 and iron amendments were increased, whereby the treatment efficiency in the sandy soil was increased slightly. With both H_2O_2 and $KMnO_4$, treatment efficiency increased to varying degrees as

FIGURE 2-11. Oxidation of TCE and PCE in soil slurries using KMnO₄ or H₂O₂(average [TCE]₀ = 76.5 mg/kg and [PCE]₀ = 18.8 mg/kg. KMnO₄ loadings were low = 1.5 and high = 15.0 g/kg; H₂O₂ loadings were low = 4.5 and high = 25.5 g/kg) (after Gates et al. 1995)

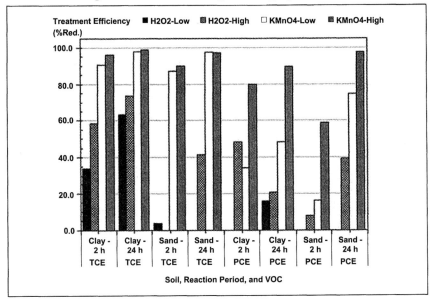

FIGURE 2-12. Oxidation of TCE and PCE in soil slurries using KMnO₄ versus H₂O₂ or Fenton's regent (KMnO₄ loading = 15.0 g/kg; H₂O₂ = 25.5 g/kg; Fenton's reagent = 25.5 g/kg H₂O₂ plus 5 mM FeSO₄) (after Gates et al. 1995).

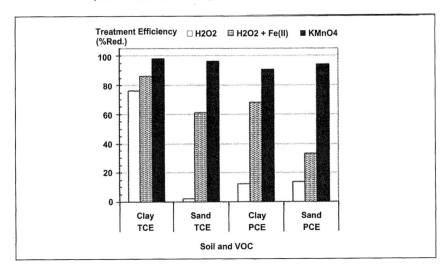

the oxidant loading rate (g/kg) and reaction time (h) were increased. Multiple oxidant additions or surfactant addition were not found to have any significant effect on VOC treatment efficiency. Also, very limited TCA treatment was observed with either $KMnO_4$ or H_2O_2. These experiments also provided insight into $KMnO_4$ oxidation effectiveness for sorbed (and also NAPL phase) organics. Prior to oxidant addition, 8 and 11% of the TCE mass (total = 76.5 mg/kg) and 19 and 25% of the PCE mass (total = 18.8 mg/kg) was present in the sorbed phase in silty clay and sand soils, respectively. Nevertheless, high treatment efficiencies (>95% destruction of starting total mass) were still observed during 2- and 24-hr reaction periods (Figure 2-11 and 2-12).

DNAPL Phase Organics

A theoretical basis for chemical oxidation of chlorinated ethenes when they are present in a DNAPL phase is supported by experimental research as noted in this section. An increased rate of DNAPL degradation (defined as dissolution with oxidative destruction) can be attributed to chemical reaction that causes the aqueous phase oxidant and DNAPL concentration gradients to increase and result in an increase in the dissolution rate and oxidant mass flux towards the interface (Schnarr et al. 1998, Hood et al. 1999, Urynowicz 2000). The conceptual model shown in Figure 2-13 is based on mass transfer via diffusion through a stagnant film boundary layer of fluid (Sherwood et al. 1975, Pfannkuch 1984, deZabala and Radke 1986, Miller et al. 1990).

Mass transfer can involve several steps, one or more of which may be considered rate limiting. For mass transfer across fluid-fluid interfaces, diffusion of the chemical species across the stagnant film boundary layer between the interface and the bulk aqueous phase is usually the rate-determining step (Pfannkuch 1984, deZabala and Radke 1986). Fick's first law was developed from the assumption that a substance diffusing across a given cross-section area per unit time is proportional to the concentration gradient (equation 2.21).

$$J = -D\frac{dC}{dx} \tag{2.21}$$

where, J = mass flux of the solute (g cm^{-2} s^{-1}), D = diffusion coefficient (cm^2 s^{-1}), x = distance (cm), and C = concentration (g cm^{-3}).

Nernst performed much of the early work on mass-transfer and is credited with formulating the concept of diffusion through a stagnant film

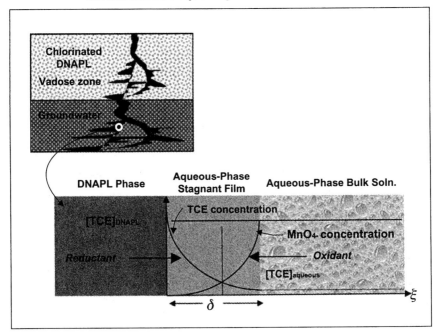

FIGURE 2-13. Conceptual model for ISCO of DNAPL in groundwater due to MnO₄⁻ oxidant transport to the DNAPL-water interface which increases interphase mass transfer due to oxidation reactions in the film and bulk aqueous phase.

boundary layer of fluid (Miller et al. 1990). The conceptual model, which is referred to as the stagnant film model (Sherwood et al. 1975), assumes steady-state diffusion across the boundary layer. The steady-state diffusion approximation requires that the volume of the boundary layer is relatively small when compared to the adjacent fluids and that both fluids are well mixed. In the stagnant film model illustrated schematically in Figure 2-13, a nonaqueous phase liquid (e.g., TCE) is in contact with a well-mixed aqueous liquid. At the liquid-liquid interface, the dissolved liquid is in thermodynamic equilibrium with the non-aqueous phase liquid (C_S). At the other side of the stagnant film boundary layer, the concentration of dissolved phase liquid is equal to the bulk aqueous concentration (C). Equation 2.22 shows how the stagnant film model can be related to Fick's first law.

$$J = -D\frac{dC}{dx} = -\left(D\frac{dC_f}{d\xi}\right) = \frac{D}{\delta}(C_s - C) \tag{2.22}$$

where, C_f = concentration in the stagnant film (g cm^{-3}), ξ = local coordinate system within the stagnant film boundary layer normal to the surface of the liquid-liquid interface (cm), δ = stagnant film boundary layer thickness (cm), and C_S = aqueous phase concentration that corresponds to a condition of thermodynamic equilibrium with the non-aqueous phase (g cm^{-3}).

A mass-transfer model can also be used to describe steady-state diffusion across a stagnant film (equation 2.23).

$$J = -K(C_s - C) \qquad (2.23)$$

where , K = mass-transfer coefficient (cm s^{-1}). As shown in equation 2.24, the diffusion and mass transfer models can be considered equivalent for liquid-liquid interface mass-transfer.

$$K = \frac{D}{\delta} \qquad (2.24)$$

Dissolution or interface mass transfer is often considered the rate-limiting step for the remediation of DNAPL contaminated sites. Chemical reaction can substantially increase (e.g., orders of magnitude) the rate of interface mass transfer by reducing the concentration of the reactant in the aqueous phase, thereby, increasing the concentration gradient and flux (Cussler 1997). Chemical reaction can take place in solution (homogeneous) or at the interface (heterogeneous). Because homogeneous chemical reaction occurs in both the bulk aqueous phase and the stagnant film boundary layer the concentration gradient across the boundary layer is no longer linear (See Figure 2-13). The mass-transfer coefficient for interface mass transfer with first-order homogeneous chemical reaction is shown in equation 2.25 (Cussler 1997).

$$K = (Dk_1)^{\frac{1}{2}} \coth \left[\left(\frac{k_1}{D} \right)^{\frac{1}{2}} \delta \right] \qquad (2.25)$$

where: k_1 = first- or pseudo first-order reaction rate constant (s^{-1}). Unfortunately, the film thickness (δ) is unknown. However, the correction to mass transfer caused by chemical reaction can be determined by combining equation 2.24 with equation 2.25 to yield equation 2.26.

$$\frac{K}{K_o} = \left[\frac{Dk_1}{(K^o)^2}\right]^{\frac{1}{2}} \coth\left\{\left[\frac{Dk_1}{(K^o)^2}\right]^{\frac{1}{2}}\right\} \qquad (2.26)$$

where, K^o = mass-transfer coefficient without chemical oxidation ($cm\ s^{-1}$). When chemical reaction is relatively slow (half-lives on the order of hours), equation 2.26 reduces to equation 2.24. When chemical reaction is relatively fast (half lives on the order of seconds), equation 2.26 reduces to equation 2.27 (Cussler, 1997).

$$K = (Dk_1)^{\frac{1}{2}} \qquad (2.27)$$

Based on the magnitude of the second-order rate constant for MnO_4^- oxidation of TCE, MnO_4^- would be predicted to have a significant effect on the rate of degradation of TCE_{DNAPL}, especially at higher oxidant concentrations. Assuming a TCE_{DNAPL} with a $K^o = -1.42\ \mu l\ day^{-1}$ (the interface mass transfer coefficient determined from control experiments with no oxidant present, Figure 2-15) and a pseudo first-order reaction rate constant equal to the $[MnO_4^-]_0$ times the second-order rate constant ($0.89 \pm 0.07\ mol^{-1}\ dm^3\ s^{-1}$), the correction to mass transfer caused by chemical reaction can be predicted. As shown in Figure 2-14, at low oxidant concentrations ($[MnO_4^-]_0 = 0.0016\ mol\ dm^{-3}$) MnO_4^- is only calculated to have a slight impact on the rate of DNAPL degradation. However, at relatively high concentration ($[MnO_4^-]_0 = 0.1583\ mol\ dm^{-3}$), chemical reaction with MnO_4^- should increase the rate of TCE_{NAPL} degradation by an order of magnitude.

The first controlled experimental work regarding DNAPL degradation by permanganate oxidation was completed at CSM by Urynowicz (Urynowicz 2000, Urynowicz and Siegrist 2000). A series of integrated experiments were deemed necessary to elucidate the chemical oxidation of DNAPLs in an groundwater system. Results from micro-reaction vessel and syringe reactor experiments demonstrated that the chemical oxidation of TCE with MnO_4^- is a second-order reaction ($0.88 \pm 0.07\ mol^{-1}\ dm^3\ s^{-1}$) and that an irreversible second-order kinetic model (e.g., eqn. 2.17) could effectively describe TCE concentrations with time. From a series of TCE_{DNAPL} degradation experiments, permanganate was shown to significantly increase the initial rate of degradation of TCE_{DNAPL} (Figure 2-15 and 2-16). As predicted, the initial rate of TCE_{DNAPL} degradation observed experimentally was increased at the high MnO_4^- concentrations as a result of chemical reaction. However, the rate decreased rapidly with time. It appears TCE_{DNAPL} degradation in the

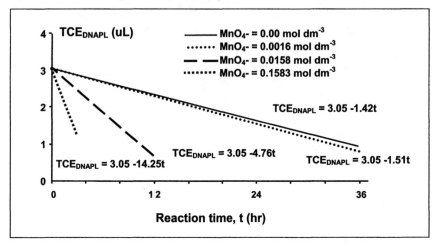

FIGURE 2-14. Calculated volume of TCE_{NAPL} versus time as affected by homogeneous chemical reaction (after Urynowicz 2000). $[TCE]_o = 3.05\mu L$; $[MnO_4^-]_o = 0, 0.0016, 0.0158, 0.1583$ mol dm^{-3}.

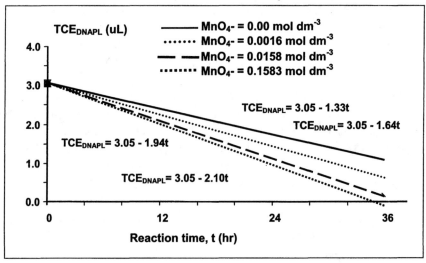

FIGURE 2-15. Experimentally observed volume of TCE_{NAPL} versus time during permanganate oxidation in an aqueous system under static conditions (after Urynowicz 2000). $[TCE]_o = 3.05 \pm 5.0\% \mu L$; $[MnO_4^-]_o = 0, 0.0016, 0.0158, 0.1583$ mol dm^{-3}; pH ~ 6.9; I = 0.05 M.

FIGURE 2-16. **Predicted versus experimentally measured DNAPL degradation (dissolution plus oxidation) (Urynowicz 2000). K = mass transfer coefficient with chemical reaction, K^o = mass transfer coefficient without chemical reaction, k_1 = pseudo first-order reaction rate constant, D = diffusion coefficient; $[TCE]_o = 3\mu L$; pH ~ 6.9; I = 0.05 M. Individual data points represent $[MnO_4^-]_o$ = 0.0016, 0.0158 and 0.1583 mol dm^{-3}.**

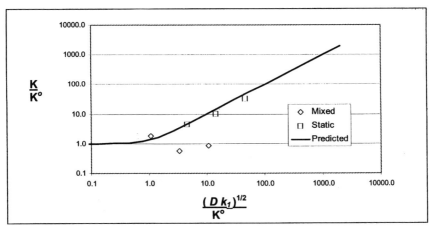

presence of MnO_4^- is subject to interfacial resistance caused by the local deposition of MnO_2 solids produced as a result of chemical oxidation. The interfacial resistance appeared to have increased slightly under mixing conditions as opposed to quiescent conditions (Figure 2-16). From flow-through TCE_{DNAPL} degradation experiments, the rate of TCE_{DNAPL} degradation was shown to decrease at the high MnO_4^- concentration and eventually cease altogether after a period of 12 hours as a result of interfacial resistance caused by an MnO_2 film. Aqueous-phase flow velocity was shown to have little or no effect on the rate of TCE degradation within the experimental range. Results from a series of interfacial MnO_2 deposition experiments revealed that the interfacial deposition of MnO_2 solids occurred very rapidly. The permanganate concentration had little effect on the rate of MnO_2 solids deposition. The MnO_2 film was shown to consist of plate-like structures.

The work of Urynowicz suggests that a predicted enhancement to DNAPL degradation due to homogeneous chemical reaction in the stagnant film may or may not occur depending, in large part, on the mass and distribution properties of the DNAPL. In low residual saturations where the DNAPL is present as films on media surfaces or dispersed ganglia, the oxidation-enhanced degradation may destroy the DNAPL before interfacial mass transfer resistance develops and slows the degradation process. On the other hand, if the DNAPL is present in pools or extensive connected ganglia, then development of mass transfer resistance could slow down or even preclude complete destruction of the DNAPL by permanganate. An alternative perspective on the latter situation is that the DNAPL would be 'stabilized' by a coating of MnO_2 solids and the rate of dissolution could be reduced such that the aqueous phase concentration might remain below a risk-based cleanup concentration.

Other applied experiments have explored the degradation of DNAPLs in soil and groundwater. These lab and field experiments were completed to determine if DNAPLs originally present in a soil and groundwater system could be degraded by permanganate oxidation. Drescher et al. (1999) performed batch and column experiments to evaluate the chemical oxidation of DNAPL residual in a homogeneous sandy soil. Complete destruction of the DNAPL mass was attained after treatment with several pore volumes of a 1000 mg/L potassium permanganate ($KMnO_4$) solution. McKay et al. (1998) performed a pilot study to evaluate the feasibility of in situ chemical oxidation with MnO_4^- for the treatment of TCE in low-permeability, vadose zone soils. The presence of a DNAPL residual was suspected based on elevated TCE concentrations. Increased chloride (Cl^-) and decreased TCE concentrations were observed following treatment, nevertheless elevated TCE concentrations remained. It was concluded that larger volumes or higher concentrations of MnO_4^- (applied as $KMnO_4$) were required for complete cleanup.

Schnarr et al. (1998) conducted in situ chemical oxidation field experiments at the Borden research site using $KMnO_4$ as the oxidant. The experiments were performed within a 7.5 m^3 cell isolated from the surrounding aquifer by sheet piles and artificially contaminated with TCE and tetrachloroethene (PCE). Results from the field test cell experiments and a companion laboratory column study revealed that the removal of TCE and PCE residual in porous soil could be accelerated with high concentrations of $KMnO_4$ solution ($\geq$ 10 g L^{-1}).

West et al. (1998a,b) completed a full-scale field application of in situ chemical oxidation using horizontal wells (two parallel wells 200 ft. each and 100 ft. apart at 30 ft. bgs) to deliver 2 to 4 wt.% $KMnO_4$ for in situ treatment of DNAPL contaminated groundwater. DNAPL residual was inferred from groundwater concentrations that were well above 10% of saturation (up to 820 mg/L). Over 30 days, TCE concentrations in the groundwater were reduced from 820 mg/L to ≤ 5 ug/L in 13 of 17 wells and an estimated 300 kg of TCE were destroyed within a 10^6 L region of groundwater.

Struse (1999) reported on a controlled laboratory study where intact cores of silty clay soil were contaminated with pure-phase TCE and then a 5000 mg/L solution of $KMnO_4$ was allowed to diffuse through the core for a period of a month or longer (Struse 1999, Struse and Siegrist 2000). Based on a post-treatment extraction of the entire core, near complete destruction of the TCE was observed.

2.4.   OXIDATION OF OTHER ORGANIC CONTAMINANTS

As described in the previous sections of this chapter, much is known about the permanganate oxidation of chlorinated ethenes. However, there are other organic COCs that have been studied and for which valuable information is available. This section describes information related to chlorinated alkanes, BTEX compounds, phenols, polyaromatic hydrocarbons, polychlorinated biphenyls, and high explosives.

Chlorinated Alkanes

Under conditions common for ISCO applications, permanganate is not a very effective oxidant for degradation of chlorinated alkanes such as 1,1,1-trichloroethane (Gates et al. 1995, Tratnyek et al. 1998, Gates-Anderson et al. 2001). Limited treatment of 1,1,1-TCA is expected, because chemical oxidation with $KMnO_4$ (as well as H_2O_2) appears to be favored in compounds with pi bonds or other electrophilically favorable chemical structures. A saturated aliphatic compound, such as 1,1,1-TCA, has no readily available electron pairs and is not as easy to chemically oxidize as unsaturated compounds such as TCE and PCE. Experiments have confirmed the low treatment efficiency achieved with permanganate oxidation of chlorinated alkanes. For example, Tratnyek et al. (1998) investigated the feasibility of treating mixtures of VOCs in site and laboratory water by sequential reduction and oxidation. Although sequential reduction and oxidation of VOCs proved problematic, the experiments demonstrated that MnO_4^- effectively oxidized chlorinated ethenes but had little effect on halogenated alkanes. Gates et al. (1995) reported similar findings for treatment of 1,1,1-TCA in soil slurries.

BTEX Compounds

Experimental evidence regarding the oxidation of BTEX compounds by permanganate during ISCO suggests that substituted aromatics could be degraded. This is supported by information regarding the potential viability of permanganate oxidation based upon research and experimentation in organic chemistry and characterization of NOM. Permanganate oxidation of alkylbenzenes has been shown to occur, yielding benzoic acids with intermediate phenones (Arndt 1981). The oxidation of toluene and substituted toluenes (e.g., NO_2, Cl,) also leads to benzoic acids along with a small amount of ring cleavage (Arndt 1981). However, benzene appears relatively stable in the presence of permanganate (Verschueren 1983).

Phenols

Potassium permanganate has been used for chemical oxidation of phenolic compounds during wastewater treatment. The stoichiometric equation for the oxidation of phenol by $KMnO_4$ is given by eqn. 2.28:

$$3C_6H_5OH + 28KMnO_4 + 5H_2O \rightarrow 18CO_2 + 28KOH + 28MnO_2 \quad (2.28)$$

According to Vella et al. (1990) the oxidation of phenol proceeds through organic acid intermediates (oxalic acid, tartaric acid, formic acid) before yielding carbon dioxide. Mineralization of phenol consumes a relatively large quantity of permanganate, 15.7 g of $KMnO_4$ per g of phenol (eqn. 2.28).

Vella et al. (1990) examined the $KMnO_4$ oxidation of phenolic wastewater in the laboratory and at two industrial wastewater treatment plants. The laboratory treatability studies demonstrated that $KMnO_4$ can oxidize phenol by 90% or more in 60 min (Figure 2-17). Based on field studies, $KMnO_4$ well in excess of stoichiometric requirements effectively reduced the phenol concentrations to <20 ug/L. Vella et al. (1990) did observe that the $KMnO_4$ resulted in a colored effluent due to MnO_2, but that it could be minimized through the use of coagulant and flocculant addition.

Polyaromatic Hydrocarbons

Polyaromatic hydrocarbons such as naphthalene, phenanthrene, and pyrene can be oxidized by permanganate. Cleavage of one of the aromatic rings can occur. For example, oxidative cleavage of naphthalenes in acidic solutions leads to phthalic acid. Ring cleavage can also occur with

FIGURE 2-17.   **Oxidation of phenol in wastewater versus permanganate dose (reaction period = 60 min at pH 8.5 and temperature of 22°C) (after Vella et al. 1990).**

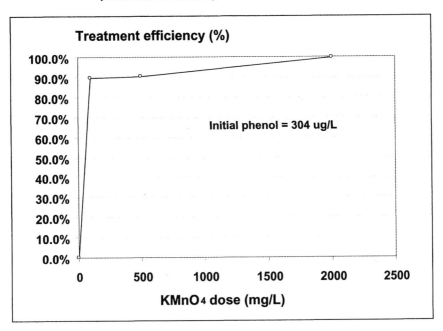

oxidation to quinone. Oxidation of phenanthrene in neutral solutions can lead to phenanthroquinone and biphenyl-2,2'-dicarboxylic acid as a byproduct (Arndt 1981). Equations for the mineralization of these compounds are shown in equations 2.29 to 2.31.

Naphthalene
$$16KMnO_4 + C_{10}H_8 + 16H^+ \rightarrow 16MnO_2 + 10CO_2 + 16K^+ + 12H_2O \qquad (2.29)$$

Phenanthrene
$$22KMnO_4 + C_{14}H_{10} + 22H^+ \rightarrow 22MnO_2 + 14CO_2 + 22K^+ + 16H_2O \qquad (2.30)$$

Pyrene
$$74KMnO_4 + 3C_{16}H_{10} + 74H^+ \rightarrow 74MnO_2 + 48CO_2 + 74K^+ + 52H_2O \qquad (2.31)$$

The pH of the system can affect the reaction as shown for the oxidation of naphthalene, which yields different products under acidic and basic conditions as shown in Figure 2-18.

FIGURE 2-18. Oxidation of naphthalene under acidic versus alkaline conditions (after Arndt 1981).

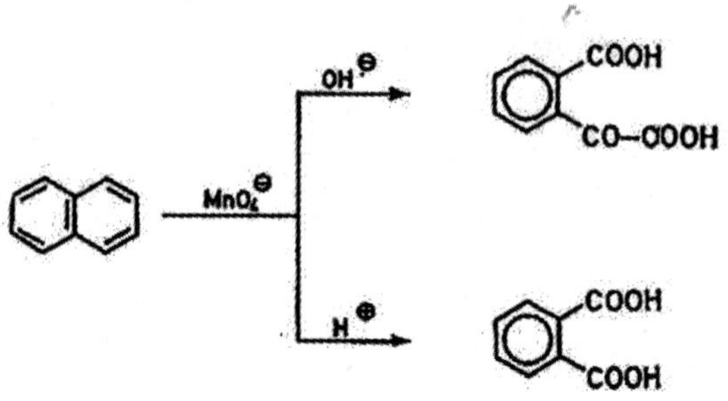

Experimental work has been completed with soil slurry reactors to compare the treatment effectiveness of KMnO₄ compared to H₂O₂ (alone or with FeSO₄ added and/or pH adjustment) (Gates et al. 1995). In these experiments, sand and silty clay soil were contaminated with a mixture of three PAHs (naphthalene, phenanthrene, and pyrene). The relative treatment effects of soil type, oxidant loading rate as well as the use of surfactant or iron amendments and pH adjustment were examined using batch experiments. The initial concentrations of PAHs were: naphthalene = 260-337 mg/kg, phenanthrene = 248-341 mg/kg, and pyrene = 226-331 mg/kg. The PAH distribution between phases prior to oxidation addition was estimated using a fugacity-based partitioning model (Dawson 1997), and it was estimated that 90 to 99 wt.% would be in the NAPL phase with only limited mass in the aqueous phase and sorbed to the soil solids. For both soil types, the KMnO₄ treatment efficiencies for naphthalene, phenanthrene, and pyrene were >90% at a 20 g/kg oxidant loading (Figure 2-19). In the clay soil, comparable treatment efficiencies of >95% were achieved with H₂O₂ amended with iron, but in the sand soil the treatment efficiencies were negligible. Treatment efficiencies with H₂O₂ only (i.e., without iron addition) were comparatively lower in the clay soil (30 to 75% range) and near zero in the sand soil. Of the three PAHs treated, naphthalene appears to be the most easily oxidized at low KMnO₄ loadings. This could be due to the fact that it has the fewest number of aromatic rings that need to be cleaved. As all three of the PAHs had a substantial percent of the starting mass present as NAPL, the high treatment efficiency suggests that there is potential for oxidation of the NAPL compounds.

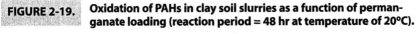

FIGURE 2-19.   Oxidation of PAHs in clay soil slurries as a function of permanganate loading (reaction period = 48 hr at temperature of 20°C).

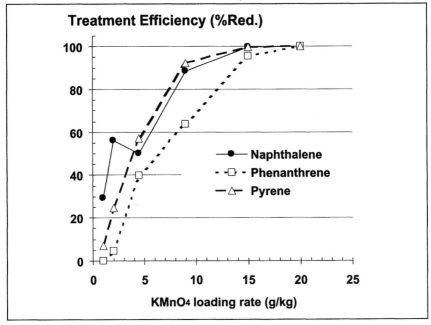

Polychlorinated Biphenyls

A limited study was conducted at Oak Ridge National Laboratory to investigate the ability of the potassium permanganate and Fenton's reagent to degrade polychlorinated biphenyls in artificially spiked soil and aqueous/organic matrices. The experiments were conducted on a sandy soil (91 wt.% sand, 9 wt.% silt; 1840 mg/kg TOC, pH = 8.7) and a 20% acetone and 80% water solution. The PCBs used for this study included 4-chlorobiphenyl (4-CB) and 2,5,2'-trichlorobiphenyl (2,5,2'-TCB). Liquid and soil samples were spiked with 2.5 mg of 4-CB and 1.2 mg of 2,5,2'-TCB, respectively. Oxidant applications for the soil samples were either 0.15 g (0.95 mmoles) or 0.3 g (1.9 mmoles) of $KMnO_4$, 0.255 g (4 mmoles) or 0.51 g (8 mmoles) of H_2O_2. Oxidant applications for the liquid samples were either 0.25 g of $KMnO_4$ or 0.425 g of H_2O_2. Specific oxidant solutions used were a 5% $KMnO_4$ solution, and 8.5% H_2O_2. The H_2O_2 applications were followed with a solution of $FeSO_4$ to initiate Fenton's reagent reactions. Note that the oxidant to PCB mole ratios were on the order of 1000:1. These experiments revealed that Fenton's reagent was effective in degrading 4-CB within an hour of oxidant application (Figure 2-20). On the other hand, $KMnO_4$ was not as effective. These results are supported with observations in the soil experiments (Figure 2-21 and 2-22). Minimal change in PCB concentrations was observed in the $KMnO_4$ treatments, while a significant reduction in congener concentration was observed within 5 minutes of H_2O_2 application.

High Explosives

High explosives (HE) have contaminated soil and water at numerous government installations. High explosives are toxic and mutagenic and classified as environmental hazards and as priority pollutants by the USEPA (Alnaizy and Akgerman 1999). More than 60 highly explosive compounds have been developed including those that are predominantly used for military purposes. These compounds include 2,4,6-trinitroltoluene (TNT), cyclotrimethylenetrinitroamine (cyclo-1,3,5-trimethylene-2,4,6-trinitramine or royal demolition explosives [RDX]), cyclotetramethylenetetranitramine (cyclo-1,3,5,7-tetramethylene-2,4,6,8-tetranitramine or high melting explosives [HMX]), and 2,4-dinitrotoluene (2,4-DNT).

FIGURE 2-20.   **Concentration of 4-chlorobiphenyl vs reaction time in a 20% acetone and 80% water solution containing either 5% KMnO₄ or 8.5% H₂O₂.**

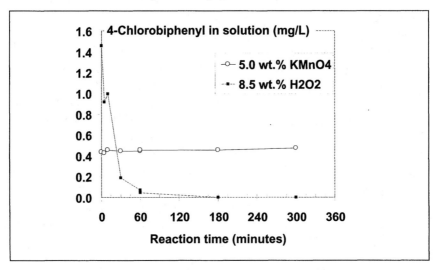

FIGURE 2-21.   Oxidation of 2,5,2′ TCB on sandy soil with 5 wt.% KMnO₄.

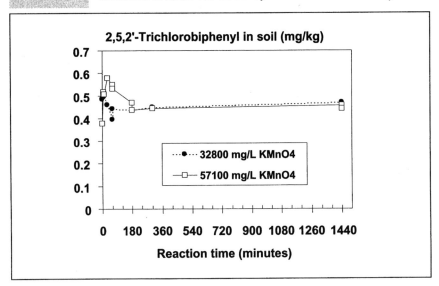

FIGURE 2-22.   Oxidation of 2,5,2′ TCB on sandy soil with 8.5 wt.% H₂O₂.

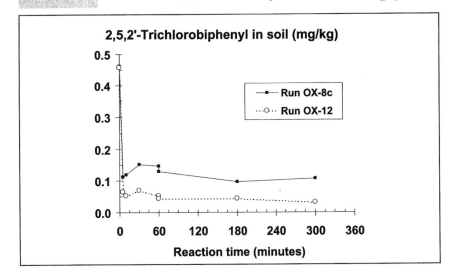

Oxidation of HE compounds has been demonstrated in the laboratory primarily using Fenton's reagent and other advanced oxidation processes (e.g. ultraviolet [UV], ozone [O_3], and various combinations) (Bier et al. 1999, Bose et al. 1998a,b, Mohanty 1993). Use of $KMnO_4$ for chemical oxidation of HEs is still emerging. Some of the first studies were recently completed by Clayton et al. at the IT Group (IT Corp. and SM Stollar Corp. 2000). Both laboratory and field scale testing have been conducted to determine the feasibility of HE oxidation using $KMnO_4$. As noted below, this testing indicated that $KMnO_4$ could effectively treat the HE compounds of interest (HMX, RDX, TNT, 2,6-DNT and 2,4-DNT) to below regulatory criteria. Laboratory testing included bench-scale batch tests using continuously mixed slurry systems at various $KMnO_4$ concentrations ranging from 400 to 20,000 mg/L $KMnO_4$. Monitoring of the tests was designed to determine HE degradation rate and extent, oxidant demand, by-product formation, and microtoxicity.

During laboratory testing, the extent of HE degradation was determined by measuring the concentration of the HE compounds of interest at various time points within each experimental set-up (Figure 2-23). The greatest mass degradation for each compound of interest occurred at higher permanganate concentrations. The TNT mass appeared to have completely degraded. The compounds of interest were successfully treated with $KMnO_4$ at virtually any concentration, with the $KMnO_4$ concentration dictating the degradation kinetics. Preliminary data were collected to evaluate potential transformation products of RDX treatment. Mono-, di-, and tri-nitrosoamines (MNX, DNX, TNX) were produced by the oxidation of RDX, however, short-lived and subsequently oxidized by the permanganate. Two unknown peaks were detected and were present in the background system. These unknowns either remained constant or decreased with time during treatment.

The reaction kinetics of HE-permanganate reactions were confirmed to be second order, with pseudo first-order RDX half-lives of 0.5 to 35 days (Figure 2-24). These fairly slow reaction kinetics may be favorable for subsurface oxidant delivery and transport. Permanganate consumption was shown to relate strongly to the permanganate concentration applied with lower concentrations of solution exerting lesser oxidant demand. Total permanganate consumption ranged from approximately 0.5 to 1 g $KMnO_4$ per kg of soil treated. While TOC comprised most of the oxidant demand, it could be minimized by decreasing the $KMnO_4$ concentration.

FIGURE 2-23.   RDX destruction due to KMnO₄ based on laboratory testing (IT Corporation and SM Stoller Corporation 2000).

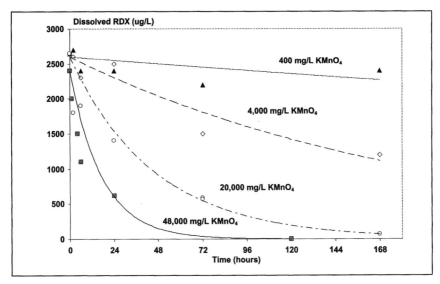

FIGURE 2-24.   RDX and HDX reaction rates based on laboratory testing (IT Corporation and SM Stoller Corporation 2000).

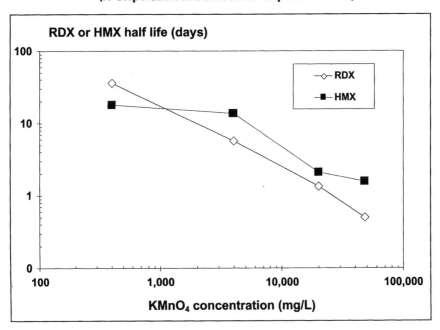

Microtox bioassay was performed by CSM to screen the changes in aqueous toxicity with the chemical oxidation of HE. The results from the $KMnO_4$ control showed that permanganate itself imposes a degree of biotoxicity that may be difficult to separate from other effects. The fact that the toxicity observed for the samples from the treatability study with $KMnO_4$ were equal to or lower than those associated with the reference samples with $KMnO_4$ and clean water suggest that treatment of the HE did not increase the biotoxicity apart form the effect of the oxidant itself. Within the active treatment zone, there may be biotoxicity, but in a typical groundwater system, the oxidant will be depleted and bio-adverse conditions will dissipate with time.

Based on the favorable results of the laboratory testing, a field scale treatability test was conducted (IT Corp and SM Stollar Corp, 2000). Field testing included a series of single-well push-pull tests to obtain proof of concept in the field and assess treatment design parameters. The test was restricted to a 3 day treatment time. Significant degradation of all the HE compounds of interest (RDX, HMX, TNT, and DNT) was observed with no evidence of transformation products. The RDX half-life was determined to be ~4 days at ~7,000 mg $KMnO_4$/L of groundwater, which is slightly slower than measured in the laboratory tests. This is an expected result since subsurface mass transfer and diffusion of $KMnO_4$ and the contaminants of interest typically limit the overall treatment rate. While permanganate consumption rates were not determined, permanganate persisted in the test wells for several weeks after the test.

2.5. SUMMARY

Permanganate has a long history of use as a chemical oxidant in organic chemistry and the study of NOM structure and composition. It has also been widely used for treatment of water and wastewaters. During the 1990s it has emerged as an oxidant for in situ remediation of contaminated sites. As of this writing, reaction stoichiometry, pathways, and reaction kinetics are well understood for MnO_4^- (either as $KMnO_4$ or $NaMnO_4$) oxidation of chlorinated ethenes (e.g., TCE, PCE) dissolved in aqueous environments. The effects of key environmental conditions such as temperature, pH and ionic composition are also established. Oxidative degradation of chlorinated ethenes in porous media in the sorbed or DNAPL phases is feasible, though the rate and extent of reaction for a given MnO_4^- dose is highly dependent on contaminant properties and matrix interactions.

2.6. REFERENCES

Almendros, G., F.J. Gonzalez-Vila, and F. Martin (1989). Room temperature alkaline permanganate oxidation of representative humic acids. *Soil Biology & Biochemistry*, 21:481-486.

Alnaizy, R. and A. Akgerman (1999). Oxidative treatment of high explosives contaminated wastewater. *Water Resources*, 33(9):2021-2030.

Arndt, D. (1981). Manganese Compounds as Oxidizing Agents in Organic Chemistry. Open Court Pub. Co., LaSalle, IL.

Bier, E.L., J. Singh, Z. Li, S.D Comfort, and P.J, Shea (1999). Remediating hexa-hydro-1,3,5-trinitro-1,2,5-trazine-contaminated water and soil by Fenton oxidation. *Environmental Toxicology and Chemistry*, 18(6):1078-1084.

Bose, P., W. H. Glaze, and D. S. Maddox (1998a). Degradation of RDX by various advanced oxidation processes: i. reaction rates. *Water Resources*, 32(4):997-1004.

Bose, P., W.H. Glaze, and D.S. Maddox (1998b). Degradation of RDX by various advanced oxidation processes: ii. organic by-products. *Water Resources*, 32(4):1005-1018.

Case, T.L. (1997). Reactive permanganate grouts for horizontal permeable barriers and in situ treatment of groundwater. M.S. Thesis, Colorado School of Mines, Golden, CO.

Cherry, A.K. (1962). Use of potassium permanganate in water treatment. *J. Am. Water Works Assoc.*, 54:417-424.

Cline, S.R., J.M. Giaquinto, M.K. McCracken, D.L. Denton, R.C. Starr (1999). Laboratory evaluation of in situ chemical oxidation for groundwater remediation, Test Area North, Operable Unit 1-07B, Idaho National Engineering and Environmental Laboratory. Oak Ridge National Laboratory Report, ORNL/TM-13711/V1, Oak Ridge, TN.

Colthurst, J.M. and P.C. Singer. (1982). Removing trihalomethane precursors by permanganate oxidation and manganese dioxide precipitation. *J. Amer. Water Works Assoc.*, February, pp. 78-83.

Cussler, E.L. (1997). Diffusion. Cambridge University Press.

Dawson, H.E. (1997). Screening Level tools for modeling fate and transport of NAPLs and trace organic chemicals in soil and groundwater: SOILMOD, TRANS1D, NAPLMOB. Colorado School of Mines, Office of Special Programs and Continuing Education. Golden, CO.

deZabala, E.F., and C.J. Radke (1986). A non-equilibrium description of alkaline water flooding. *Society of Petroleum Engineering Journal*, 26(1):29-43.

Drescher, E.A., R. Gavaskar, B.M. Sass, L.J. Cumming, M.J. Drescher, and T. Williamson (1999). Batch and column testing to evaluate chemical oxidation of DNAPL source areas. Proc. Intern. Conf. on Remediation of Chlorinated and Recalcitrant Compounds. Monterey, Calif. pp. 425–432.

Fatiadi, A.J. (1987). The classical permanganate ion: still a novel oxidant in organic chemistry. *Journal of Synthetic Organic Chemistry*, (2):85-206.

Freeman, F. (1976). Postulated intermediates and activated complexes in the permanganate ion oxidation of organic compounds. *Reviews on Reactive Species in Chemical Reactions*, 1:179-226.

Gates, D.D., R.L. Siegrist, R.L., and S.R. Cline (1995). Chemical oxidation of contaminants in clay or sandy soil. Proceedings of ASCE National Conference on Environmental Engineering. Am. Soc. of Civil Eng., Pittsburgh, PA.

Gates-Anderson, D.D., R.L. Siegrist and S.R. Cline (2001). Comparison of potassium permanganate and hydrogen peroxide as chemical oxidants for organically contaminated soils. *J. Environmental Engineering*, 127(4):337-347.

Griffith S.M., and M. Schnitzer (1975). Oxidative degradation of humic and fulvic acids extracted from tropical volcanic soils. *Canadian Journal of Soil Science*, 55:251-267.

Hanninen, K. (1992). Cupric oxide oxidation of peat and coal humic acids. *The Science of the Total Environment*, 117/118:75-82.

Hood, E.D., N.R. Thomson, D. Grossi, and G.J. Farquhar (1999). Experimental determination of the kinetic rate law for the oxidation of perchloroethene by potassium permanganate. *Chemosphere*.

Huang, K., G.E., Hoag, P. Chheda, B.A. Woody, and G.M. Dobbs (1999). Kinetic study of oxidation of trichloroethene by potassium permanganate. *Environmental Engineering Science*, 16(4):265-274.

IT Corporation and SM Stollar Corporation (2000). Implementation report of remediation technology screening and treatability testing of possible remediation technologies for the Pantex perched aquifer. October, 2000. DOE Pantex Plant, Amarillo, Texas.

Khan, S.U., and M. Schnitzer (1971). Further investigation on the chemistry of fulvic acid, a soil humic fraction. *Canadian Journal of Chemistry*, 49:2302-2309.

Khan, S.U. and M. Schnitzer (1972). Permanganate oxidation of humic acids, fulvic acids, and humins extracted from ah horizons of a black chernozem, a black solod, and a black solonetz Soil. *Canadian Journal of Soil Science*, 52:43-51.

Khan, S.U. and M. Schnitzer (ed.) (1978). Soil Organic Matter. Elsevier Scientific Publishing Company, New York pp. 27-58.

Ladbury, J.W., and C.F. Cullis (1958). Kinetics and mechanism of oxidation by permanganate. *Chem Rev,* 58, p. 403.

Lee, D.G. (1980). The Oxidation of Organic Compounds by Permanganate Ion and Hexavalent Chromium. Open Court Publishing Company, LaSalle, Illinois.

Logan, S.R. (1996). Fundamentals of Chemical Kinetics. Longman Group Limited, Essex, England.

Lowe, K.S., F.G. Gardner, R.L. Siegrist, and T.C. Houk (1999). Field pilot test of in situ chemical oxidation through recirculation using vertical wells at the Portsmouth Gaseous Diffusion Plant. EPA/625/R-99/012. U.S. EPA ORD, Washington, DC. pp. 42-49.

McKay, D., A. Hewitt, S. Reitsma, J. LaChance, and R. Baker (1998). In Situ oxidation of trichloroethene using potassium permanganate: Part 2. pilot study. Proc. Intern. Conf. on Remediation of Chlorinated and Recalcitrant Compounds. Monterey, Calif. pp. 377-382.

Makino T, et al. (1998). Determination of optimal chromium oxidation conditions and evaluation of soil oxidative activity in soils. *Journal of Geochemical Exploration,* 64:435-441.

Miller, C.T., M.M. Poirier-McNeill and A.S. Mayer (1990). Dissolution of trapped nonaqueous phase liquids: mass transfer characteristics. *Water Resources Research,* 26(11):2783-2796.

Mohanty, N.R. and I.W. Wei (1993). Oxidation of 2,4-Dinitrotoluene using Fenton's reagent: reaction mechanisms and their practical applications. *Hazardous Waste & Hazardous Materials,* 10(2):171-183.

Nriagu, J.E. (ed.) (1976). Environmental Biogeochemistry, Volume 1, Carbon, Nitrogen, Phosphorus, Sulfur, and Selenium Cycles. Ann Arbor Science Publishers Inc., Ann Arbor, Michigan.

Ortiz De Serra M. and M. Schnitzer (1972). Extraction of humic acid by alkali and chelating resin. *Canadian Journal of Soil Science,* 52:365-374.

Ortiz De Serra M.I. and M. Schnitzer (1973). The chemistry of humic and fulvic acids extracted from Argentine soils-permanganate oxidation of methylated humic and fulvic acids. *Soil Biology and Biochemistry,* 5:287-296.

Pfannkuch, H.O. (1984). Determination of the contaminant source strength from mass exchange processes at the petroleum groundwater interface in shallow aquifer systems. Proceedings of the NWWA Conference on Petroleum Hydrocarbons and Organic Chemicals in Groundwater, National Well Association, Dublin, Ohio. pp. 111-129.

Pohlman A.A. and J.G. McColl (1989). Organic oxidation and manganese and aluminum mobilization in forest soils. *Soil Science Society of America Journal*, 53:686-690.

Schnarr, M.J. and G.J. Farquhar (1992). An in situ oxidation technique to destroy residual DNAPL from soil. Subsurface Restoration Conference, Third International Conference on Ground Water Quality, Dallas, Texas, Jun 21-24, 1992.

Schnarr, M., C. Truax, G. Farquhar, E. Hood, T. Gonully, and B. Stickney (1998). Laboratory and controlled field experimentation using potassium permanganate to remediate trichloroethene and perchloroethene DNAPLs in porous media. *Journal of Contaminant Hydrology*, 29:205-224.

Sherwood, T.K., R.L. Pigford, and C.R. Wilke (1975). "Mass Transfer." McGraw-Hill, New York.

Siegrist, R.L., K.S. Lowe, L.C. Murdoch, T.L. Case, and D.A. Pickering (1999). In situ oxidation by fracture emplaced reactive solids. *J. Environmental Engineering*, 125(5):429-440.

Siegrist, R.L., M.A. Urynowicz, and O.R. West (2000). An overview of in situ chemical oxidation technology features and applications. EPA/625/R-99/012. U.S. EPA ORD, Washington, DC. pp. 61-69.

Singer, P.C., J.H. Borchardt, and J.M. Colthurst (1980). The effects of permanganate pretreatment on trihalomethane formation in drinking water. *J. Am. Water Works Assoc.*, 72(10):573.

Spicher, R.G., and R.T. Skrinde (1963). Potassium permanganate oxidation of organic contaminants in water supplies. *Jour. Amer. Water Works Assoc.*, Sept, pp. 1174-94.

Steel, E.W., and T.J. McGhee (1979). Water Supply and Sewerage, Fifth Edition. McGraw-Hill, New York, NY.

Stewart, R. (1965). In: K. Wiberg (Editor). Oxidation in Organic Chemistry: Oxidation by Permanganate. Academic Press, New York, NY.

Struse, A.M. (1999). Mass transport of potassium permanganate in low permeability media and matrix interactions. M.S. Thesis, Colorado School of Mines, Golden, CO.

Struse, A.M. and R.L. Siegrist (2000). Permanganate transport and matrix interactions in silty clay soils. In: Wickramanayake, G.B., A.R. Gavaskar, and A.S.C. Chen (ed.). Chemical Oxidation and Reactive Barriers. Battelle Press, Columbus, OH. pp. 67-74.

Tratnyek, P.G., T.L. Johnson, S.D. Warner, H.S. Clarke, and J.A. Baker (1998). In situ treatment of organics by sequential reduction and oxidation. Proc. First Intern. Conf. on Remediation of Chlorinated and Recalcitrant Compounds. Monterey, Calif. pp. 371–376.

Urynowicz, M.A. (2000). Reaction kinetics and mass transfer during in situ oxidation of dissolved and DNAPL trichloroethene with permanganate. Ph.D. dissertation, Environmental Science & Engineering Division, Colorado School of Mines. May.

Urynowicz, M.A. and R.L. Siegrist (2000). Chemical degradation of TCE DNAPL by permanganate. In: Wickramanayake, G.B., A.R. Gavaskar, and A.S.C. Chen (ed.). Chemical Oxidation and Reactive Barriers. Battelle Press, Columbus, OH. pp. 75-82.

Vella, P.A., G. Deshinsky, J.E. Boll, J. Munder, and W.M. Joyce (1990). Treatment of low level phenols with potassium permanganate. *Res. Jour. Water Pollution Cont. Fed.*, 62(7):907-14.

Vella, P.A. and B. Veronda (1994). Oxidation of Trichloroethene: A comparison of potassium permanganate and Fenton's reagent. 3rd Intern. Symposium on Chemical Oxidation. In: In Situ Chemical Oxidation for the Nineties. Vol. 3. Technomic Publishing Co., Inc. Lancaster, PA. pp. 62-73.

Verschueren, K. (1983). Handbook of Environmental Data on Organic Chemicals. Van Nostrand Reinhold Company, New York. p.239, p.1105, p.1194.

Wagner, G. and J. Russ (1895). Phys.-Chem. Soc., 27, 219.

Weber, W.J. (1972). Physicochemical Processes for Water Quality Control, Wiley-Interscience, A Division of John Wiley & Sons, Inc., New York.

West, O.R., S.R. Cline, W.L. Holden, F.G. Gardner, B.M. Schlosser, J.E. Thate, D.A. Pickering, and T.C. Houk (1998a). A full-scale field demonstration of in situ chemical oxidation through recirculation at the X-701B site. Oak Ridge National Laboratory Report, ORNL/TM-13556.

West, O.R., Cline, S.R., Siegrist, R.L., Houk, T.C., Holden, W.L., Gardner, F.G. and Schlosser, R.M. (1998b). A field-scale test of in situ chemical oxidation through recirculation. Proc. Spectrum '98 International Conference on Nuclear and Hazardous Waste Management. Denver, Colorado, Sept. 13-18, pp. 1051-1057.

Wiberg, K.B., K.A. Saegebarth (1957). The mechanisms of permanganate oxidation: IV. hydroxylation of olefins and related reactions. *J. Am. Chem. Soc.* 79:2822-2824.

Yan, Y. and F.W. Schwartz (1996). In situ oxidative dechlorination of trichloroethene by potassium permanganate. Proc. Third International Conference on AOTs. October 26-29, Cincinnati, Ohio.

Yan, Y.E., and F.W. Schwartz (1998). Oxidation of chlorinated solvents by permanganate. Proc. Intern. Conf. on Remediation of Chlorinated and Recalcitrant Compounds, Battelle Press, Columbus, Ohio. pp. 403-408.

Yan, Y.E. and F.W. Schwartz (1999). Oxidative degradation and kinetics of chlorinated ethylenes by potassium permanganate. *Journal of Contaminant Hydrology*, 37:343-365.

CHAPTER 3

Permanganate Effects on Metals in the Subsurface

Metals can exist naturally in the subsurface or be present as anthropogenic contaminants. In addition, metals can be added to the subsurface during in situ chemical oxidation (ISCO) as impurities in permanganates (see Table 2-3). The metals of interest and potential concern include lead, chromium, arsenic, zinc, cadmium, copper, and mercury as these are found as primary or co-contaminants at contaminated sites (Figure 3-1). ISCO using permanganate can impact metal behavior in the subsurface through several mechanisms including: the alteration of system pH and Eh, changes in ionic composition, removal or the alteration of natural organic matter, and production of MnO_2 solids. It is important to understand the effects of permanganate oxidation on metals so that successful reduction in risk due to destruction of an organic contaminant does not lead to an increased risk from a metal co-contaminant (Figure 3-2). This Chapter describes the potential impact of permanganate oxidation on subsurface metal mobility, including governing principles and the results of laboratory and field studies.

3.1. METALS AND THEIR CHEMICAL BEHAVIOR IN THE SUBSURFACE

Metal Behavior

The behavior of metals in the soil and groundwater environment depends on the form of the metal, which in turn is affected by soil and groundwater chemistry. Reactions that influence the mobility of metals

FIGURE 3-1. Metal contaminants most commonly found at Superfund Sites.

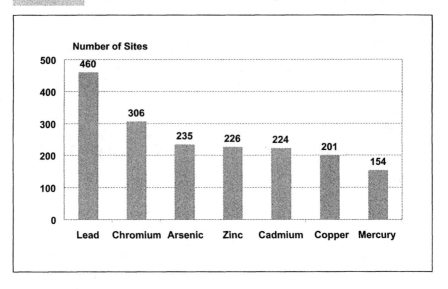

FIGURE 3-2. Illustration of potential oxidation-induced changes in subsurface geochemistry and mobility.

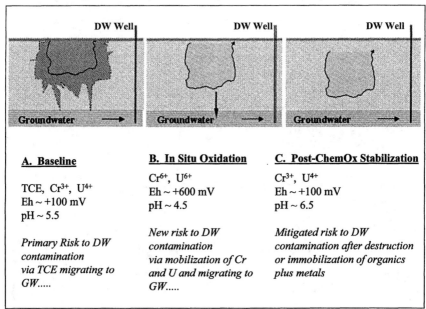

A. Baseline

TCE, Cr^{3+}, U^{4+}
Eh ~ +100 mV
pH ~ 5.5

Primary Risk to DW contamination via TCE migrating to GW.....

B. In Situ Oxidation

Cr^{6+}, U^{6+}
Eh ~ +600 mV
pH ~ 4.5

New risk to DW contamination via mobilization of Cr and U and migrating to GW.....

C. Post-ChemOx Stabilization

Cr^{3+}, U^{4+}
Eh ~ +100 mV
pH ~ 6.5

Mitigated risk to DW contamination after destruction or immobilization of organics plus metals include acid/base, precipitation/dissolution, oxidation/reduction, sorption, or ion exchange. Those reactions depend, to varying degrees, on factors such as pH, Eh, and solution composition and chemistry. For example, Figure 3-3 illustrates the influence of pH on the sorption of cations (metals) and anions onto iron oxide. Table 3-1 lists the common metals found at contaminated sites and highlights their chemical forms and general factors affecting their mobility and fate.

Soil pH values are typically between 4.0 and 8.5. Under low pH conditions, buffering is provided by aluminum, and under high pH conditions buffering is due to calcium carbonate. Inorganic ions, such as carbonate, phosphate, and sulfide, in the subsurface can influence the ability of soil to fix metals. These anions can form insoluble complexes with the metal ions, causing them to desorb from the soil or to precipitate. At low pH

FIGURE 3-3. **Metal adsorption to hydrous iron oxide gels (Evanko and Dzombak 1997).**

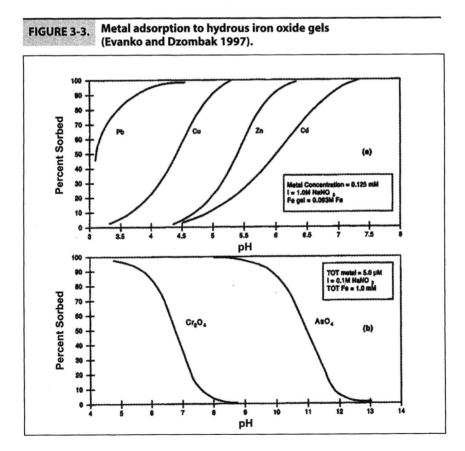

TABLE 3-1.   Metal fate and mobility at contaminated sites (after Evanko and Dzombak 1997).

Metal	Natural forms	Waste forms	Effects of pH and other environmental factors	Environmental behavior
Lead	Pb^0 PbO	Pb^0 PbO $Pb(OH)_2$	At pH > 6, lead forms lead carbonate In solutions with high sulfide concentrations, lead precipitates as PbS, which is converted to lead hydroxide, carbonate, or sulfate when sulfur is oxidized Low solubility compounds are formed by complexation with inorganic and organic ligands	Most lead released to environment is retained in soil where adsorption, ion exchange, precipitation, and complexation with organic matter limit transport In surface and ground waters, a significant fraction of lead exists as undissolved precipitates ($PbCO_3$, Pb_2O, $Pb(OH)_2$, $PbSO_4$)
Chromium	Occurs only as compounds (e.g., $FeCr_2O_4$)	Cr(VI) Cr(III)	Cr(III) is the dominant form at pH < 4 The leachability of Cr(VI) increases as soil pH increases Cr(VI) → Cr(III) through reduction by soil organic matter, S^{2-}, and Fe^{2+} under anaerobic conditions Chromate and dichromate are common species that precipitate in presence of metal cations, and can also sorb onto soil surfaces (metal oxides)	Usually particle-associated and will ultimately be deposited Cr(VI) is the most common form found at waste sites, and is more toxic and mobile than Cr(III) Cr(III), dominant at low pH, forms complexes with NH_3, Cl^-, OH^-, F^-, CN^-, SO_4^{2-}, and soluble organic ligands Cr(III) mobility is decreased by adsorption to clays and oxide minerals below pH 5, and has low solubility above pH 5 due to formation of $Cr(OH)_3(s)$

continued

TABLE 3-1. (continued)

Metal	Natural forms	Waste forms	Effects of pH and other environmental factors	Environmental behavior
Arsenic	As_2O_3	$As(V)$, AsO_4^{3-} $As(III)$ AsO_3^{3-} $(CH_3)AsO_2H_2$ $(CH_3)_2AsO_2H$	Arsenic mobility increases with increasing pH. $As(V)$ is dominant in aerobic environments. $As(V)$ coprecipitates with, or adsorbs onto iron oxyhydroxides under acidic and moderately reducing conditions. Arsenate (AsO_4^{3-}) precipitates when metal cations are present. $As(III)$ dominates, as arsenite (AsO_3^{3-}), under reducing conditions. Arsenite can adsorb or co-precipitate with metal sulfides and has a high affinity for other sulfur compounds	As compounds sorb strongly to soils and are transported only short distances in surface and ground water. Sorption and coprecipitation with iron oxides are the most important removal mechanisms under most environmental conditions. Becomes mobilized under reducing, alkaline, and saline conditions; in the presence of ions that compete for sorption sites; and in the presence of organic compounds that form complexes with As. $As(V)$ coprecipitates are immobile under acidic and moderately reducing conditions

continued

TABLE 3-1. (continued)

Metal	Natural forms	Waste forms	Effects of pH and other environmental factors	Environmental behavior
Zinc	Extracted from mineral ores to form ZnO	$Zn(II)$ $Zn(OH)_2(s)$ $ZnCO_3(s)$ $ZnS(s)$ $Zn(CN)_2(s)$	At higher pH values, forms carbonate and hydroxide complexes that control solubility Precipitates under reducing conditions, and may coprecipitate with hydrous oxides of Fe or Mn Sorption of Zn increases as pH increases or salinity decreases	Present as soluble, mobile compounds at neutral and acidic pH values Sorption to sediments, suspended solids (hydrous Fe or Mn oxides), clay minerals, and organic matter is the primary fate
Cadmium	CdS CdCO₃	Cd^{2+} $Cd(OH)_2$ Cd-Cn compl.	At high pH, $Cd(OH)_2$ and $CdCO_3$ dominate At pH < 8, Cd^{2+} and $CdSO_4$ dominate If reducing with sulfur, CdS(s) is formed Possible precipitation with phosphate and other anions	Mobile in water as hydrated ions and at lower pH may form complexes with chloride and sulfate Removed from water via sorption onto minerals at pH>6 Higher removal at higher pH and with higher CEC media Removal as CdS under reducing conditions

continued

TABLE 3-1. (continued)

Metal	Natural forms	Waste forms	Effects of pH and other environmental factors	Environmental behavior
Copper	Copper sulfide and oxide ores	Cu^{2+} $CuCO_3$ $CuOH^+$ $Cu(OH)_2$	$CuCO_3$ is the dominant soluble species in aerobic, alkaline systems The affinity of Cu for humates increases as pH increases and ionic strength decreases In anaerobic environments with sulfur present, CuS(s) forms	Forms strong solution complexes with humic acids Mobility decreases by sorption to mineral surfaces over a wide range of pH values
Mercury	Sulfide ore— cinnabar	Hg^{2+} Hg_2^{2+} Hg^0 Methyl/ethyl mercury	Acidic conditions (pH < 4) favor formation of methyl mercury; higher pH values favor precipitation of HgS(s) Sorption increases as pH increases Hg_2^{2+} and Hg^{2+} dominant forms under oxidizing conditions With mildly reducing conditions, organic or inorganic mercury may be reduced to elemental mercury, which may then be converted to alkyllated forms	Sorption to soils, sediments, and humic materials is an important removal mechanism Can be removed from solution by coprecipitation with sulfides Hg(II) forms strong complexes with a variety of inorganic and organic ligands, making it very soluble in oxidized aquatic systems

values, metal cations are most mobile, while metal anions would tend to sorb to oxide minerals. At higher pH values, cations precipitate or sorb to mineral surfaces, while the anions are mobilized. As such, the presence of metal oxides of iron, aluminum, and manganese can strongly influence metal concentrations because, even when initially sorbed to the oxides under certain subsurface conditions, ion exchange, specific adsorption, and surface precipitation are possible with any change in conditions. Sorption of metal cations onto metal oxides increases sharply with increasing pH, while sorption of metal anions is greatest at low pH and will decrease with increasing pH. The cation exchange capacity (CEC) of the soil dictates the ability of soils to uptake metal cations. Natural organic matter (NOM) will also influence the sorption of metals to mineral surfaces. It can complex metals and impact their concentrations in solution. The particle size distribution of the soil is another influencing factor. Smaller particles have a greater surface area than coarser materials, thus are able to sorb significantly greater amounts of metal.

ISCO Effects on Metals

ISCO by use of permanganate can affect sorption and ion exchange in the subsurface to varying extents depending on the oxidant dose and the subsurface conditions prior to application. Altered mobility of metal contaminants as well as naturally occurring metals can potentially occur during ISCO with permanganate. For example, in situ permanganate reactions can yield low pH (e.g., pH 3) and high Eh conditions (e.g., +800 mV) depending on the oxidant dose and pre-treatment environmental conditions (e.g., buffering capacity and reductants present). As illustrated in Figure 3-2, ISCO could successfully reduce the mass of a target organic like TCE such that the risk of groundwater contamination could be lessened. However, a metal that was present and sorbed to the soils under ambient conditions (e.g., reducing conditions and neutral to alkaline pH), could conceivably be mobilized during permanganate oxidation due to ISCO-induced changes in NOM content (decreased), pH (decreased), Eh (increased), and ionic composition (increased). It is important to note that the oxidizing conditions presented with ISCO treatment can have different influences on different metals. However, the production of MnO_2 solids as a byproduct of the permanganate oxidation could yield a strong sorbent that could be retained in the subsurface as coatings on mineral surfaces or as particles, and the sorption capacity of the MnO_2 could dominate and offset the above changes. Mobilization effects may be attenuated additionally by natural processes once ISCO is complete (e.g., return to naturally present reducing conditions).

3.2 MANGANESE OXIDE GENESIS DURING PERMANGANATE OXIDATION

Manganese Oxide Production by ISCO

In situ treatment involving chemical oxidants can produce particles by shearing off fragments of natural soil or by yielding reaction products (e.g. iron or manganese oxides). Unique to permanganate compared to other oxidants such as Fenton's reagent or ozone, is the production of MnO_2 solids as a reaction product (e.g., see eqn. 2.9).

$$2MnO_4^- + C_2HCl_3 \rightarrow 2MnO_2(s) + 2CO_2 + 3Cl^- + H^+ \qquad (2.9)$$

Not surprisingly, observations made during a wide range of experiments with soil and groundwater media, low to high concentrations of oxidizable organics (chlorocarbons or NOM), and low to high permanganate doses, have revealed that particles are inevitably generated. Figures 3-4 and 3-5 are photographs of manganese oxide particles produced during two field applications of ISCO with permanganate. The manganese oxides were observed to be micron-sized particles (West et al. 1998a,b, Siegrist et al. 2000a,b).

It is known that manganese oxides are produced when MnO_4^- reacts with reductants, whether they are organic contaminants, like chlorinated solvents (e.g., eqn. 2.9), NOM, or reduced metal species. However, permanganate also can react with water to form manganese oxides and oxygen (Rees 1987):

$$4MnO_4^- (aq) + 2H_2O \rightarrow 3O_2(g) + 4MnO_2(s) + 4OH^- \qquad (3.1)$$

Pure potassium permanganate reacts very slowly with water, but the reaction shown in eqn. 3.1 is catalyzed by solid MnO_2 in the system. This can result in an apparent decomposition of permanganate and the production of MnO_2 even when reductants such as target organic chemicals or NOM are absent.

Recent and ongoing research has provided insight into the mass of particles generated by ISCO with permanganate and the factors affecting their behavior. At CSM, a duplicated, 2^5 factorial design was recently

FIGURE 3-4.   **Scanning electron micrograph (SEM) of MnO$_2$ solids in groundwater collected from a horizontal extraction well used in a paired well delivery of 2 to 4 wt.% KMnO$_4$ to treat DNAPL contaminated groundwater (West et al. 1998a).**

FIGURE 3-5.   **Photomicrograph of solids retained on a 1-um cartridge filter in the above-ground oxidant amendment and reinjection piping after passage through a 5-um filter (mag. = 5000x) (Siegrist et al. 2000a).**

completed to examine the mass of particles produced as affected by oxidant type, oxidant concentration, TCE concentration, reaction time, and
presence of silt/clay particles (Siegrist et al. 2000b). The results of this
experiment revealed that the quantity of the particles produced during
chemical oxidation was impacted by several conditions (Figure 3-6).
Higher TCE concentrations increased the mass of filterable particles produced by 150 to 200% regardless of whether there were any added
silt/clay particles in the groundwater. However, increasing the silt/clay
particle content in the groundwater dramatically increases (300 to 700%)
the mass of particles produced when there is TCE and oxidant present.
The oxidant form (either Na vs. K) and oxidant concentration appeared to
have only a minimal effect on particle formation. Increasing the reaction
time from 15 min to 300 min had no marked effect on the production of
filterable particles. Since the experiments were carried out in 10 mL of
groundwater, the net mass (mg) of particles produced multiplied by 100
yields the mg/L concentration of particles in the groundwater. Under
some conditions, such as higher TCE, MnO_4^-, and silt/clay, there can be
a net increase of 300 to 400 mg/L of >0.45 um particles in groundwater.

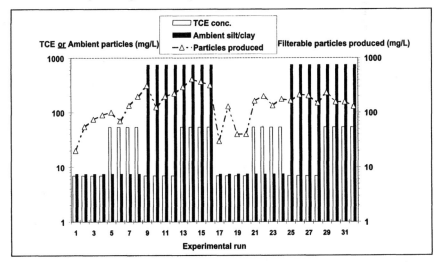

FIGURE 3-6. Filterable solids (>0.45 um) produced (mg/L) as affected
by process conditions during chemical oxidation with
permanganate (Siegrist et al. 2000a).

A series of multiple regression analyses were performed to determine if the filterable particles produced and the oxidant consumption could be described as a function of the conditions examined (Siegrist et al. 2000b). The general form of the relationship is shown in eqn. 3.2.

$$Y = \beta_o + \beta_1 X_1 + \beta_2 X_2 + \beta_3 X_3 \qquad (3.2)$$

where Y = particles generated (or oxidant consumed) (mg/L), X_1 = oxidant dose (mg/L), X_2 = TCE concentration (mg/L), X_3 = ambient particles in the groundwater matrix (mg/L), and β_o, β_1, β_2, and β_3 are fitting parameters. After exploring different forms of equation 3.2, the regression equation that made physical sense and accounted for the most variance in the particle generation data (for both oxidant forms plus both reaction times combined) was that shown in eqn. 3.3 ($R^2 = 0.85$).

Particles = 0.0089 (MnO_4^- dose)+ 2.14 (TCE conc.)+ 0.197 (ambient filterable particles)   (3.3)

where all concentrations are in mg/L. The constant, 2.14, compares with the stoichiometric data of eqn. 2.9 which indicates that for TCE oxidation 1.32 g of MnO_2 are produced when 1.81 g of MnO_4^- reacts with 1 g of C_2HCl_3 (per eqn. 2.9). This relationship suggests that the presence of ambient filterable particles in the groundwater containing TCE and to which permanganate is added results in an increased net production of filterable particles beyond that contributed by the production of MnO_2 alone. This is speculated to be due to the interaction of the MnO_2 solids with other ambient but nonfilterable solids (<0.45 um) such that they are incorporated into and become part of a MnO_2-based filterable particle. Confirming the central role of MnO_2 in the formation of filterable particles, under the conditions studied, the filterable particles created by permanganate oxidation were readily dissolved by addition of a solution of peracetic acid.

Manganese Oxide Genesis and Properties

The genesis, properties, and effects of MnO_2 particles produced by ISCO with permanganate have not yet been thoroughly studied but, as noted below, some insight can be gained from the literature. When the Mn^{+2} in permanganate is reduced (e.g., during oxidation of TCE), manganese oxides are produced. Manganese oxides are often represented as

MnO_2 in stoichiometric equations. Naturally occurring manganese oxides can include a number of minerals such as birnessite, lithiophorite, hollandite and pyrolusite (Table 3-2 and 3-3) (McKenzie 1970). The system Eh and pH can play a key role in determining which oxidation state and form the Mn will be in (Figure 3-7). The presence of sulfur has little effect on the predominance region for MnO_2 while the absence of carbonate yields a much broader pH range in which they exist. Manganese oxide minerals exist in a number of crystal forms, exist in various states of hydration, and contain a variety of foreign ions (Pisarcyzk 1995) (Figure 3-8). In this chapter, the manganese oxides produced during chemical oxidation with permanganate will be represented simply as MnO_2.

There are questions regarding generated MnO_2 mass and chemical characteristics, their size and ultimate fate, and what, if any, effects they may have on remediation system function and performance that have not yet been fully examined. Manganese oxides can be generated ex situ (e.g., during above-ground oxidant amendment of contaminated groundwater before reinjection) or in situ (e.g., during oxidation in the subsurface). If the quantities are large enough and if precautions are not taken to manage particles, they could plug well screens, filter packs, or pores in the formation and thereby reduce permeability (discussed further in Chapter 4). In addition, they could pose problems if they enable colloidal transport of metals in an uncontrolled fashion. However, if the particles are managed (removed from recirculated groundwater or dispersed/ retained in the

TABLE 3-2. Properties of natural pyrolusite (β-MnO_2) (CRC 1993).

Property	Value
Molecular wt.	86.94 g
Specific Gravity	5.026 g/cm³
Melting point	535C
Solubility	Insoluble in hot and cold water Soluble in HCl but insoluble in HNO_3
Physical state	Brown-black solid

TABLE 3-3. **Manganese oxide minerals and selected properties.**

Mineral name [1]	Other names [1]	Composition [1]
Pyrolusite	β-MnO$_2$, polianite	MnO$_2$
Ramsdellite	-	MnO$_2$
Nsutite	γ-MnO$_2$, ρ-MnO$_2$	variable
Birnessite	δ-MnO$_2$	variable
Hollandite	α-MnO$_2$	Ba$_2$Mn$_8$O$_{16}$
Coronadite	α-MnO$_2$	Pb$_2$Mn$_8$O$_{16}$
Manganite	γ-MnOOH	MnOOH
Partidgeite	α-Mn$_2$O$_3$	Mn$_2$O$_3$
Hausmannite	-	Mn$_3$O$_4$

Form	Description	Common mineral	pH$_{PZC}$
δ-MnO$_2$	MnOx, x=1.95-2.00, 5-40% bound water	birnessite	1.5
α-MnO$_2$	MnOx, x=1.88-1.95	hollandite gp.	4.5
γ-MnO$_2$	hydrated MnO$_2$	nsutite	5.5
β-MnO$_2$	MnOx, x=1.95-2.00	pyrolusite	7.3
Mn(II)Manganite	–	–	1.8

Sources: [1]Dixon et al. 1977. [2]Pisarcyzk 1995. [3]Murray et al. 1967.

formation), they could sorb and remove high levels of heavy metals such as Pb, Ni, Cr, Zn, Cd (e.g., 0.1 to 1.0 wt.%) (McKenzie 1977, Lu et al. 1991). Also, the slow reduction of Mn^{+4} in these particles to Mn^{+2} may also facilitate oxidation of intermediate organic acids or other groundwater NOM (Stone and Morgan 1984a,b).

The manganese oxides generated through permanganate oxidation reactions from a target organic like TCE or NOM can exist in colloidal or particulate form depending on the matrix conditions presented in Table 3-4. It has been demonstrated that the MnO$_2$ can be present in the form of micron-sized particles (e.g., Figure 3-4 and 3-5) (West et al. 1998a,b), but no attempts to specifically characterize the settling effects

FIGURE 3-7.	Predominance area diagram for manganese compounds. Note: at 25C with total dissolved carbonate at $10^{-1.4}$ (Duggan 1993).

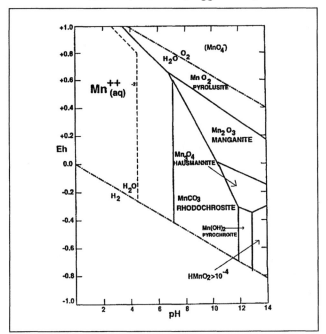

FIGURE 3-8.	Crystal structures of manganese oxides (Pisarcyzk 1995).

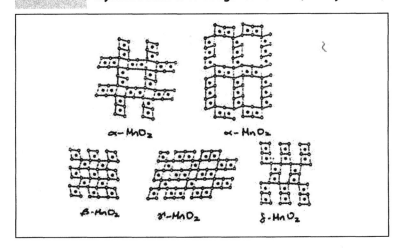

TABLE 3-4. **Conditions impacting manganese oxide colloid stability.**

Matrix Condition	Impact	Reference(s)
Increased permanganate conc.	Decreased stability*	Perez-Benito et al., 1990
Phosphate, and potentially other anions	Increased stability	Insausti et al., 1992; Perez-Benito and Arias, 1991; Doona and Schneider, 1993
Cations, specifically multivalent	Decreased stability	Morgan and Stumm, 1964; Perez-Benito et al., 1989; Perez-Benito et al., 1990; Doona and Schneider, 1993; Chandrakanth and Amy 1996
Decreased pH	Decreased stability	Morgan and Stumm, 1964; Perez-Benito, 1990
Protective colloids	Increased stability	Insausti et al., 1992; 1993; Perez-Benito et al., 1989; Perez-Benito et al., 1990; Perez-Benito and Arias, 1992; Doona and Schneider, 1993; Chandrakanth and Amy, 1996

*Stability refers to the ability of the colloid to remain in colloidal phase as opposed to settling.

of these particles in the context of permanganate-based remediation have been encountered in the literature. Some studies have reported that the particles are initially colloidal in size but under quiescent conditions can coagulate/flocculate and grow in size. Ongoing research at CSM is examining the character, including the particle size, of the MnO_2 generated through permanganate oxidation of organic contaminants and natural organic matter. By following the evolution of the generated particles spectrophotometrically at 418 nm (the wavelength of maximum absorbance for manganese oxides), the rates of particle generation, growth (coagulation/flocculation), and settling appear to be dependent on organic material concentration, oxidant dose, and matrix conditions such as pH and cation (calcium) content (Figure 3-9). A higher absorbance at 418 nm may correspond with a higher colloidal MnO_2 concentration, and a

FIGURE 3-9. Absorption data (λ = 418 nm) for a solution under varied matrix conditions (Crimi 2001). (a) Ca vs. no Ca at 50 mg/L KMnO$_4$ and pH=7, (b) 25 mg/L vs. 50 mg/L KMnO$_4$ and pH=7 vs. pH=3, no Ca.

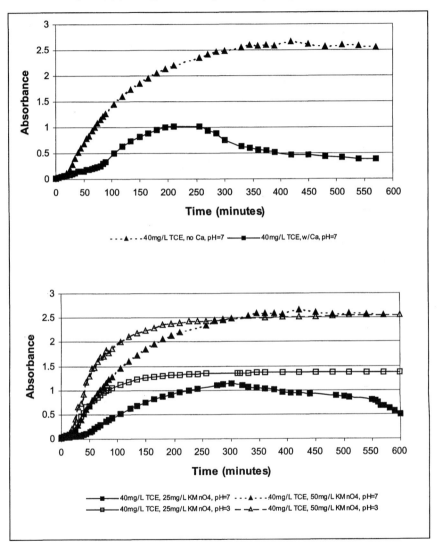

subsequent decrease in absorbance may correspond with increasing particle size and settling. This rise then fall in 418 nm absorbance data has been observed under most reaction conditions studied thus far.

An increase in the concentration of permanganate in the groundwater matrix will drive the oxidation reaction to an increase in manganese oxides production, resulting in a higher percentage present in the particulate vs. colloidal phase (Table 3-4). This is due to the increased interactions between the MnO_2 particles as concentration increases. This occurrence has only been reported in systems where the permanganate was the limiting reagent. Even if the target reductant is the limiting reagent of the reaction, though, that could still be expected based on the decomposition of excess permanganate catalyzed by manganese oxides (eqn. 3.1). Phosphate has been noted to increase colloidal manganese oxides stability, thereby preventing aggregation. Phosphate ions can sorb onto negatively charged MnO_2 (i.e., pH > pH_{pzc}) (Perez-Benito and Arias 1991, Perez-Benito and Brillas 1992) through surface complexation (Sposito 1989, Stumm 1992). With the negative ions sorbed onto the surface of the colloidal manganese oxides, repulsive forces prevent interaction and aggregation. When cations such as calcium are present in solution, there is a strong attractive force that neutralizes and destabilizes the manganese oxides, leading to particle growth. This is supported in the data collected in the aforementioned CSM preliminary studies comparing systems with and without the presence of calcium in the reaction matrix (Figure 3-9). The effect of pH on stability, as indicated by the literature, would be similar; at lower pH the colloids are destabilized and will aggregate. Preliminary CSM research data (also included in Figure 3-9), though, indicates an opposite effect. Based on spectrophotometer data collected at 418 nm, it appears that the colloidal manganese oxides are more stable in pH 3 solution as compared with pH 7. A visual inspection of the pH 3 systems also demonstrated no apparent particle settling, which would have indicated colloid destabilization and aggregation. Ongoing studies will clarify if colloidal stabilization is the case, or if the manganese has been converted to a soluble form and is not present as suspended MnO_2. Additionally, a protective colloid can sorb onto the surface of colloidal manganese oxides and protect them from interaction with cations. This effect leads to an increased stabilization of the colloidal manganese oxides.

Turbulence, or mixing, is another important factor in the particle growth of colloidal MnO_2 (Amirtharajah and O'Melia 1990). Increasing the degree of mixing increases interparticle collisions, promoting aggregation and floc formation. However, highly vigorous mixing can cause floc break-up through development of turbulent shear forces. Mixing is important in field permanganate systems where manganese oxides are generated because a high degree of mixing within the subsurface may lead to particle growth and deposition of manganese oxides. That is significant, especially in recirculation systems where colloidal manganese oxides are reinjected, because deposition in the subsurface can lead to pore clogging (discussed in greater detail in Chapter 4).

Stumm (1990, 1997) indicates that mineral dissolution rates are affected by the same factors controlling the stability of colloidal MnO_2 (Table 3-4). Stability, with regard to the particulate phase, refers to hindrance of dissolution. The degree to which conditions affect dissolution of particulate vs. colloidal manganese oxides can be expected to differ. Since the degree to which conditions can actually impact stability depends on the surface available, meaning the processes are surface controlled (Stumm and Wollast 1990), the impact of the factors in Table 3-4 on particulate manganese oxides is expected to be less than for the colloidal manganese oxides. As manganese oxides come together during coagulation, there are fewer surface sites "open" as the particle size gets larger.

Specifically, increased permanganate concentration in solution may increase particulate stability (i.e., hinder dissolution) because it is a more reactive oxidant than manganese oxides. In other words, if a reductant is available in solution, permanganate will preferentially oxidize it, even though manganese oxides can act as an oxidant for some reductants under some conditions. Phosphate ions, and other anions, in solution have a similar effect on both colloidal and particulate manganese oxides. The phosphate will protect reactive surface sites. Protection from reductants that would promote dissolution is one example. Protective colloids, that are not reductants, would have a similar effect. Cations, sorbed to particulate manganese oxides, would promote stability, again by protecting reactive surface sites. Additionally, cations can serve to link manganese oxides, making the oxides even larger with even fewer remaining reactive surface sites. Solution pH has a potentially significant impact on particulate stability. At the point of zero charge (pH_{pzc}), the dissolution rate typically is minimized, since surface protonation and deprotonation is minimized (Stumm 1997). Since the pH_{pzc} of manganese oxides is relatively low (ranging from 1.5–5), a pH range typical in soil systems would be expected to result in some mineral dissolution, although Murray (1973) demonstrated no *hydrous* manganese dioxide dissolution is observed for samples examined in solutions above pH 3.5 (upper limit of pH tested is not given). With respect to MnO_2 formation, it has been demonstrated that pH, Eh, cations, anions, turbulence, NOM and organic contaminant presence and form are all influencing factors. Many of these same properties, as well as the ultimate properties of the MnO_2, will impact the sorption of metals onto the MnO_2.

Metal Sorption onto Manganese Oxides

As shown in Table 3-3, there are numerous forms of manganese oxide. The pH_{pzc} is different for the different forms and it ranges from 1.5 to 5.0 (Murray 1973). This is important as it affects the surface chemistry and the adsorption of metals by the oxides as well as the affinity for the oxides to other soil minerals. The chemical and electrochemical reactivity of the oxides results from the presence of cation vacancies in the crystal lattice (Pisarcyzk 1995). The adsorption of some divalent cations has been explained by ion-exchange mechanisms. A selectivity series developed for δ-MnO_2 was reported by Murray (1973) as:

$$Co \geq Mn > Zn > Ni > Ba > Sr > Ca > Mg$$

The ionic radius of the sorbed cation will affect its degree of sorption (as demonstrated above); the smaller the crystalline radius, the less likely it will sorb.

Manganese oxides can contribute significantly to the adsorption of trace metals. In soils, significant amounts of copper, cobalt, zinc, nickel, lead, silver, and cadmium have been found to be associated with manganese oxides (Fu et al. 1991). At low concentrations of metal cations, pH exerts a great effect on the adsorptive properties of the oxides. At low concentrations, an increase in pH (above the pH_{pzc}) increases the amount of metal adsorbed. At high metal concentrations, pH exerts little or no effect. Typically, MnO_2 have not been expected to sorb anions because of their high negative surface charge in the range of typical soil pH. However, recent studies suggest that these oxides adsorbed selenite in a manner similar to cation adsorption (Saeki et al. 1995). There was high adsorption at low to moderate pH (pH 3 to 7) but rapidly diminishing sorption above pH 7.

As a demonstration of the sorption potential of generated MnO_2, the following example is provided based on the manganese oxides generated through permanganate oxidation and investigated by Fu et al. (1991). An adsorption capacity of the generated MnO_2 for cadmium of 1.08 mmol/g at pH 7 is reported. Based on reaction stoichiometry (eqn. 2.9), the oxidation of 50 mg of TCE by permanganate could produce 51 mg of manganese oxides, resulting in the potential sorption of 6 mg Cd. One could expect an approximate one-to-one mass of oxidized TCE to manganese oxides generated with deviations attributed to permanganate decomposition, and a one-to-8.5 mass ratio of sorbed cadmium to manganese oxides (or approximately 0.12 mg-Cd per mg manganese oxides).

Comparing the relative importance of generated MnO_2 and naturally present soil materials in terms of sorption potential, McLaren and Crawford (1973) and He and Singh (1993) report in their soil studies with copper and cadmium, respectively, that manganese (and/or iron) oxides in the soils studied contained the highest fraction of bound metal; over organic matter and other mineral components. That can lead to the expectation of a high degree of metals retention by the manganese oxides. Godtfredsen and Stone (1994) demonstrated in their research with manganese dioxide-bound copper that influx of NOM brought on metal release caused by complexation of Cu by soluble NOM. Petrovic et al. (1998) found an opposite effect: upon addition of humic acids (HA) to systems with mineral-bound metals, an increase in sorption was demonstrated. Presumably, this was due to sorption of the metal onto the HA that had sorbed onto mineral surfaces.

The presence and/or influx of low molecular weight organic compounds (e.g., excess TCE) into a system with particulate manganese oxide-bound metal can either (1) compete with manganese oxides for the bound metal or (2) react with the manganese oxides, resulting in their dissolution. The rate of reduction of the manganese oxides by molecular material was demonstrated to be a function of the molecular weight of the organic material. Lower-molecular weight reductants (e.g., oxalate) resulted in faster dissolution than higher-molecular weight compounds (e.g., pyruvate, NOM-rich water samples). In sample mixtures with high- and low-molecular weight reductants, the higher-molecular weight compounds act as a shield and protect the manganese oxides from reduction by the lower-molecular weight compounds. This was attributed to sorption of the high-molecular weight compounds onto the surfaces of the manganese oxides, blocking reaction sites from the smaller, easier-to-oxidize compounds (Godfredtsen and Stone 1994).

Based on information presented in Sections 3.1 and 3.2, a great deal of site-specific information regarding the impact on the mobility of natural and contaminant metals may be needed before ISCO is used for organic contaminant destruction at some sites. Not only is it important to know those conditions and to understand their influence before ISCO treatment begins, it is essential to know and understand the impact of ISCO on those conditions. Table 3-5 summarizes the site-specific information that will influence ISCO decisions with regard to metal mobility, as well as their potential impact on metals pre- and post-oxidation.

3.3. IMPACT OF PERMANGANATE ISCO ON METAL MOBILITY

Several laboratory and field investigations have examined the impact of permanganate on metal mobility. Observations have documented both mobilization and immobilization of metals following ISCO operations. In a laboratory study to investigate problems that could impact a permanganate oxidation scheme, Li and Schwartz (2000) examined the effect of permanganate on dissolved metals concentrations resulting from reaction with samples of aquifer materials including alluvium, glacial till, glacial outwash, and carbonate sand. Fifteen grams of each sample were mixed with 15 mL of 0.5 g/L of potassium permanganate. The solution then was covered, kept in the dark, and shaken once a day. After 10 days, when most of the permanganate had been consumed, 5mL of the solution was removed, filtered, and measured by ICP-MS. Of 12 elements analyzed, chromium, selenium, and rubidium generally demonstrated an increase in concentration (as converted to pore water concentration and compared to controls with no permanganate); vanadium, zinc, cadmium, and lead decreased in concentration. The magnitude of change in concentration varied with the type of soil reacted. Titanium, arsenic, molybdenum, cesium, and mercury demonstrated no significant change in concentration (Figure 3-10).

At a former industrial site in California, Moes et al. (2000) reported increased detection of chromium and selenium following ISCO, where the naturally occurring metals were apparently oxidized by permanganate to a more mobile state. At this site, TCE concentrations as high as 260 mg/L were treated with 550 kg of $KMnO_4$ in a pilot-scale study in March of 1999. Samples from one of the site's monitoring wells were filtered and analyzed for dissolved metals pre- and post-treatment yielding the results presented in Table 3-6. The higher levels of chromium and selenium were expected because it is known these metals are more soluble in their higher oxidation state and also because chromium (and other metals) exist as impurities in technical grade potassium permanganate and can

| TABLE 3-5. | Conditions impacting ISCO decisions with respect to metal mobility. |

Condition	Impact
pH, Eh	• Low pH = high metal cation mobility; metal anions sorb to mineral oxides • High pH = high metal anion mobility; metal cations sorb to mineral oxides
NOM content	• Competes with mineral oxides for metals sorption; sorption potential will change under ISCO conditions due to changes in NOM • Will react with permanganate to exert a demand on the oxidant and will generate manganese oxides as a by-product • Can slowly reduce generated MnOx
Mineral oxides content	• Competes with NOM for metals sorption; sorption potential can change under ISCO conditions due to changes in pH, Eh
CEC	• Dictates metal cation sorption potential
Particle size distribution	• Smaller particles generally contain highest metals contaminant concentrations
Naturally present metals	• Redox sensitive metals (e.g., Cr, As) will be impacted by ISCO; can be mobilized/immobilized
Cation content	• Compete with metals for sorption sites • Will influence size and stability of generated MnOx
Inorganic anion content	• Can form insoluble complexes with metals to cause desorption or precipitation • Will influence the stability of generated MnOx
Target organic contaminant phase	• Influences oxidant demand and build-up/deposition of generated MnOx impacting subsurface permeability • Impacts the ability of natural soil constituents and generated MnOx to sorb metals
Buffer capacity	• Dictates initial and final pH
Hydraulic conductivity, porosity, tortuosity	• Influences flow through the subsurface and will dictate the mobility of metals and generated particles

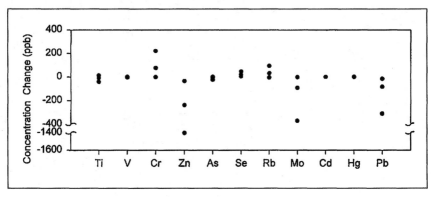

FIGURE 3-10. Average, maximum and minimum concentrations for the elements in pore water after treatment with 0.5 g/L KMnO$_4$ for ten days (Li and Schwartz 2000).

TABLE 3-6. Dissolved metals (mg/L) detected[a] in groundwater at MW-1 (Moes et al. 2000).

Date	Cd	Cr(tot)	Cr^{+6}	Hg	Ni	Se	Ag	Zn
12/10/98	0.13	<0.02	<0.01	<0.0002	<0.02	<0.08	0.051	0.058
4/22/99	0.11	0.48	0.53	0.0032	0.098	0.57	0.097	0.056
5/27/99	0.048	0.38	0.35	0.0026	0.11	0.53	0.11	0.047
6/23/99	0.11	0.48	0.41	0.0027	0.14	0.63	0.11	0.024

(a) Metals analyzed for but not detected included Sb, As, Be, Pb, and Th

contribute to the dissolved metals concentrations. Also, the metals at this site were expected to revert eventually to their reduced states, of lower solubility, as the groundwater Eh decreased post-treatment and as the groundwater migrated into areas of more reducing potential. The authors indicate that the initial deleterious side effect of increased metal mobility may be acceptable, in terms of risk, because the occurrence of increased oxidized metals may be temporary as compared to the potential lifespan of leaving the organic contaminants untreated. Also, the metals effect may be acceptable (as demonstrated by regulatory approval at the California site) if the treated groundwater is not used for drinking-water supply because there would be no negative impact on human health risk.

Studies by Crimi et al. at CSM have examined the impact of MnO_2 generation during oxidation of organic materials on the mobility of metals (Crimi 2001). In one study, batch experimental systems were used to compare cadmium sorption before and after permanganate was added to the system. A characterized, silty-clay field soil (7.5 g), with organic carbon of 500-1,500 mg/kg and clay content of 7-8 wt.%, was reacted for 24 hours with 150 mL of simulated groundwater matrix (Struse 1999) containing 5 mg/L of Cd as cadmium chloride. The permanganate concentration used was 5,000 mg/L, the potassium chloride concentration used was 2,359 mg/L (for equal potassium ion concentration), and the TCE concentration was 5 mg-TCE/kg-soil. Duplicates of six experimental conditions were explored, as shown in Table 3-7. The initial pH of the systems ranged from 4.0 - 4.4, and changed less than 1 pH unit following reaction. In the systems with no $KMnO_4$ or KCl (systems 1 and 2), approximately 65% of added Cd was sorbed onto the soil. When $KMnO_4$ was present, regardless of the presence of TCE, virtually all of the available Cd was sorbed (systems 3 and 4). The increased sorption may be attributable to the formation of MnO_2 from reaction of MnO_4^- with the silty-clay matrix. When KCl was present in non-MnO_4^- systems, much

TABLE 3-7.	Cadmium sorption in batch experimental systems.			
System	Cd concentration added (mg-Cd/kg dry soil)	Cd concentration sorbed (mg-Cd/kg dry soil)		% of initial Cd sorbed
	Rep. 1	Rep. 2	Average	
1. Cd	100	64.8	65.0	64.9
2. Cd + TCE	100	64.8	64.8	64.8
3. Cd + $KMnO_4$	100	99.7	100	99.9
4. Cd + TCE + $KMnO_4$	100	99.5	100	99.8
5. Cd + KCl	100	15.8	22.8	19.3
6. Cd + TCE + KCl	100	21.8	16.8	19.3

less of the available Cd was sorbed (systems 5 and 6). This indicates that the impact of MnO_4^- oxidation on Cd sorption may outweigh the displacement effects of the added K^+ ion as an exchangeable cation.

Additional studies were conducted at CSM in support of ISCO operations at the DOE Portsmouth Gaseous Diffusion Plant in Piketon, Ohio. Laboratory experiments were conducted to determine the possible cause of injection well pressure build-up during the first three weeks of operation of a vertical well used for $NaMnO_4$ (3,000 mg/L) injection into a DNAPL contaminated aquifer (Siegrist et al. 2000a). It was speculated that the deposition and build-up of manganese oxides resulting from the reaction of permanganate with TCE and NOM, and which potentially coagulated/flocculated with clay colloids, caused permeability reduction at the well screen (more discussion in Chapter 4). A characterization of Cd sorption was conducted to gain insight into the composition of aquifer sediments and any pore-filling agents. For this study, samples included media from a filter cake taken from an inline cartridge filter (5 um and 1 um), as well as aquifer sediments at 6 locations from 2 ft. to over 100 ft. from the injection well. Samples of the subsurface solids (2 g) from each location were reacted for 72 hrs. with 40 mL of simulated groundwater (Struse 1999) and 5 mg/L of Cd as cadmium chloride. The sorption of Cd by the filter cake was nearly 100% while sorption at the closest distances (2-5 ft.) from injection was >90% (Figure 3-11). The furthest distance represents background media, where permanganate treatment was not detected. Results of MnO_2 content with distance from the injection well correspond with the Cd sorption data (Figure 3-12). Since manganese oxides can strongly sorb metal ions, the sorption data support the presence of MnO_2 formed by reaction and suggest that permanganate treatment of the TCE in the region could result in altered metal mobility. The data demonstrate that samples that experienced just a 0.70–0.90 $mg\text{-}MnO_2$ increase per gram of media also demonstrated an almost 50% increase in Cd sorption. Calculations based on the data indicate a possible sorption capacity of 0.20-0.30 mg-Cd per $mg\text{-}MnO_2$ per gram of media, an even higher sorption potential than expected. This suggests that the generation of even small quantities of MnO_2 through permanganate oxidation can have a significant impact on the sorption of metal co-contaminants.

To further support the oxidation field application at the DOE Portsmouth Gaseous Diffusion Plant mentioned above, column studies were conducted at CSM with three operational modes; (C1) groundwater,

FIGURE 3-11. Cd sorption with distance from the oxidant injection well.

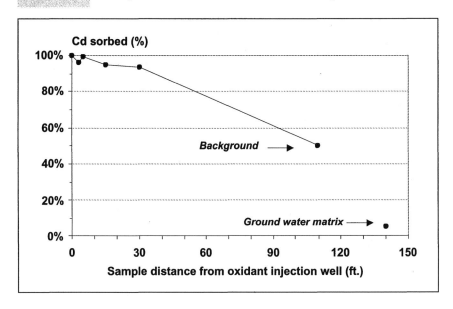

FIGURE 3-12. Manganese oxide content with distance from oxidant injection well.

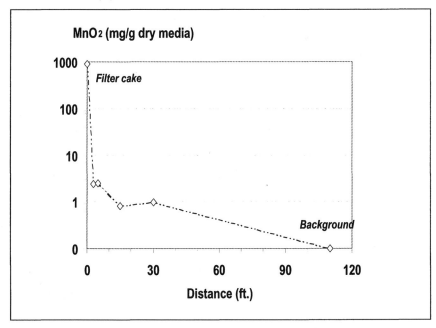

TCE, and permanganate; (C2) groundwater only; (C3) groundwater and permanganate. The soil used was untreated soil from the site. The columns received a series of varied groundwater throughputs: (1) groundwater only, (2) TCE addition to column 1 and additional groundwater to the others, (3) 3,000 mg/L of $KMnO_4$ to columns 1 and 3, with groundwater in column 2, (4) groundwater plus Cd, and (5) particle-laden groundwater in column 1. Relevant to the impact of treatment on Cd-retention, simulated groundwater with 50 mg/L of cadmium chloride was passed through each of the columns for approximately 100 pore volumes (referred to as groundwater throughput condition (4) above). Effluent samples were taken periodically, filtered (0.45um), and measured for Cd content (Figure 3-13). The addition of permanganate resulted in increased retention of Cd in the system as evidence by Cd breakthrough behavior and by the fact that 74% of Cd was retained in C1 and 79% in C3 but only 49% in C2. The cadmium sorption study was followed by a particle

FIGURE 3-13.   **Cadmium breakthrough in columns C1, C2, and C3.**

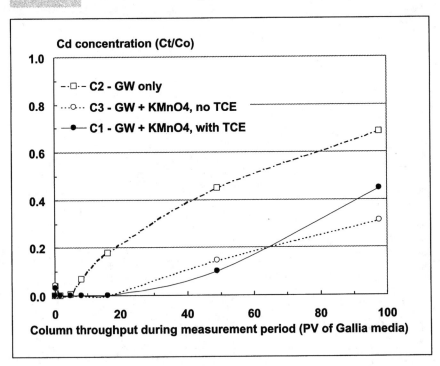

genesis operational mode in column C1 to study pressure build-up attributable to MnO_2 generation in the system. This was followed by addition of a peracetic acid solution for 58 pore volumes to determine if permeability could be recovered through dissolution of the MnO_2. Both Cd and Mn were released during peracetic acid treatment, in a fashion that corresponded with a release of filterable solids (Figure 3-14). The release of manganese suggests dissolution of MnO_2, and the corresponding release of Cd is consistent with a Cd-retention by the MnO_2. A common theme of the experimental results from the studies conducted at CSM is that

FIGURE 3-14. Metal leaching from permanganate treated silty sand soil with groundwater alone followed by peracetic acid solution.

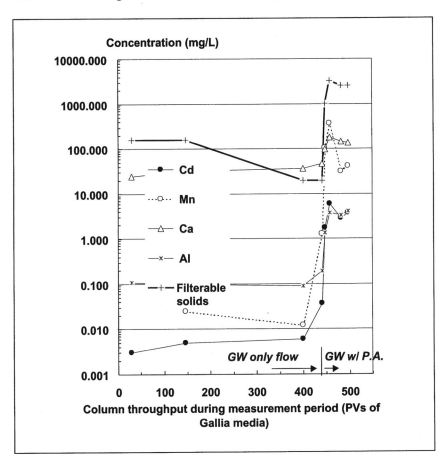

introduction of permanganate to the soil and groundwater systems resulted in increased Cd retention. It is believed that this increased retention is due to the Cd-sorption properties of MnO_2 generated through permanganate oxidation of TCE and NOM.

Studies examining metal mobility following permanganate oxidation have also focused on post-ISCO attenuation (other than from sorption onto generated MnO_2) of the increased concentrations of metals (such as chromium, selenium, and rubidium) that can result from oxidation of the less soluble reduced form to the more soluble oxidized form (e.g., $Cr^{+3} \rightarrow Cr^{+6}$). Chambers et al. (2000a,b) completed laboratory and pilot-scale studies to evaluate use of $KMnO_4$ for in situ remediation of VOCs in groundwater at a former electronics manufacturing plant in Sunnyvale, CA (Figure 3-15 and 3-16). A focus of the studies was to examine the impact of the oxidant on the speciation of naturally occurring chromium, present in soils at levels of approximately 70 mg/kg. Neither Cr^{+3} nor Cr^{+6} were detectable in groundwater before oxidant injection, but there were several reasons for concern over higher levels: (1) the high level of naturally present reduced chromium, (2) the presence of chromium as an

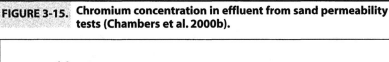

FIGURE 3-15. Chromium concentration in effluent from sand permeability tests (Chambers et al. 2000b).

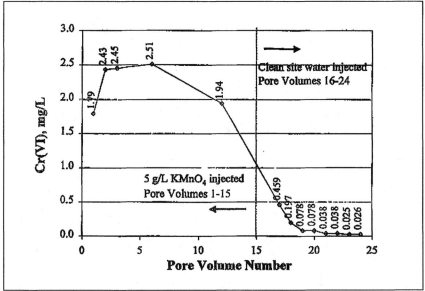

FIGURE 3-16. Trends in chromium concentrations in B-zone groundwater (Chambers et al. 2000a).

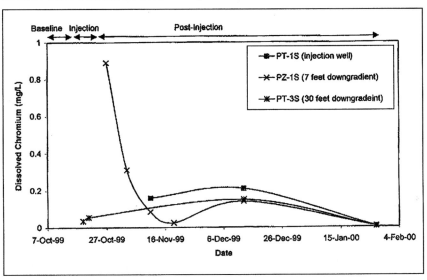

impurity in potassium permanganate at levels of approximately 30 to 60 mg/kg, (3) the known greater solubility of the higher oxidized form of chromium, and (4) California's maximum contaminant level (MCL) for chromium of 0.05 mg/L and the federal MCL of 0.10 mg/L. In the laboratory studies, the groundwater and soil oxidant demands were first determined, followed by soil permeability tests and an evaluation of post-oxidation chromium levels. In the chromium studies, equal amounts of soil and potassium permanganate were mixed, centrifuged, then filtered and colorimetrically analyzed for chromium. Column effluent, resulting from the permeability studies, was also analyzed for chromium. These actions were followed by examination of the ability to attenuate the oxi-dized chromium provided by three reductants: ferrous iron, ascorbic acid, and molasses. Natural attenuation by site groundwater also was tested in a column study.

In the batch studies of Chambers et al. (2000a,b), the control samples (non-oxidized) resulted in non-detectable levels of chromium, while all three of the oxidized samples resulted in concentrations of chromium of approximately 0.8 mg/L for each of the soils exposed (sand, silt, and clay). Some of the elevated chromium is attributed to the impurities in the $KMnO_4$, which contained 63 mg-Cr per kg-$KMnO_4$. That was not enough to account for the majority of the detected chromium which, therefore was attributed to oxidation of the naturally present Cr^{+3} to Cr^{+6} by the permanganate. From the permeability sand column effluent (Figure 3-15), chromium was detected in concentrations ranging from 1.79 mg/L to 2.51 mg/L during flushing with the $KMnO_4$, but decreased to 0.459 mg/L after flushing with about 2 pore volumes of clean site water. After 5 to 6 pore volumes of flushing, levels decreased to below the MCL of 0.05 mg/L. The results of the examination of reductants demonstrated that of the three examined, ferrous iron most affected chromium concentrations, while the molasses had little effect as compared to site groundwater treatment. In all cases, though, the Cr^{+6} concentration was less than 0.05 mg/L within 24 hours of reductant introduction. It is to be noted that, even with no reductant addition, the Cr^{+6} concentration decreased with time, which suggests that field natural attenuation is likely. The study of natural attenuation in a column of simulated subsurface conditions indicated a decrease in Cr^{+6} concentrations from 1.15 mg/L in soil pore water to 0.68 mg/L (41% reduction) within seven days.

Chambers et al. (2000a,b) followed their laboratory studies with a pilot-scale evaluation of ISCO for this site. The evaluation focused on the natural capacity of soil to reduce chromium to the more stable Cr^{+3} form. For the study, 4,000 gallons of 40 to 50 mg/L of $KMnO_4$ was injected at a rate of 1-2 gpm for a one-week period; a three-month monitoring period followed. Prior to injection, no chromium was detected in the site groundwater, but it was known that the soil contained approximately 70 mg/kg of naturally present chromium and that the $KMnO_4$ contained approximately 30-60 mg/kg of chromium as an impurity. Maximum dissolved chromium concentrations of approximately 2,000 ug/L were detected in shallow groundwater within five feet of injection approximately one week after injection. The chromium concentrations rapidly attenuated to concentrations of about 100 ug/L within 2 to 3 weeks. Three months after injection, the dissolved chromium concentrations were less than 5 to 10 ug/L in all groundwater samples. This demonstrated the significant capacity for the soil to reduce the Cr^{+6} to Cr^{+3}, presumably by natural organic matter, ferrous iron, and other site substances (Figure 3-16).

Clayton et al. (2000) report a similar metal mobilization, followed by attenuation effect, from field data of multiple projects using potassium or sodium permanganate to treat organic contaminants. They demonstrate that metals such as chromium, uranium, vanadium, selenium, and molybdenum, which can exist in a non-detectable reduced state naturally, become more soluble and thus mobile in their oxidized state. That could present a problem at sites with high levels of naturally occurring metal concentrations or where metals contamination had initially been attenuated through natural reduction. The oxidation of Cr^{+3} to Cr^{+6} is the most commonly reported mobilization effect, as demonstrated in Table 3-8. Of note is the fact (demonstrated in Figure 3-17 from Site B listed in

TABLE 3-8.	Lab and field data on chromium liberation and attenuation (Clayton et al. 2000).		
Site	Lab or Field	Pretreatment Total Cr in soil (mt/kg)	Maximum Dissolved Cr^{+6} Liberated (mg/l)	Cr^{+6} Attenuation Observed
A	Lab	368	105 mg/L	Yes, 40mg/l per pore volume of soil contact
B	Field	65	3 mg/L	Yes, to 0.007 mg/l in the field after 45 days
C	Field	28-94	1.5 mg/L	Yes, to 0.15 mg/l after 23 days

FIGURE 3-17. Time-series plot of dissolved Cr (VI) attenuation after permanganate treatment (Clayton et al. 2000).

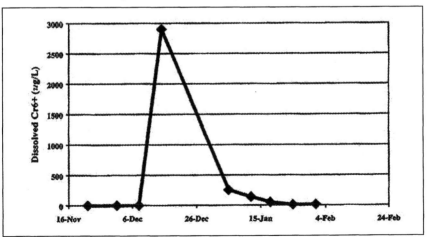

Table 3-8) that although the permanganate oxidation initially resulted in elevated levels of chromium as Cr^{+6}, the chromium levels attenuated over time. This attenuation is attributed to sorption and chemical reduction of the chromium. Sorption can occur by such materials as iron hydroxides, organic carbon, or even the manganese oxides generated through the permanganate oxidation reaction with organic contaminants and natural organic materials. The authors also report that sites with background chromium concentrations in soils of less than about 5 mg/kg have not demonstrated this effect.

3.4. SUMMARY

At sites where ISCO may be an option, there are many site-specific considerations with respect to metal mobility that must be examined with respect to both pre- and post-treatment conditions (e.g., Tables 3-1 and 3-5). While those considerations must not be oversimplified, there are some site conditions that provide insight into whether or not there could be a resulting problem with metals (Table 3-9). The oxidant dose concentrations and total loading delivered to the subsurface must be considered. If relatively low concentrations of $KMnO_4$ (e.g., 1000 mg/L) are delivered for a short period (e.g., 30 days), then concern over any impurities in the permanganate and longer-term release of Mn^{+2} will be low. Also, there would be few metal-related concerns at sites with pre-ISCO conditions that include: a high pH and buffering capacity, highly oxidizing conditions, high natural CEC, and the presence of metals that may be less mobile under oxidative conditions (e.g., arsenic). The same would be true if local groundwater was not used for drinking water or agricultural purposes or if the flow velocity was high and flushing were to dissipate metals concentrations relatively quickly.

In contrast, potential concern about metals would be higher at sites where high concentrations of permanganate are to be used (e.g., >1000 mg/L) for longer time periods (e.g., 90 days or more). These conditions could raise concern over any impurities added in the permanganate as well as the longer-term behavior of the added Mn. In addition, due consideration would be warranted at sites characterized by a low pH and low buffering capacity, low CEC, high inorganic anion concentrations, and presence of natural metals or co-contaminant metals (especially those that are redox sensitive) (Table 3-10 includes typical background concentrations of some common metals). In addition, concerns could exist at a site

TABLE 3-9.	Contaminated site conditions and level of concern over ISCO and metal mobility.		
Site and ISCO characteristic	Attribute	Conditions indicating a *lower* level of concern for post-ISCO metal mobility	Conditions indicating a *higher* level of concern for post-ISCO metal mobility
Oxidant loading	Dose concentration (e.g., mg/L or wt.%)	Low	High
	Oxidant loading (e.g., g/kg of media)	Low	High
Pre-ISCO site conditions	pH	Neutral to alkaline	Acidic
	Buffer capacity	High	Low
	Eh	Oxidizing	Reducing
	Porous media permeability	Low	High
	Porous media CEC	High	Low
	Inorganic ion concentrations	Low	High
	Natural metals	Low levels	High levels
	Co-contaminant metals	No redox metals	Redox sensitive metals present (e.g., Cr)
Groundwater features	Local uses	Not used for drinking water or agricultural purposes	Used for drinking water or agricultural purposes
	Flow velocity and flushing	High	Low

Metal	Common range for soils (ppm)	Selected average for soils (ppm)
Arsenic (As)	1 - 50	5
Cadmium (Cd)	0.01 - 0.70	0.06
Chromium (Cr)	1 - 1,000	100
Lead (Pb)	2 - 200	10
Mercury (Hg)	0.01 - 0.3	0.03
Barium (Ba)	100 - 3,000	430
Boron (B)	2 - 100	10
Copper (Cu)	2 - 100	30
Manganese (Mn)	20 - 3,000	600
Nickel (Ni)	5 - 500	40
Selenium (Se)	0.1 - 2	0.3
Silver (Ag)	0.01 - 5	0.05
Tin (Sn)	2 - 200	10
Zinc (Zn)	10 - 300	50

TABLE 3-10. Representative metal content typical of soils (USEPA 1995).

where the local groundwater is used for drinking water supply or agricultural purposes and particularly where the groundwater velocity is low and elevated metal levels could persist for some time period.

For sites where metals might become a potential problem during or after ISCO, simple batch and/or column studies using site soil and groundwater with permanganate should be performed to determine the potential impact of ISCO on metal mobility. Additionally, contaminant-specific properties (e.g., the impact of sulfur on Cu as demonstrated in Table 3-1) and the properties of manganese oxides that are generated through oxidation of target organic contaminants and NOM must be considered. At some sites, field pilot studies may be necessary to accurately mimic redox conditions and ISCO effects on metal mobility and fate.

Not only must site conditions, such as soil, groundwater, and contaminant properties be considered, so must other important factors, including: (1) engineering options for generated particle control, (2) strategies for mitigating deleterious metals effects, (3) ultimate planned use for the site (i.e., potential for human exposure), and (4) risk management/assessment. Risk-based decisions should include comparisons of pre- and post-treatment risk from both the contaminants of concern and from potential secondary effects (e.g., metal mobility), options to mitigate effects of ISCO on naturally occurring metals or on metal co-contaminants, and the long-term sustainability of anticipated ISCO effects—both positive and negative.

3.5. REFERENCES

Amirtharajah, A. and C.R. O'Melia (1990). Coagulation processes: destabilization, mixing, and flocculation. Water Quality and Treatment, 4th ed. 269-365.

Chambers J., A. Leavitt, C. Walti, C.G. Schreier, and J. Melby (2000a). In-Situ destruction of chlorinated solvents with $KMnO_4$ oxidizes chromium. In: G.B. Wickramanayake, A.R. Gavaskar, A.S.C. Chen (ed.). Chemical Oxidation and Reactive Barriers: Remediation of Chlorinated and Recalcitrant Compounds. Battelle Press. Columbus, OH. pp. 49-56.

Chambers J., A. Leavitt, C. Walti, C.G. Schreier, J. Melby, and L. Goldstein. (2000b). Treatability study—fate of chromium during oxidation of chlorinated solvents. In: G.B. Wickramanayake, A.R. Gavaskar, A.S.C. Chen (ed.). Chemical Oxidation and Reactive Barriers: Remediation of Chlorinated and Recalcitrant Compounds. Battelle Press. Columbus, OH. pp. 57-66.

Chandrakanth, M.S. and G.L. Amy (1996). Effects of ozone on the colloidal stability and aggregation of particles coated with natural organic matter. *Environ. Sci. Technol.*, 30(2):431-442.

Clayton, W.S., B.K. Marvin, T. Pac, and E. Mott-Smith (2000). A multisite field performance evaluation of in situ chemical oxidation using permanganate. 2000. In: G.B. Wickramanayake, A.R. Gavaskar, A.S.C. Chen (ed.). Chemical Oxidation and Reactive Barriers: Remediation of Chlorinated and Recalcitrant Compounds. Battelle Press. Columbus, OH. pp. 101-108.

CRC (1993). Handbook of Chemistry and Physics. 74th Ed. CRC Press, Boca Raton, FL.

Crimi, M. (2001). Ph.D. Dissertation proposal. Colorado School of Mines, Golden, CO.

Dixon et al. (1977). Minerals in Soil Environments. *Soil Sci. Soc. America.* pp. 181-193.

Doona C.J. and F.W. Schneider (1993). Identification of colloidal Mn(IV) in permanganate oscillating reactions. *J. Am. Chem. Soc.*, 115:9683-9686.

Duggan et al. (1993). Abatement of manganese in coal mine drainages through the use of constructed wetlands. Bureau of Mines research report, U.S. Department of Interior.

Evanko, C.R. and D.A. Dzombak (1997). Remediation of metals-contaminated soils and groundwater. Technology Evaluation Report, TE-97-01. Ground Water Remediation Technologies Analysis Center, Pittsburgh, PA.

Fu, G., H.E. Allen, and C.E. Cowan (1991). Adsorption of cadmium and copper by manganese oxide. *Soil Science*, 152(2):72-81.

Godtfredsen K.L. and A.T. Stone (1994). Solubilization of manganese dioxide-bound copper by naturally occurring organic compounds. *Env. Sci. Tech.*, 28(8):1450-1456.

He Q.B. and B.R. Singh (1993). Effect of organic matter on the distribution, extractability and uptake of cadmium in soils. *Journal of Soil Science*. 44:641-650.

Insausti, M.J., F. Mata-Perez, and P. Alvarez-Macho (1992). Permanganate oxidation of glycine: influence of amino acid on colloidal manganese dioxide. *International Journal of Chemical Kinetics*, 24(5):411-419.

Insausti, M.J., F. Mata-Perez, and P. Alvarez-Macho (1993). UV-VIS spectrophotometric study and dynamic analysis of the colloidal product of permanganate oxidation of a-amino acids. *React. Kinet. Catal. Lett.*, 51(1):51-59.

Li, X.D. and F.W. Schwartz (2000). Efficiency problems related to permanganate oxidation schemes. In: G.B. Wickramanayake, A.R. Gavaskar, A.S.C. Chen (ed.). Chemical Oxidation and Reactive Barriers: Remediation of Chlorinated and Recalcitrant Compounds. Battelle Press. Columbus, OH. pp. 41-48.

McKenzie, R.M. (1970). The reaction of cobalt with manganese dioxide minerals. *Aust. J. Soil. Res.*, 8:97-106.

McKenzie, R.M. (1977). Manganese oxides and hydroxides. In: Dixon, J.B. and S.B. Weed (ed.). Minerals in Soil Environments. Soil Society of America. Madison, WI. pp. 181-193.

McLaren, R.G. and D.V. Crawford (1973). Studies on soil copper; II. The specific adsorption of copper by soils. *Journal of Soil Science*, 24(4):443-452.

Moes M., C. Peabody, R. Siegrist, and M. Urynowicz (2000). Permanganate injection for source zone treatment of TCE DNAPL. In: G.B. Wickramanayake, A.R. Gavaskar, A.S.C. Chen (ed.). Chemical Oxidation and Reactive Barriers: Remediation of Chlorinated and Recalcitrant Compounds. Battelle Press. Columbus, OH. pp. 117-124.

Morgan, J.J. and W. Stumm (1964). Colloid-chemical properties of manganese dioxide. *Journal of Colloid Science*, 19:347-359.

Murray (1967). The adsorption of aqueous metal on colloidal hydrous manganese oxide. Adsorption from Aqueous Solution.

Murray (1973). The surface chemistry of hydrous manganese dioxide. *J. Col. and Int. Sci.* 46(3):357-370.

Perez-Benito, J.F., E. Brillas, and R. Pouplana (1989). Identification of a Soluble form of colloidal manganese (IV). *Inorganic Chemistry*, 28:390-392.

Perez-Benito, J.F., C. Arias, and E. Brillas (1990). A kinetic study of the autocatalytic Permanganate oxidation of formic acid. International *Journal of Chemical Kinetics*, 22:261-287.

Perez-Benito, J.F. and C. Arias (1991). A kinetic study of the permanganate oxidation of triethylamine. Catalysis by Soluble Colloids. International *Journal of Chemical Kinetics*, 23:717-732.

Perez-Benito, J.F. and C. Arias (1992a). A kinetic study of the reaction between soluble (colloidal) manganese dioxide and formic acid. *Journal of Colloid and Interface Science*, 149(1):92-97.

Perez-Benito, J.F. and C. Arias (1992b). Occurrence of colloidal manganese dioxide in permanganate reactions. *Journal of Colloid and Interface Science*, 152(1):70-84.

Petrovic M., M. Kastelan-Macan, A.J.M. Horvat (1998). Interactive sorption of metal ions and humic acids onto mineral particles. *Water, Air, and Soil Pollution,* 111:41-56.

Pisarczyk, K. (1995). Manganese Compounds. In: Encyclopedia of Chemical Technology, 4th Ed., John Wiley & Sons, Inc. pp. 1031-1032.

Rees, T. (1987). The stability of potassium permanganate solutions. *J. Chemical Education,* Volume 64, p. 1058.

Saeki, et al. (1995). Selenite adsorption by manganese oxides. *Soil Sci.,* 160(4):265-272.

Siegrist, R., M.A. Urynowicz, M. Crimi, and A. Struse. (2000a). Particle genesis and effects during in situ chemical oxidation of trichloroethene in groundwater using permanganate; Final Report for Oak Ridge National Laboratory, Grand Junction, Colorado/Oak Ridge, Tennessee.

Siegrist, R., M.A. Urynowicz, O.R. West, M. Crimi, A.M. Struse, K.S. Lowe (2000b). In situ chemical oxidation for remediation of contaminated soil and groundwater. Water Environment Federation: Annual Conference. October, 2000. Anaheim, CA.

Sposito, G. (1989). The Chemistry of Soils. Oxford University Press. New York.

Stone, A.T. and J.J. Morgan (1984a). Reduction and dissolution of manganese (III) and manganese (IV) oxides by organics. 1. Reaction with hydroquinone. *Environ. Sci. Technol.* 18(6):450-456.

Stone, A.T. and J.J. Morgan (1984b). Reduction and dissolution of manganese (III) and manganese (IV) oxides by organics: 2. Survey of the reactivity of organics. *Environ. Sci. Technol.* 18(8):617-624.

Struse, A.M. (1999). Mass transport of potassium permanganate in low permeability media and matrix interactions. M.S. Thesis, Colorado School of Mines, Golden, CO.

Stumm W. (1992). Chemistry of the Solid-Water Interface: Processes at the Mineral-Water and Particle-Water Interface in Natural Systems. John Wiley & Sons, Inc. New York.

Stumm, W. (1997). Reactivity at the mineral-water interface: dissolution and inhibition. Colloids and Surfaces A: Physicochemical and Engineering Aspects, 120:143-166.

Stumm, W. and R. Wollast (1990). Coordination chemistry of weathering: kinetics of the surface-controlled dissolution of oxide minerals. *Reviews of Geophysics*, 28(1):53-69.

U.S. Environmental Protection Agency (1995). Contaminants and remedial options at selected metal-contaminated sites. Office of Research and Development, Washington, D.C., EPA 540-R-95-512.

West, O.R., S.R. Cline, W.L. Holden, F.G. Gardner, B.M. Schlosser, J.E. Thate, D.A. Pickering, and T.C. Houk (1998a). A Full-scale field demonstration of in situ chemical oxidation through recirculation at the X-701B Site. Oak Ridge National Laboratory Report, ORNL/TM-13556.

West, O.R., Cline, S.R., Siegrist, R.L., Houk, T.C., Holden, W.L., Gardner, F.G. and Schlosser, R.M. (1998b). A Field-scale test of in situ chemical oxidation through recirculation. Proc. Spectrum '98 International Conference on Nuclear and Hazardous Waste Management. Denver, Colorado, Sept. 13-18, pp. 1051-1057.

CHAPTER 4

Permanganate Effects on Subsurface Conditions

In situ treatment of contaminated soil and groundwater by permanganate requires careful consideration of the potential effects that permanganate can have on subsurface conditions. As described already in Chapter 3, permanganate can impact metal behavior and this must be carefully considered. In this chapter, additional potential secondary effects are described that can be beneficial and/or deleterious depending on the context. As described below, these *potential* effects conceivably include: (1) altered behavior of organic COCs, (2) changes in permeability due to particle genesis and gas evolution, (3) geochemical and microbial perturbations, and (4) introduction of toxic COCs from impurities and reaction byproducts.

4.1. ALTERED BEHAVIOR OF ORGANIC COCS

Altered mobility of metal co-contaminants potentially can occur during in situ chemical oxidation (ISCO) as described in Chapter 3. For example, in situ permanganate reactions can yield low pH/high Eh conditions depending on the pre-treatment environmental conditions (e.g., buffering capacity and reductants present). In some cases, this can alter ion exchange chemistry. This can affect the mobility of metals such as Zn, Cd, Cr, and As (see Chapter 3). If altered mobility is not planned for, it can yield potentially adverse risk consequences. Alternatively, if the

125

mobility is anticipated and adequate controls are put in place, it is conceivable that metals could be mobilized and captured.

The mobility of untreated organic COCs also must be considered. Mobilization of organics dissolved in groundwater beyond the boundary of the subsurface region to be treated by ISCO can result if adequate hydraulic controls are not put in place (see Chapter 5). For example, if an injection well is used to deliver clean water amended with oxidant to a groundwater zone, this can create hydraulic gradients that can push groundwater containing COCs out of the treated region before ISCO can occur. If pressures are high enough, there is some chance that DNAPLs could be mobilized, however this will be highly site specific and likely not of great concern in most settings.

As noted in Chapters 2 and 3, during permanganate oxidation of organic COCs or other reductants, MnO_2 solids can be produced as a reaction product. For example, during the mineralization of TCE, 1 mole of MnO_2 solids is produced for each mole of permanganate consumed (e.g., see eqn. 2.9).

$$2MnO_4^- + C_2HCl_3 = 2MnO_2(s) + 2CO_2 + 3Cl^- + H^+ \qquad (2.9)$$

Not surprisingly, observations made during ISCO with permanganate have revealed that manganese oxide (MnO_2) solids are generated. These solids can have varying effects on subsurface conditions depending on the pre-ISCO site conditions and the design of the permanganate ISCO (Figure 4-1). For example, the solids can be discrete and agglomerated micron-sized particles that can be mobile in some settings (West et al. 1998a,b, Siegrist et al. 2000, West et al. 2001). Alternatively, the MnO_2 solids may interact with other solids in the soil and groundwater system. For example, the MnO_2 solids may coat porous media surfaces and provide a strong sorbent for immobilizing metals (see Chapter 3). The solids may grow in size and become pore-filling agents that cause a loss in permeability.

Relevant to the discussion of altered behavior of organic COCs, MnO_2 solids can coat organics and reduce interphase mass transfer. With DNAPLs, that can dramatically slow down the rate of DNAPL dissolution and oxidative degradation under some conditions (Urynowicz 2000). Figure 4-2 presents a series of photographs illustrating the formation of an interfacial coating or film on a TCE DNAPL after chemical oxidation with permanganate. Development of such a film has been shown to occur

FIGURE 4-1. Illustration of particle genesis processes during ISCO and the potential for development of interfacial deposits and coatings and/or pore plugging agents.

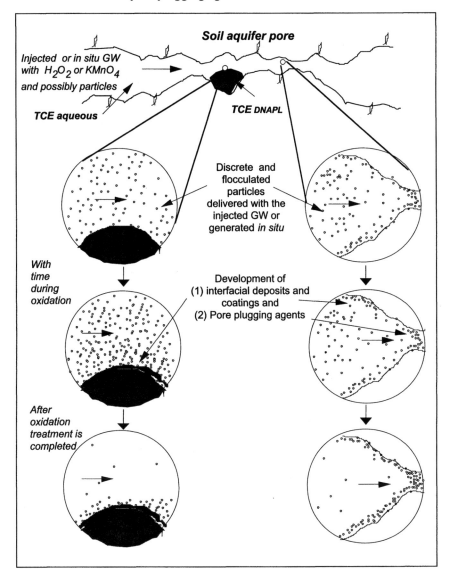

FIGURE 4-2.   **MnO$_2$ film formation at the DNAPL-water interface explaining changes in DNAPL degradation rate with time (Urynowicz 2000).**

(a) TCE$_{DNAPL}$ droplet suspended in phosphate buffered de-ionized water from the tip of a syringe needle prior to chemical oxidation with MnO$_4^-$. Needle diam. = 0.8 mm; pH 6.9; I=0.3 M.

(b) TCE$_{DNAPL}$ droplet suspended in phosphate buffered de-ionized water approximately 3 hours after chemical oxidation with MnO$_4^-$.

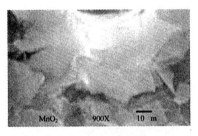

(c) MnO$_2$ film residual left following chemical oxidation with MnO$_4^-$ and complete dissolution of he TCE$_{DNAPL}$ droplet.

(d) Photomicrograph of plate-like MnO$_2$ film left following chemical oxidation with MnO$_4^-$ and complete dissolution of the TCE$_{DNAPL}$ droplet (magnification = 900x).

during laminar flow conditions representative of groundwater flow regimes in aqueous flow-through cells (Urynowicz 2000), as well as 1-D and 2-D cells containing porous media (Li and Schwartz 2000, Reitsma and Marshall 2000). As the film develops, it can reduce the oxidative degradation of DNAPLs (Urynowicz 2000). That behavior could be detrimental in settings with DNAPL pools or high levels of residuals if the performance goal were to degrade all of the DNAPL mass. However, if the goal were to reduce the concentration of dissolved COCs in the mobile groundwater, then this film formation might yield a form of chemo-stabilization that could enable achieving that goal. Long-term behavior of the 'stabilized' DNAPL is obviously of importance for this approach to be effective.

4.2. PERMEABILITY CHANGES DUE TO PARTICLE GENESIS AND GAS EVOLUTION

Particulate Generation

In situ treatment involving chemical oxidants can produce particles by shearing off fragments of natural soil or by yielding reaction products (e.g., iron or manganese oxides). Unique to permanganate, compared to other oxidants such as Fenton's reagent or ozone, is the production of MnO_2 solids as a reaction product (e.g., see eqn. 2.9). While manganese oxide is often represented as MnO_2 in stoichiometric equations, it is more appropriately described as $MnO_{1.7-2.0}$ because the oxide contains varying percentages of lower valent manganese. Those particles can be of concern with regard to their effects on permeability of a subsurface formation or on the delivery system (e.g., well screen and filter pack). The MnO_2 solids themselves can react with other dissolved and colloidal solids and form agglomerations of mixed composition. These particles can be filterable at 1 um or less and thus can be considered suspended solids.

At CSM, a duplicated, 2^5 factorial design was recently completed to examine the mass of particles produced as affected by oxidant type, oxidant concentration, TCE concentration, reaction time, and presence of silt/clay particles (Siegrist et al. 2000) (see Chapter 3). The results of this experiment revealed that the quantity of the particles produced during chemical oxidation was impacted by several conditions. The regression equation that made physical sense and accounted for the most variance in the particle-generation data (for both oxidant forms plus both reaction times combined) was that shown previously in eqn. 3.1 ($R^2 = 0.85$) which is reproduced here:

$$\text{Particles} = 0.0089 \, (MnO_4^- \text{ dose}) + 2.14 \, (\text{TCE conc.}) + 0.197 \, (\text{ambient filterable particles}) \quad (3.1)$$

where all concentrations are in mg/L. This relationship is applicable when MnO_4 is in excess of TCE demand. Eqn. 3.1 suggests that the presence of

ambient filterable particles in the groundwater increases the net production of particles beyond that of just the MnO_2 alone. Under some conditions, such as higher TCE (54 mg/L), MnO_4^- (3000 mg/L) and 750 mg/L
ambient silt/clay, there can be a net increase of 300 to 400 mg/L of >0.45
um particles in groundwater (Siegrist et al. 2000).

The mass and chemical characteristics of the MnO_2 formed, their size
and ultimate fate, and what, if any, effects they may have on remediation
system function and performance depend on environmental conditions,
contaminant levels, and oxidant loading and ISCO design. Permeability
effects observed during laboratory studies and field applications are discussed further below.

Gas Evolution

With peroxide-based oxidants, substantial gas and heat can be evolved
during exothermic reactions including autodecomposition and organic
chemical oxidation. This can cause concern over pressure gradients and
gas migration yielding fugitive emissions as well as gas entrapment with
concomitant permeability loss (Wiesner et al. 1996). However, with permanganates, gas-evolving autodecomposition reactions do not occur and
the oxidation reactions are slower and only weakly exothermic.
Nevertheless, CO_2 is a product of the mineralization of organics like TCE
(as shown in eqn. 2.9). There have been some reports of gas evolution
during permanganate ISCO in porous media with DNAPLs present and of
problems with permeability as noted below (e.g., Li and Schwartz 2000).
However, fugitive emissions as a result of gas evolution are not anticipated to be a problem with permanganate ISCO.

Permeability Effects

Laboratory and field studies have provided results which indicate that
subsurface permeability may or may not be markedly impacted by in situ
permanganate oxidation. Manganese oxides and carbon dioxide gas
generated through permanganate oxidation of organic materials
conceivably can impact flow patterns in the subsurface. The degree to
which they can impact permeability appears to be related to the amount
of contaminant in the reaction zone, as well as the reaction rate, which are
inter-related. As noted previously, the reduction of MnO_4^- during oxidation of target organics and NOM leads to the production of MnO_2 solids.
Depending upon the system design, contaminant characteristics, and

groundwater chemistry, MnO_2 may be formed ex situ or in situ. The MnO_2 are initially colloidal in size and therefore could be transported out of the treatment zone. Excessive MnO_2 production, possibly coagulated/flocculated, however, may lead to pore clogging within the formation and/or at a well screen/filter pack and thereby cause reduced groundwater flow. Described below are observations made during several experiments and field applications that have provided insight into the potential effects of permanganate ISCO on system permeability, including delivery points (e.g., wells) and groundwater formations.

A set of ISCO applications in Ohio revealed the types of problems that can be created with particles that are generated ex situ and not removed before vertical well injection. The first application involved groundwater flushing using a 5-spot vertical well recirculation network to deliver 250 mg/L $NaMnO_4$ to treat 1 to 2 mg/L TCE in groundwater (see Site 2 in Chapter 7). As part of this application, pre- and post-treatment hydraulic conductivity evaluations were made using single well slug tests (Lowe et al. 2000). Results of slug tests at 10 wells indicated that there was no loss in formation permeability due to ISCO under the conditions of the application (i.e., low levels of TCE and low oxidant dose) (Table 4-1). A second application, based in part on the success of the first, was conducted in the same aquifer formation but at another location at the same facility (near Site 1 as described in Chapter 7). This application also employed a 5-spot vertical well network (four perimeter wells for extraction at 55 ft. radius from one center well for reinjection) but the concentrations of TCE were much higher (up to 600 mg/L) and the oxidant dose was also higher (3000 mg/L). This application was focused on treatment of a groundwater plume that had developed in the aquifer formation, which typically occurs at a depth of 23 to 30 ft. bgs and is a 3- to 8-ft. thick layer. The aquifer formation is comprised of a saturated silty-sandy gravel unit of higher permeability overlying a lithified unit of coarse angular gravel in a silty clay matrix (41 to 79 dry wt. % levels of silt plus clay). The hydraulic conductivity is variable in the range of 30 to 300 ft/day. During operation, groundwater was extracted from the four perimeter wells, supplemented with 3,000 mg $NaMnO_4$ per L of extracted groundwater, passed through 5- and 1-um in-line cartridge filters, and then injected into a central injection well at a maximum rate of about 6 gpm. After approximately 5 days of operation, increasing injection well pressures (up to 18 psig) caused reduced recirculation rates (down to 4 gpm). Redevelopment of the injection well recovered the well efficiency, and operations were

TABLE 4-1	Hydraulic conductivities measured during single well aquifer tests (Lowe et al. 2000).[1]		
Well Number	Average pre-oxidation (ft/day)	Average post-oxidation (ft/day)	Ratio of post- to pre-oxidation
X770-EW02	26.5	26.7	1.00
X770-EW03	86.5	87.3	1.00
X770-EW04	44	54.3	1.23
X770-MW01	148	94	0.64
X770-MW03	297.5	271.7	0.91
X770-MW05	691.5	532.3	0.77
X770-MW07	202.5	187.7	0.93
X770-MW08	121	143	1.18
X770-MW10	145	113.7	0.78
X770-MW12		366.3	–
X770-MW14	601.5	337	0.56

[1]Refer to Site 2 in Chapter 7 for further site information.

resumed, but at reduced recirculation rates (4 gpm). Increasing injection pressures and reduced recirculation rates again were rapidly observed resulting in attempts to redevelop the injection well on day 10 and 13. On day 15, surface breaching of the injected fluids at a nearby monitoring well resulted in system shutdown.

Siegrist et al. (2000) completed laboratory experiments to understand the nature and cause of the permeability problems observed in the field as noted for the application just described. The experimental work included (1) characterization of field samples collected from boreholes made along a transect from the oxidant injection well to a background location, (2) batch experiments using a 2^5 factorial design to examine oxidant consumed and particles produced as affected by oxidant type, oxidant concentration, TCE concentration, reaction time, and presence of silt/clay particles, and (3) flow-through column experiments. Analyses of soil cores from the ISCO field site revealed that MnO_2 solids were present in the subsurface near the injection well for $NaMnO_4$, but at low levels (2.3-2.5 mg/g dry wt. media) amounting to <1% v/v of the porosity. Batch tests revealed that the mass of filterable particles (>0.45 um) produced was increased during MnO_4^- oxidation at higher TCE concentrations (54 vs. 7 mg/L) and in the presence of groundwater particles (750 mg/L vs. 7.5 mg/L). Under otherwise comparable conditions, increasing the MnO_4^- dose did not markedly affect the particle production, but it did markedly increase the oxidant consumption. The oxidant form ($NaMnO_4$ vs. $KMnO_4$) or reaction time (15 vs. 300 min) had little effect on oxidant consumption or filterable particle production. Flow-through column experiments revealed that permeability loss was possible during ISCO but only under conditions with very high MnO_2 particle production (e.g., high oxidant loading and high reductant levels). The filterable particles produced during ISCO with MnO_4^- could be dissolved by a weak peracetic acid solution suggesting a possible method for permeability recovery.

West et al. (1998a, 1998b, 2000) observed substantial reductions in hydraulic conductivities as a result of an ISCO application where 2 to 4 wt.% of $KMnO_4$ was used to treat TCE at 100 to 800 mg/L concentrations in groundwater (see Site 1 in Chapter 7). Hydraulic conductivities were measured 14 months prior to ISCO in some monitoring wells in the vicinity of the test site. Re-measurement of hydraulic conductivities in these wells 10 months after completion of the ISCO field test showed order of magnitude decreases in four wells (73G, 74G, 77G, and 81G) (see Table 4-2). The most dramatic decrease was observed in well 74G that had been used for vertical well oxidant injection into groundwater with high TCE present. The clogging was judged to be caused by precipitation of MnO_2, which also could have been aggravated by the mobilization of colloidal aquifer material (West et al. 2000). Total suspended solid (TSS) levels were significantly higher than background levels in wells where MnO_4^- had been detected during and subsequent to oxidant injection (Figure 4-3). Elevated TSS levels were also measured in wells that are downstream of the test site but where MnO_4^- had never been detected, indicating that the particles, either created or mobilized by the ISCO treatment, were being carried out of the treatment zone by ambient groundwater flow.

Li and Schwartz (2000) conducted 1-D column and 2-D test cell studies to examine mass removal rates and related flushing efficiencies resulting from reaction of permanganate with some typical aquifer materials (Table 4-3). Columns were packed with a medium silica sand, and 1 mL of TCE was added to the opening of the upside-down columns

Well no.	Pre-treatment K March 1997 (ft/day)	Post treatment June 1998 (ft/day)	Post treatment June 1998 Trial 2 (ft/day)	Ratio of post-treatment average pre-treatment value
TABLE 4-2 Results of formation hydraulic conductivity tests before and after in situ oxidation with 2 to 4 wt.% KMnO₄ to treat high levels of TCE in groundwater (West et al. 1998a,b).				
73G	39.6	1.0	1.7	0.040
74G	182.8	0.6	0.8	0.004
76G	24.2	57.7	55.8	2.34
77G	411.1	93.6	139.8	0.284
78G	65.5	153.8	142.4	2.26
79G	142.6	34.0	20.1	0.19
80G	31.7	63.7	57.0	1.90
81G	60.2	5.5	7.6	0.109

[1]Refer to Site 1 in Chapter 7 for additional site description. TCE concentrations in groundwater were as high as 800 mg/L suggesting the presence of DNAPL residual.

FIGURE 4-3.   Total suspended solids in background and MnO₄⁻ impacted wells. Samples were collected 10 months after the ISCOR field test (West et al. 2000).

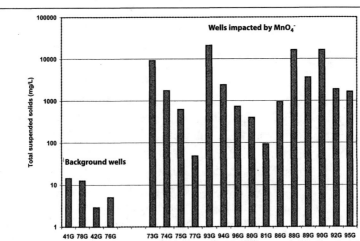

Note: Well group A consists of wells that are upstream of the horizontal injection well and where MnO₄⁻ was never detected; these wells are assumed to represent background conditions. Well Group B are wells where MnO₄⁻ was still detected 10 months after ISCO. Well Group C are wells where MnO₄⁻ was detected 2 weeks after the ISCO, but not detected 10 months after ISCO. Wells 9 and 21G (well group D) are ~50 and ~400 ft downstream of the injection horizontal well; MnO₄⁻ was not detected in either well through 10 months after ISCO.

TABLE 4-3	Experimental design for 1-D column and 2-D flow experiments (Li and Schwartz 2000).				
Type	Apparatus dimensions (mm)	Porous medium	Flow rate (mL/min)	KMnO$_4$ (g/L)	Porosity
1-D	605 long by 50 I.D.	Silica sand	1.2	1.0	0.385
2-D	305 long by 50 high by 3 thick	Glass beads	0.5	0.2	0.42

to create an even zone of residual DNAPL. Columns were returned to their upright position, and fluid was pumped upward through the column. Column effluent was collected and analyzed for permanganate, TCE, and chloride (by-product of the oxidation reaction) every 12 hours for two weeks. The study was concluded when TCE effluent concentrations reached below 5 ug/L. Following the experiment, several samples of the sand were taken along the column, treated with thiosulfate to dissolve the MnO$_2$, then analyzed by ICP-MS for Mn to quantify the MnO$_2$. The distribution of manganese oxides in the column (Figure 4-4) indicates that

FIGURE 4-4.   Distribution of Mn along a 1-D silica sand column following upflow by 1 g/L KMnO$_4$ for ISCO of TCE (Li and Schwartz 2000).

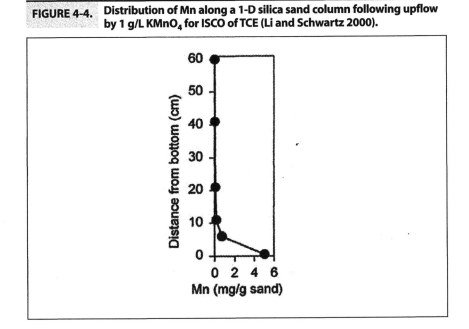

the majority of the Mn was located close to or at the DNAPL zone. These precipitates tended to plug the column, and flushing became more difficult as the experiment progressed.

In the 2-D studies (Table 4-3), a small tank was filled with transparent glass beads with a mean grain size of 1mm. Recharge and discharge wells were designed on each end of the tank. The tank was saturated with water, then two pumps running at the same steady state maintained inflow and outflow rates. One mL of TCE was added to the top of the tank to form a zone of residual DNAPL saturation across the vertical depth of the tank, and a small DNAPL pool was formed at the bottom. For two weeks, the tank was flushed with potassium permanganate. Effluent samples were taken three times a day and were analyzed in the same manner as those of the 1-D columns. Results were similar to those of the 1-D column experiments, except that removal rates were much smaller. Based on visual observation, once the permanganate came into contact with the residual DNAPL, there was a tendency for the permanganate flood to bypass the zone with the highest DNAPL saturation, moving through the less saturated zone toward the top of the tank. Over time, precipitation of manganese oxides reduced the permeability within the tank. The MnO_2 formed a precipitation rind above the DNAPL pool, and toward the end of the experiment a greater injection pressure was required to maintain the flow. The experiment was ended when the manganese oxides plugged the tank nearly completely, to the point that injection was not possible. Examination of the tank indicated the presence of tiny bubbles, presumed to be CO_2 produced from the oxidation. The gas bubbles were presumed to reduce the permeability and flow in the system. A conclusion of the study was that the tendency for preferential flow paths to develop will be promoted as manganese oxides precipitate in the zones of higher DNAPL saturation or as carbon dioxide bubbles are trapped in the porous medium. The rapid rate of TCE oxidation and relatively slow mass transfer rate from the DNAPL to the aqueous phase means that the manganese oxides would tend to precipitate at or immediately adjacent to the DNAPL. This would result in a zone of more concentrated precipitates around the zones of greatest saturation. This may cause the flow to bypass these zones to follow a more permeable flow path, which would cause the DNAPL oxidation process to become diffusion-controlled.

Reitsma and Marshall (2000) also conducted 2-D experimental studies to examine the processes that occur during permanganate ISCO of pooled NAPL in porous media. A tank was packed with a porous fine sand medium (porosity = 0.44), with a coarse sand lens (porosity = 0.49) containing a DNAPL pool in the center of the tank. An initial flush of water (3.0 L/d) was used to determine the baseline mass transfer rates from the pooled DNAPL. Then, $KMnO_4$ was introduced at a uniform rate from one end of the tank, and effluent samples were taken from the opposite end. There were additional sampling collection points established around the DNAPL pool to determine the dissolved DNAPL concentration distributions. The permanganate flush was then followed by an additional water flush to determine post-treatment mass transfer rates. Three different experiments were conducted in the tank: (1) 5.0 mL of TCE injected into the coarse sand lens, with 10 g/L $KMnO_4$ flow-through, (2) 5.0 mL of PCE, with 5 g/L $KMnO_4$, and (3) 2.0 mL of PCE with 1 g/L $KMnO_4$. The experiments were designed so that the reaction rate of experiment 1 was 200 times the rate of experiment 2, and 1,000 times the rate of experiment 3.

Experiments 1 and 2 conducted by Reitsma and Marshall (2000) provided similar results. Initially, most of the flow was diverted around the DNAPL due to the low permeability of the coarse lens caused by the presence of the NAPL. Visual inspection for manganese oxides provided an indicator of where the reaction was taking place since the MnO_2 were precipitating almost immediately after reaction and were not being transported. Early on, small amounts of MnO_2 were observed in the first few centimeters of the DNAPL, indicating that some of the permanganate solution had entered the coarse lens. Most of the reaction took place within the first millimeter of this zone, which was indicated by heavy build-up of manganese oxides. This type of build-up also was observed on the upper and lower bounds of the lens but not at the posterior end. There were thin areas of manganese dioxides, which coincided with mixing of dissolved TCE from its down-gradient plume with $KMnO_4$ from its surrounding zone. As the experiment continued, the production of CO_2 gas was rapid, and led to nearly complete desaturation of DNAPL from the coarse sand lens, followed by reduction of permeability of the sand to $KMnO_4$. As the gas volume increased, the bubbles traveled upwards vertically above the coarse sand lens. Manganese dioxide formation was observed several centimeters above the lens and coincided with the zones of CO_2-induced desaturation. This suggested that CO_2 gas may have transported TCE vapor from the lens into the sand above. It may also have enhanced overall mass transfer by providing an additional transport mechanism from the TCE pool. In experiment 3, reaction rates were relatively much slower than in experiments 1 and 2, and that resulted in a greater permanganate front through the DNAPL in the coarse sand lens (Figure 4-5). As the experiment progressed, gas began to desaturate the coarse sand lens (Figure 4-6). The shape of the precipitated manganese oxides plume in experiment 3 was different than in the prior two experiments. Heavy precipitation occurred throughout the DNAPL zone and fanned out down-gradient from the pool. Precipitation was greatest within the actual coarse sand lens, unlike the other experiments. Due to the slower reaction rate in experiment 3, permanganate was able to pass through the entire DNAPL zone and continue to react downstream. After approximately 12 pore volumes of flow-through, the MnO_2 plume became twice the thickness of the coarse sand lens (Figure 4-7). This was due to upward movement of carbon dioxide gas near the front of the pool. Experiment 3, as compared to the other two experiments, demonstrated that even at slow reaction rates carbon dioxide gas and manganese dioxides precipitation remain an issue. An overall conclusion of the experimental studies was that manganese dioxides and carbon dioxide

FIGURE 4-5. Front view of a sand tank during a 1.0 g/L $KMnO_4$ flush through a PCE contaminated coarse sand lens surrounded by fine sand (Reitsma and Marshall 2000).

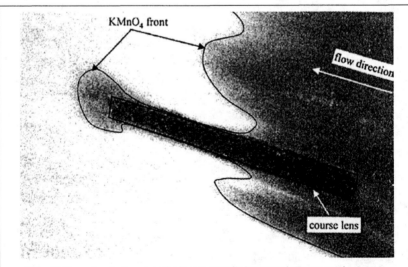

KMnO$_4$ front

flow direction

course lens

Note: Oblique view is shown as the lens is actually horizontal. Coarse sand n = 0.49, fine sand n = 0.44. PCE in coarse sand = 2.0 mL, Q = 3.0 L/d with 1.0 g/L $KMnO_4$.

FIGURE 4-6.	Front view of a sand tank after 1.0 g/L KMnO$_4$ had been flushed through it for 110 hours and CO$_2$ had evolved and MnO$_2$ precipitation was observed (Reitsma and Marshall 2000).

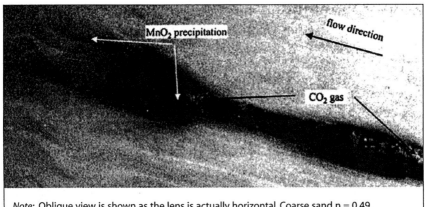

Note: Oblique view is shown as the lens is actually horizontal. Coarse sand n = 0.49, fine sand n = 0.44. PCE in coarse sand = 2.0 mL, Q = 3.0 L/d with 1.0 g/L KMnO$_4$.

FIGURE 4-7.	Front view of a sand tank after completion of a 1.0 g/L KMnO$_4$ flush and the post treatment water flush (Reitsma and Marshall 2000).

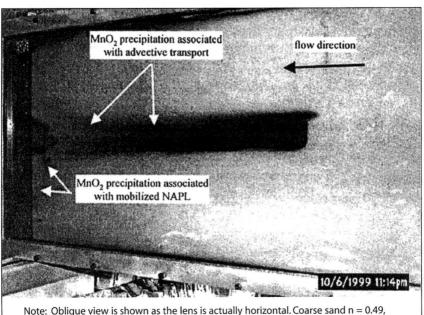

Note: Oblique view is shown as the lens is actually horizontal. Coarse sand n = 0.49, fine sand n = 0.44. PCE in coarse sand = 2.0 mL, Q = 3.0 L/d with 1.0 g/L KMnO$_4$.

production are important because of their potential for changing flow patterns. The behavior of the permanganate delivery is affected by the NAPL zone permeability, which is impacted by the CO_2 and MnO_2. This effect likely would be even more apparent with greater system heterogeneity.

While the studies just described demonstrate the possible deleterious effects of permanganate ISCO on system permeability and degradation of DNAPLs, there are other studies that indicate no reduction in permeability (or treatment effectiveness) during permanganate ISCO. The study described above by Lowe et al. (2000), where 250 mg/L of $NaMnO_4$ was used to flush a groundwater zone with 1 to 2 mg/L of TCE, is one example (see Site 2 in Chapter 7). Another site with low levels of chlorinated solvent contamination is described by Chambers et al. (2000). These investigators performed column studies, including standard permeability tests, on sandy soil in support of field operations at a Sunnyvale, CA, site that contained up to 8 mg/L of cis-DCE and TCE in groundwater. A standard permeability test (ASTM D 5084) was performed with clean water before ISCO to determine permeability and porosity values for the sand. Then the test was modified to examine post-treatment permeability and porosity with $KMnO_4$ as the permeant flushed through at 1, 5, and 25 g/L for 15 pore volumes. Effluent was collected after each pore volume of flowthrough. The results indicated no permeability reduction due to permanganate treatment (Table 4-4). In fact, the hydraulic conductivity was increased by almost an order of magnitude after application of the 25 g/L potassium permanganate treatment. MnO_2 was measured in column effluent, and varied from 37 to 75 mg/L, but with no obvious trend.

TABLE 4-4	Hydraulic conductivities of sand (cm/s) after flushing with different concentrations of $KMnO_4$ solutions (Chambers et al. 2000).		
	$KMnO_4$ = 1 g/L	$KMnO_4$ = 5 g/L	$KMnO_4$ = 25 g/L
Before[1]	3.2×10^{-5}	2.2×10^{-5}	3.2×10^{-5}
After[2]	3.4×10^{-5}	2.5×10^{-5}	$11. \times 10^{-5}$
% change	+6.2	+13.6	+240

[1]Conductivity measured with water.
[2]Conductivity measured after 15 pore volumes of flushing with $KMnO_4$ solution.

A field study of ISCO of DNAPL at a Cape Canaveral, Florida, site also demonstrated very little permeability reduction due to permanganate treatment of DNAPL residuals (Mott-Smith et al. 2000). This site had three primary lithologic units: (1) Upper Sand Unit (USU) containing fine sands, (2) Middle Fine Grained Unit (MFGU) containing silty fine sand with lenses of sandy clay, and (3) Lower Sandy Unit (LSU) with inter-bedded layers of shell hash, fine sand, silty fine sand, and sandy clay (Figure 4-8). The USU, MFGU, and LSU contained 846 kg, 1,048 kg, and 4,228 kg of TCE respectively, with TCE contamination starting at about 4.6 m below ground in the USU. The average hydraulic conductivity of the soil varied from 1.8 to 4.6 ft/day and varied significantly within each lithologic unit. The treatment system included an automated $KMnO_4$ feed system delivering solution at 0.1 to 3 wt.% into the subsurface through injection probes installed using a direct push method. The maximum throughput of $KMnO_4$ was approximately 159 kg/hr with flow rates at each point ranging from 0.1 to 6.1 gpm at wellhead pressures of 15 to 55 psig. Results indicated significant mass destruction of TCE with no evident significant formation plugging. Injection flow rates decreased less than 10% during the first two phases of operation at the site.

FIGURE 4-8. **Lithology of the LC-34 site at Cape Canaveral where injection probes were used to deliver $KMnO_4$ for in situ treatment of TCE DNAPL residual (Mott-Smith et al. 2000).**

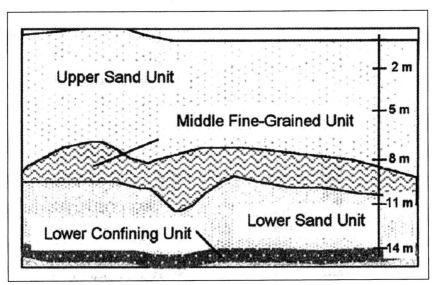

As evidenced by the above studies, ISCO with permanganate may or may not contribute to a localized reduction in delivery point or formation permeability. When low concentrations of organics are present in porous media of average density and porosity, calculations based on the density of generated MnO_2 indicate very low hypothetical pore volume reduction due to the MnO_2. For example, Struse (1999) measured MnO_2 concentrations in silty clay soil after diffusion of MnO_4^- from a source (5000 mg/L $KMnO_4$) for 40 days and found only low levels of MnO_2 ($\leq$ 2.2 mg/g) compared to 0.13 mg/g in untreated media (Table 4-5). Initially, Struse hypothesized that the deposition of MnO_2 as a by-product of the oxidation reaction might clog or partially block soil pores, thereby increasing the tortuosity of the porous media system. However, it was not possible to identify any manganese, and therefore MnO_2, using scanning electron microscopy (SEM). Based on the chemical analysis, the MnO_2 accounted for only 0.082% of the total pore volume in core 1 and 0.086% in core 2.

With very high organic contaminant concentrations (e.g., 10,000 mg/kg), permeability reductions could be quite significant. For example, Schnarr and Farquhar (1992) noted a 5.6% decrease in pore volume after

TABLE 4-5.	Changes in soil constituent concentrations after diffusive transport of $KMnO_4$ through intact cores of silty clay soil (after Struse 1999).	
Distance from influent chamber	**Core 1 TOC ($\mu g/mg$)**	**Core 2 TOC ($\mu g/mg$)**
TOC measured at 0.64 cm	0.73	0.50
TOC measured at 1.91 cm	1.07	0.64
TOC in untreated duplicate core	1.09	0.65
Loss in TOC at 0.64 cm	0.34	0.14
Loss in TOC at 1.91 cm	0.02	0.01

Core number / cm from influent end	Extractable constituents								
	K^+ mg/g	Ca^{+2} mg/g	Mg^{+2} mg/g	Mn^{+2} mg/g	Fe^{+2} mg/g	Br^- mg/L	Cl^- mg/L	MnO_4^- mg/L	MnO_2 mg/g
1 0.64	2.06	0.35	0.22	0.08	0.00	71.0	59.8	1886	2.15
1.91	1.54	1.30	0.54	0.15	0.00	45.9	50.2	840	1.14
2 0.64	2.9	0.54	0.26	0.28	0.11	23.0	69	1727	2.0
1.91	1.2	0.72	0.36	0.03	0.05	8.6	40	169	1.5

injecting potassium permanganate into PCE-contaminated soil, as compared to the 1% volume change attributed to the PCE itself.

Concern over permeability loss appears greatest for sites with high concentrations of organic COCs (e.g., DNAPLs) in groundwater to which high concentrations of permanganate (wt.%) are delivered using horizontal flushing from vertical or horizontal wells. Permeability loss appears to be due to the formation of MnO_2 solids and their deposition on well screens, filter packs, and, in some cases, the aquifer sediments. The occurrence and extent of permeability effects is dependent on the contaminant levels, environmental setting, and operational factors.

4.3. GEOCHEMICAL AND MICROBIAL PERTURBATIONS

The application of permanganate to the subsurface may have effects on the pre-existing geochemistry and microbiology of the site. This may be important to consider with respect to the overall treatment effectiveness achieved and the potential for adverse effects on the subsurface. This section provides a discussion of the following topics and the potential impacts of permanganate: (1) the fate of manganese added, (2) natural organic matter, (3) pH, (4) cation exchange, and (5) microbiology.

Fate of the Manganese added in Permanganate

The fate of the manganese added to the subsurface may be of interest and possible concern at a given site. Under appropriate geochemical conditions, manganese dioxides produced by the reaction of MnO_4^- with oxidizable materials can be dissolved with time, resulting in the release of Mn^{+2} into the groundwater. Because mineral dissolution rates depend on available surface area, the dissolution of particulate MnO_2 is expected to be slower than that of MnO_2 that remains in colloidal form.

West et al. (2000) completed an application of ISCO using $KMnO_4$ at a site where groundwater was contaminated with TCE at levels suggesting the presence of residual DNAPLs (see Site 1 in Chapter 7). As part of the long-term monitoring of this site, studies were conducted to evaluate the predominant form of residual Mn in the treatment region (West et al. 2000). Dissolved Mn levels were measured in filtered (<0.45 μm) groundwater samples collected 10 months after the ISCO field test (Figure 4-9). Very high Mn levels were measured in the MnO_4^- impacted wells where MnO_4^- was still detected at the time of sampling. However, in wells where MnO_4^- was no longer present (wells 86G, 88G, 89G, 90G,

FIGURE 4-9. **Dissolved Mn concentrations in filtered (< 0.45 μm), acidified groundwater samples collected from background and MnO₄⁻-impacted wells. Samples were collected 10 months after the ISCOR field test (West et al. 2000)[1].**

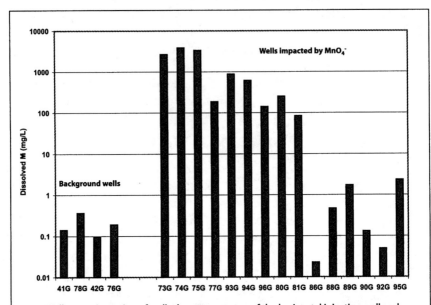

Note: Well group A consists of wells that are upstream of the horizontal injection well and where MnO_4^- was never detected; these wells are assumed to represent background conditions. Well Group B are wells where MnO_4^- was still detected 10 months after ISCO. Well Group C are wells where MnO_4^- was detected 2 weeks after the ISCO, but not detected 10 months after ISCO. Wells 9 and 21G (well group D) are ~50 and ~400 ft downstream of the injection horizontal well; MnO_4^- was not detected in either well through 10 months after ISCO.
[1]Refer to Site 1 in Chapter 7 for further site information.

92G, 95G in Figure 4-9), Mn levels had dropped to near or slightly above background levels. Mn-speciation of soil samples collected from the test site were consistent with the groundwater results (West et al. 2000). High levels of Mn extracted by a 50mM $CaCl_2$ solution corresponded to soil samples that still contained detectable levels of MnO_4^- (Figure 4-10). However, in soil samples where MnO_4^- was no longer present, very little Mn was leached out of the soil by the $CaCl_2$ solution (West et al. 2000). More Mn was extracted from the soil samples by a reducing agent (i.e., hydroquinone-ammonium acetate solution; Figure 4-9); this Mn includes exchangeable/soluble Mn^{+2}, dissolved MnO_4^- and reducible Mn-oxides. The latter probably includes natural as well as treatment-generated Mn-oxides.

FIGURE 4-10.   Results of Mn extraction and residual MnO$_4^-$ analysis in aquifer sediment samples collected from a borehole drilled near well 74G 10 months after ISCO (West et al. 2000)[1].

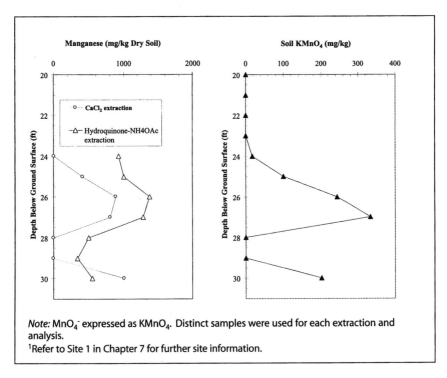

Note: MnO$_4^-$ expressed as KMnO$_4$. Distinct samples were used for each extraction and analysis.
[1]Refer to Site 1 in Chapter 7 for further site information.

Impacts of Permanganate on Natural Organic Matter

As described in Chapter 2, permanganate will oxidize many constituents in NOM, and this NOM demand for MnO$_4^-$ can be substantial. However, some fraction of the original NOM and/or products of its degradation have been shown to be resistant to complete mineralization, even in the presence of high MnO$_4^-$ levels over long time periods. Struse (1999) and Siegrist et al. (1999) conducted laboratory and field studies of KMnO$_4$ interactions with low-permeability clays and silts (K$_{sat}$ <10^{-5} cm s^{-1}). Those sediments are comprised of 70 to 95% silt (quartz and feldspars) and clay particles (illite, quartz, kaolinite, and smectite) with pH of ~6.0, cation exchange capacity of ~17.5 meq per 100g, and total organic carbon (TOC) of 500 to 1500 mg per kg (Siegrist et al. 1999). Struse (1999) used intact cores for permanganate diffusion studies in a transport cell. The oxidant demand of the uncontaminated silty clay soil also was measured during 14-day batch tests and found to be 2.8 and 10.8 mg-MnO_4^- per g soil at oxidant doses of 500 and 5000 mg L^{-1} $KMnO_4$, respectively (Struse 1999). A source concentration of 5000 mg/L of $KMnO_4$ in a simulated groundwater (pH = 5.0, Ec = 277 umhos, TDS = 160 mg/L, alkalinity = 40 mg-$CaCO_3$/L) was allowed to diffuse through the silty clay for periods of 40 days or more. Using the TOC content determined for an untreated core of silty clay as a reference value, it was estimated that the TOC was decreased through oxidation by 33% or less (Table 4-5). These results indicate that although the NOM initially present in both cores was decreased, it was not completely degraded, and substantial NOM was apparently resistant to oxidation by MnO_4^- even after about 40 days of exposure to MnO_4^- levels of 800 mg/L or more.

The laboratory results of Struse (1999) compare well with observations made during a field application of ISCO using hydraulic fractures propped with permanganate oxidative particle mixture (OPM) (see Site 5 in Chapter 7). Soil cores were obtained at the site 10 months after initial emplacement of the permanganate-filled fractures, and appreciable TOC remained even after 10 months of exposure in a highly oxidizing zone with permanganate present (Figure 4-11) (Siegrist et al. 1999).

Cline et al. completed an application of ISCO using deep soil mixing to deliver up to 5 wt.% $KMnO_4$ solutions to either 25- or 47-ft depth at a

---

FIGURE 4-11. Geochemical properties above and below a permanganate OPM-filled fracture 10 mon. after emplacement in a silty clay deposit (Siegrist et al. 1999).

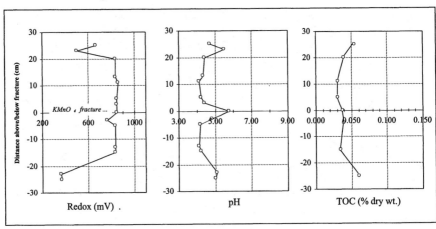

site contaminated with TCE at 1500 mg/kg or higher in silty clay soils (Cline et al. 1997, Gardner et al. 1998) (see Site 3 in Chapter 7). During the test, 6800 gal of $KMnO_4$ solution (2450 lb) was mixed into three 25-ft deep soil columns having a total soil volume of 11,425 ft^3 (average loading of 97 g $KMnO_4$ per ft^3 of soil). For application to 47-ft depth, 10,950 gal of $KMnO_4$ solution (3510 lb) was delivered (average loading of 74 g $KMnO_4$ per ft^3 soil). TCE removal was estimated at 65 and 68% for the shallow and deep treatment areas, respectively (Cline et al. 1997). Consistent with the high NOM content in the soil (1 wt.%), residual permanganate was not detected in the subsurface 3 to 5 days after oxidant injection. Prior to ISCO, the TOC content ranged from approximately 9000 mg/kg near the ground surface to 4000 mg/kg at 37 ft. depth. After treatment, the TOC content had declined and was more uniform with depth, but it remained in the range of 4000 to 7000 mg/kg throughout the permanganate treated zones.

The absence of complete mineralization of NOM, as noted above, could be due to an oxidant loading that was insufficient for the total demand of the organic COCs as well as NOM and other reductants in the system. Where adequate oxidant was provided and it persisted for a long time period, there could be formation of organic acids and other organic byproducts that are somewhat more resistant to permanganate degradation. In addition, some organics could be coated with MnO_2 solids and thereby protected from further oxidation by MnO_4^- in the surrounding aqueous phase.

Changes in Subsurface pH

Permanganate oxidation may yield acidity or alkalinity depending on the reaction chemistry for the reductants being oxidized. For example, as illustrated in equations 2.8 to 2.11 for the permanganate oxidation of PCE, TCE, DCE, and vinyl chloride (VC), respectively, the following is predicted:

PCE – 2.67 moles of H^+ per mole of PCE oxidized

TCE – 1.00 moles of H^+ per mole of TCE oxidized

DCE – 0.67 moles of OH^- per mole of DCE oxidized

VC – 2.33 moles of OH^- per mole of VC oxidized

Consistent with these predictions, observations made during permanganate oxidation of organics have revealed varied effects on pH. Extremely depressed pH (e.g., near pH 2) has been observed in aqueous systems during permanganate oxidation of high concentrations of TCE (e.g., Case 1997). Similar observations have been made during permanganate flushing experiments in sands with DNAPL residual present. Low pH may occur near the DNAPL-water interface where oxidation reactions are pronounced. However, system pH does not change dramatically in soil and groundwater systems that have high buffering capacity or in which there are lower concentrations of reductants and in which lower permanganate loadings are used. For example, Siegrist et al. (1999) found that the pH of silty clay soil with lower concentrations of TCE present was depressed from a value of pH 5.5 to 6.0 to about pH 4.5 after 10 months of exposure to permanganate (Figure 4-11). West et al. (2000) observed an increase in groundwater pH following flushing of an aquifer with 2 to 4 wt.% $KMnO_4$ to treat TCE DNAPL residuals (Figure 4-12). Background wells indicated a pH of 5.5 to 6.0 while wells in the active

FIGURE 4-12.   Groundwater pH in monitoring well samples collected 10 months after ISCO (West et al. 2000)[1].

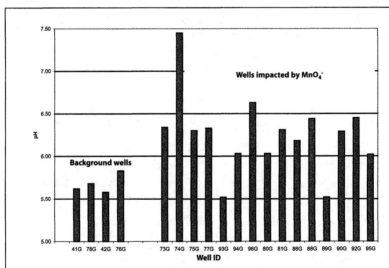

Note: Well group A consists of wells that are upstream of the horizontal injection well and where MnO_4^- was never detected; these wells are assumed to represent background conditions. Well Group B are wells where MnO_4^- was still detected 10 months after ISCO. Well Group C are wells where MnO_4^- was detected 2 weeks after the ISCO, but not detected 10 months after ISCO. Wells 9 and 21G (well group D) are ~50 and ~400 ft downstream of the injection horizontal well; MnO_4^- was not detected in either well through 10 months after ISCO.
[1]Refer to Site 1 in Chapter 7 for further site information.

ISCO treatment zone revealed a pH of about 6.25 to 6.75. Thus, based on the reaction stoichiometry and observations such as those summarized here, the effects of permanganate oxidation on subsurface pH will vary depending on site conditions, reductants present, and the oxidant loading employed.

Effects on Cation Exchange Chemistry

ISCO with permanganate can effect changes in cation ion exchange onto porous media due to changes in the balance of cations in soil and groundwater solution. In many subsurface settings, the primary cations include Ca^{+2}, Mg^{+2}, K^+, NH_4^+ Na^+, and H^+. These cations compete for exchangeable sites on porous media, where the total number of sites is indicated by the CEC in meq/100g of media. The affinity of a particular cation to the porous media is determined by its concentration and selectivity relative to the other cations in the system. The Gapon equation has been used widely for representing cation exchange processes (Bohn et al. 1979), although there are far more complicated equations available that have a better theoretical basis than the Gapon equation. For the Gapon equation, the cation exchange relationship can be represented by an equation of the following form for calcium and sodium (Bohn et al. 1979):

$$(Ca)_{1/2}X + Na^+ <=> NaX + \frac{1}{2}Ca^{2+} \qquad (4.1)$$

The resulting Gapon reaction is:

$$\frac{(NaX)}{(Ca_{1/2}X)} = K_G \frac{(Na^+)}{(Ca^{2+})^{0.5}} \qquad (4.2)$$

where X = porous media exchange site, the exchangeable cation concentrations are in meq per g (or 100 g), and soluble cation concentrations are in mmoles per liter (or moles per liter). The K_G exchange coefficients can be measured experimentally and then used to determine (by solving simultaneous equations) the distribution of sorbed vs. dissolved cations in a particular system before and after ISCO. ISCO using permanganate can alter the balance of cations in solution through the added K^+ or Na^+ ions and the production of H^+. This in turn can alter the cations on the exchange complex of the media and conceivably affect the fate of the added constituents as well as constituents already in the matrix (e.g., sorbed Ca^{+2}, Mg^{+2}, and perhaps some weakly held heavy metals).

Struse and Siegrist (2000) characterized soil constituent levels in two silty clay soil cores through which $KMnO_4$ was allowed to diffuse from a 5000 mg/L source (Table 4-5). Both soil cores 1 and 2 were also characterized to determine the concentrations of K^+, Ca^{+2}, Mg^{+2}, Mn^{+2}, Fe^{+3}, Br^-, Cl^-, MnO_4^-, and MnO_2 at 0.60 and 1.91 cm from the influent end of the core, respectively. In general, concentrations of the ions in both cores seemed to follow the same trends with cation concentrations increasing away from the MnO_4^- source and anion concentrations decreasing. Due to the much higher K^+ concentration in solution, Ca^{+2} and Mg^{+2} were probably being displaced from the soil matrix by K^+ and mobilized away from the MnO_4^- source.

In porous media that have a measurable clay fraction, careful attention also needs to be paid to the sodium adsorption ratio (SAR) to ensure that, if $NaMnO_4$ is used, the added sodium will not lead to dispersion of clay particles and a deterioration in soil structure with a reduction in permeability. Clay minerals can take up sodium and become sticky and slick, and can swell or disperse, thereby affecting the hydraulic conductivity of the media. This is of particular importance under the following conditions: (1) high $NaMnO_4$ dose, (2) measurable fraction (even low wt.% levels) of clay particles in a moderate conductivity formation, and (3) high concentrations of target chemicals and other oxidizable substances. The SAR is calculated according to eqn. 4.3:

$$SAR = \frac{(Na^+)}{(\{Ca^{+2} + Mg^{+2}\}/2)^{1/2}} \tag{4.3}$$

where all concentrations are in meq/L. The exchangeable sodium status (ESR) can be predicted from the SAR and a Gapon-type exchange equation of the form:

$$ESR = \frac{(NaX)}{(CaX + MgX)} = K_G \frac{(Na^+)}{(\{Ca^{+2} + Mg^{+2}\}/2)^{1/2}} = K_G(SAR) \tag{4.4}$$

where ESR = exchangeable sodium ratio, the exchangeable ion concentrations are in meq/g (or meq/100g), and K_G = Gapon exchange constant. The range of K_G is commonly 0.010 to 0.015 (liters/mmole)$^{1/2}$ (Bohn et al. 1979). General guidance suggests that an SAR below 10 presents little hazard of formation permeability problems while that above 18 presents greater concern. As an illustration, for a water containing 140 mg/L $CaSO_4$ and 68.8 mg/L $MgCl_2$, an SAR of 10 is reached at a concentration of 2000 mg/L $NaMnO_4$ while an SAR of 18 is reached at a concentration of 3000 mg/L.

Impacts of Permanganate on Subsurface Microbiology

The effect of permanganate oxidation on the biomass, activity, and degradation potential of microbial populations in soil and groundwater may be of interest. This is particularly relevant when bioattenuation processes are planned for treatment of any post-oxidation residual contamination in the treated region. In general, in situ chemical oxidation can affect ambient microbial populations and alter the community structure by changing biochemical conditions (e.g., electron acceptors, substrate bioavailability, temperature, biotoxicants). This could disrupt natural attenuation processes that were established prior to implementing in situ oxidation. Alternatively, oxidation can produce more available organic carbon and add electron acceptors (e.g., O_2) and stimulate post-oxidation bioactivity. This latter effect is perhaps more likely during oxidation with peroxide and ozone than with permanganate. Experiences to date regarding microbiology effects have been limited, but some information is available as noted below.

During the ISCO project completed by Cline et al. (Cline et al. 1997, Gardner et al. 1998), deep soil mixing was used to deliver up to 5 wt.% $KMnO_4$ solutions to either 25- or 47-ft depth at a site contaminated with TCE at 1500 mg/kg or higher in silty clay soils (see Site 3 in Chapter 7). Analyses for both anaerobic and aerobic bacteria revealed that neither were greatly influenced by the addition of the oxidant. For example, before ISCO the average levels of aerobic bacteria (CFU/g soil) ranged from approximately $1x10^7$ near the ground surface to $1x10^2$ at 32-ft. depth. After treatment the levels had declined by 90 to 99%, but levels remained in the range of $1x10^4$ to $1x10^6$ CFU/g.

During the horizontal well flushing of contaminated groundwater where DNAPLs were suspected, West et al. (1998a,b, 2000) examined the presence of Mn-reducing bacteria in aquifer sediment samples collected from the site (see Site 1 in Chapter 7). These samples were collected near one of the monitoring wells (92G) where MnO_4^- was detected at levels as high as 0.75 wt.% as $KMnO_4$ after the ISCO application was completed. Interest in Mn-reducing bacteria was spurred initially by the possibility of such organisms oxidizing organic compounds with the concomitant reduction of Mn-oxides and release of dissolved Mn (Lovley and Philips 1988). The reduction of Mn-oxide precipitates by Mn-reducing bacteria is also an important mechanism by which Mn can be released from solid Mn-oxide. West et al. (2000) mixed sediment samples with aqueous culture media enriched with Mn^{+4}. After approximately one month of interaction between the media and the soil, the samples were analyzed for Mn^{+2}. A control series was also prepared where the samples were autoclaved. The presence of Mn^{+2} in the Mn^{+4}-enriched samples would indicate the presence of Mn-reducing bacteria (Kostka and Nealson 1998). Analyses confirmed the presence of Mn-reducing bacteria, as indicated by elevated levels of Mn^{+2} in seven out of 10 sediment aliquots that were exposed for 1 month to Mn^{+4}-enriched culture media. A control series was also established in which soil samples were autoclaved and sterilized prior to introduction of the Mn^{+4} media, and no reduced Mn species were detected. Thus, the presence of Mn-reducing bacteria was established and it appears that these bacteria survived the exposure to high levels of MnO_4^- and concomitant strongly oxidizing conditions. The presence of Mn-reducing bacteria is relevant for two reasons: biologically facilitated reduction of Mn-oxides in the treatment zone leading to the release of Mn^{+2}, and the potential for oxidation of residual organic contaminants by these organisms (Lovley and Philips 1988).

4.4. TREATMENT-INDUCED TOXICITY

Another potential secondary effect of ISCO with permanganate involves the potential for treatment-induced toxicity to humans or other lifeforms. Permanganate contains impurities that include some heavy metals (see Table 2-3). Depending on the site conditions and oxidant dose utilized, the concentrations of metals may exceed potentially applicable regulatory limits (e.g., MCLs). In addition, depending on the target organic chemical and the environmental conditions, ISCO may produce degradation intermediates and byproducts (e.g., chlorinated organic acids or daughter products) that may be hazardous or toxic under some conditions. Experimental work to date has not documented consistent production of chlorinated daughter products, but there have been some situations where chloroethanes have been produced during oxidation of chloroethenes.

Apart from quantifying chemical components directly, there are established and emerging techniques to assess acute toxicity and carcinogenic and mutagenic potential. Acute toxicity often has been assessed by bioassays. The Microtox assay uses the luminescent bacterium *Vibrio fischeri* to measure stress to the electron transport system as measured by reduced light output during a 30-min exposure (Bulich 1979, Ross 1993). The *Daphnia magna* test uses a well-known cladoceran species to measure mortality in a 48-hr exposure (USEPA 1985). A more recently developed and more sensitive assay employs a fairy shrimp *(Thamnocephalus platyurus)* and a 24-hr. mortality test (Persoone et al. 1994). Another test involves the water flea, *Ceriodaphnia dubia*, and standard protocols commonly used to evaluate the toxicity of effluents and surface waters for compliance with water quality criteria under National Pollutant Discharge Elimination System permits (USEPA 1994).

The results of these bioassay tests often are used to estimate toxicity units (TU) associated with the system under study. For example, temporal and spatial comparisons based on Microtox tests can be made by normalizing IC_{50} values. The IC_{50} (Inhibitory Concentration - 50%) is the calculated or interpolated concentration on a dose-response curve where the biological response (in this case, luminescence) is reduced by 50%. When testing complex mixtures where concentrations of individual chemicals, or the interactions between them, are not known, TUs are calculated for the whole mixture as:

$$TU = 1 / \{mixture \; IC_{50}\} \tag{4.5}$$

where the mixture IC_{50} is the fractional dilution of the mixture that produces 50% inhibition. This approach allows definition of a baseline toxicity and then toxicity removal (i.e., fewer TUs) or addition (i.e., more TUs) as a result of in situ treatment.

Toxicity assays have been performed at several sites where permanganate has been employed for ISCO. As a component of a field application of ISCO using permanganate at a DOE site in Ohio (see Site 2 in Chapter 7), samples of TCE contaminated groundwater were obtained before and after in situ treatment by oxidant flushing (Lowe et al. 2000). At this site, groundwater containing pre-treatment TCE of 1.5 to 2.0 mg/L was reduced to the MCL of 0.005 mg/L (>95% reduction in TCE concentrations) during 10 d of flushing with 250 mg/L of $NaMnO_4$. Using a Microtox Model 500 analyzer (AZUR Environmental, Carlsbad, CA), groundwater samples were analyzed for acute toxicity to a luminescent bacteria *(Vibrio fischeri)* when exposed to 91% of the total sample concentration. No toxicity was measured for background or post-oxidant flushing samples.

West et al. (1998a,b, 2001) performed long-term monitoring at a site where ISCO was used to treat groundwater in a 5-ft thick silty gravel aquifer that was contaminated with TCE at levels that indicated the presence of residual DNAPLs (see Site 1 in Chapter 7). ISCO was implemented by use of a pair of parallel horizontal wells with 200-ft screened sections. For approximately one month, groundwater was extracted from one horizontal well while cleanwater was dosed with crystalline $KMnO_4$ (dose concentration of 2 to 4 wt.% $KMnO_4$) and injected into the other horizontal well 90 ft away. Injection via a vertical well in a hotspot also was performed during part of the 30-day period. A total of ~12,700 kg of $KMnO_4$ was delivered to the treatment region, of which 1960 kg was injected through a vertical well (well 74G). Of 206,000 gallons of oxidant solution injected, 14,000 gallons were delivered through well 74G. Post-treatment characterization showed that ISCO was effective at removing TCE in the saturated zone although aquifer heterogeneities prevented uniform delivery of the oxidant solution. At all locations where the oxidant had permeated, TCE was not detected (<5 ppb). A bioassay was completed using the water flea, *Ceriodaphnia dubia*, and groundwater collected from two wells at this site 10 months after the oxidant flushing project was completed. Well 42G was considered as a background well since MnO_4^- was never detected in it. Well 89G had levels of $KMnO_4$ as high as 2.1 wt.% after ISCO, but MnO_4^- was no longer detected in the sample collected 10 months later and used for toxicity testing. For water from well 42G (MnO_4^- never detected), reduced survival and reproduction were only observed when 100% well water was used (see Table 4-6). Water from 89G (MnO_4^- initially 2.1 wt%) appeared to be more toxic than water from well 42G, as shown by reduced survival and reproduction even in tests with 30% non-aerated well water (Table 4-6). To determine whether the toxicity of well water 89G was from dissolved TCE (483 mg/L), the toxicity tests were repeated for aerated samples and reduced reproduction was still observed (Table 4-6). The basic water quality of the 89G sample (Table 4-7) was suitable for reproduction of *Ceriodaphnia*, but alkalinity was low relative to hardness when compared with the control water. The low alkalinity and high (relative to alkalinity) hardness of full-strength water from 42G may have contributed to reduced reproduction in that sample. To evaluate the toxicity of Mn^{+2}, the tests were run on distilled water spiked with varying levels of $MnCl_2$ (see Table 4-8). Reduced reproduction was only observed at levels of Mn that were higher than that measured in the water from 89G (1.7 mg/L) and 42G (0.1 mg/L).

| TABLE 4-6 | Summary of results from the *Ceriodaphnia* toxicity tests in groundwater from an ISCO site 10 months after oxidant flushing with 2 to 4 wt.% $KMnO_4$ (after West et al. 2000). | | | |

	Well 42G[1]		Well 89G[1]	
Sample Concentration	Number of 10 animals surviving 6 d	Mean number of offspring per female (±S)	Number of 10 animals surviving 6 d	Mean number of offspring per female (±S)
		Non-aerated samples		
Control	10	25.1 ± 6.6	10	29.5 ± 2.7
0.3%	—	—	10	29.5 ± 2.7
1%	10	23.7 ± 6.3	10	23.7 ± 7.1
3%	10	21.5 ± 7.8	10	26.3 ± 4.7
10%	10	21.5 ± 7.4	10	22.9 ± 6.4*
30%	10	21.2 ± 7.5	7	16.9 ± 6.3*
100%	7	14.0 ± 6.9* [2]	—	—
		Aerated samples		
Control	—	—	10	31.3 ± 1.6
30%	—	—	9	23.9 ± 7.2*
100%	—	—	10	9.9 ± 4.6*

[1]MnO_4^- was detected in well 89G at levels as high as 2.1% as $KMnO_4$ after ISCO was completed, but was no longer detected in the sample collected 10 months later and used for toxicity testing.
[2]Asterisks (*) in the table indicate those concentrations of well water which significantly reduced survival or reproduction.

Therefore, reduced reproduction of *Ceriodaphnia* in aerated water from 89G and water from 42G does not appear to be from dissolved Mn. West et al. (2000) concluded that, precluding the effects of TCE, water from a background well and from a well where elevated levels of MnO_4^- had been detected appeared to have comparable toxicities when *Cerodaphnia* was used as the test organism.

Clayton et al. (IT Corp and SM Stollar Corp 2000) recently completed laboratory and field scale testing to determine the feasibility of destroying high explosives (HE) using $KMnO_4$. This testing indicated that

TABLE 4-7	Summary of chemical analyses of groundwater samples used for toxicity testing.				

Analyses	Units	Control	100% 89G	30% 89G	100% 42G
pH	–	8.00	5.49	7.20	6.11
Conductivity	S/cm, corrected to 25°C	201	1049	468	667
Alkalinity	mg/L as $CaCO_3$	81	66	82	26
Hardness	mg/L as $CaCO_3$	98	220	146	170

TABLE 4-8	Summary of results from the *Ceriodaphnia* toxicity test using manganese chloride.[1]		

Concentration	Number of replicates	Number of of animals surviving 6 d	Number of of offspring per female ($\pm$ S)
Control	10	10	23.5 ± 5.8
0.32	10	9	18.1 ± 5.3
0.62	10	10	19.2 ± 6.2
1.27	10	10	21.7 ± 6.3
2.51	10	10	18.7 ± 6.3
5.12	10	10	16.1 ± 6.0* [2]
9.63	10	10	5.6 ± 2.0*

[1]Water hardness = 140 mg/L as $CaCO_3$.
[2]Asterisks (*) in the table indicate those concentrations of well water which significantly reduced survival or reproduction.

$KMnO_4$ (400 to 20,000 mg/L $KMnO_4$) could effectively treat the HE compounds of interest (HMX, RDX, TNT, 2,6-DNT and 2,4-DNT) to below regulatory criteria. As part of this testing, Van Cuyk et al. at CSM determined acute toxicity using the Microtox bioassay. For the analysis, water samples of interest were serially diluted in Microtox assay matrix and light output was measured after exposure times of 5 and 15 minutes. Rapid toxicity (within 5 minutes) is an indicator of the presence of organic toxicity, while a delayed toxic effect (i.e., no toxicity until 15 minutes or more) is usually indicative of metal toxicity. Since the IC-50 is the concentration (i.e., the % of the test mixture that is from the environmental sample of interest) where the biological response is reduced by 50%, higher values of IC-50 are indicative of reduced biotoxicity. In this study, water samples were run within 24 hours of their collection. A phenol control was run each day yielding IC-50 values that corresponded to a phenol concentration between 13 and 26 mg/L, which was within the specifications for Microtox Model 500 Phenol IC-50 (5 minutes). Results from the Microtox analysis are presented in Table 4-9. The A series and the background (bgd) samples appear to have low toxicity (i.e., high IC-50 % values), while the C series samples (reportedly those treated with KMnO$_4$) have higher toxicity. Sample C2 presented an intermediate IC-50, which may be attributable to the possible quenching of this sample by a reductant added after the reaction period but before analyses were completed at CSM. There does appear to be a toxicity effect associated with the treatment using KMnO$_4$ (as seen in the C series). The reasons for this toxicity are not known, but it could be due to

TABLE 4-9	Results of analyses of samples collected during treatability studies with permanganate oxidation of high explosives.		
Sample ID	**Date** [1]	**IC-50** *5 min*	**IC-50** *15 min*
Samples received from IT Corp			
At0	2/23/00	>91% [2]	>91.1%
At0dup	2/23/00	>91%	>91.1%
C2	2/23/00	40.46%	Unable to calc
C6	2/23/00	7.84%	3.74%
C2a	2/24/00	5.72%	3.12%
Bgd (background)	2/26/00	>91%	>91%
C72	2/26/00	7.92%	3.54%
Ct168	3/1/00	7.03%	3.56%
Ct168dup	3/1/00	8.27%	3.86%
Samples prepared at CSM			
KMnO$_4$ (100 mg/L)		4.85%	1.73%
KMnO$_4$ (1000 mg/L)		0.53%	0.19%

[1] Samples were analyzed on the date received.
[2] IC-50 values are presented as % concentration of original sample.

the presence of MnO_4^- oxidant itself. Also, it could be due to the oxidant effects on the bacteria used in the test. The permanganate toxicity is supported by the high toxicity values shown in the $KMnO_4$ controls ($KMnO_4$ at 100 mg/L and 1000 mg/L in deionized water). There also may be secondary effects associated with the reaction of the added $KMnO_4$ with target contaminants of concern in the samples, such as formation of toxic intermediates or byproducts, reducing system pH, or a combination of these effects, that would result in a change in toxicity. The fact that the toxicity (IC-50 % values) observed for the samples from the treatability study with permanganate were equal to or lower than those associated with the reference samples with permanganate and clean water suggests that treatment of the sample did not increase the biotoxicity apart from the effect of the oxidant chemical itself.

As described above, the results of toxicity testing at sites treated by ISCO using permanganate have begun to provide insight into the biotoxicity that might occur with this remediation technology. Within the active ISCO treatment zone, there may be acute biotoxicity caused by the treatment process. However, in a typical groundwater system, the oxidant will be depleted and bio-adverse conditions should dissipate with time within the treatment zone and certainly with distance/time downgradient from the treatment zone.

4.5. SUMMARY

In situ treatment of contaminated soil and groundwater by permanganate requires careful consideration of the potential effects that permanganate can have on the ISCO system and subsurface conditions. As already presented in Chapter 3, permanganate can impact metal behavior, and this must be carefully considered. As described in this chapter, permanganate also can impact the subsurface conditions through several other potential processes: (1) altered behavior of organic COCs, (2) introduction of toxic COCs from impurities and reaction byproducts, (3) changes in permeability due to particle genesis and gas evolution, and (4) geochemical and microbial perturbations (Tables 4-10 to 4-12). As described in this chapter, these *potential* effects can occur and be considered beneficial and/or deleterious depending on the site conditions, level of target organic COCs, dose concentration and total oxidant loading to the subsurface, and the design of the ISCO system. Table 4-13 summarizes the types of effects that need to be considered and those conditions that represent relatively higher and lower levels of concern.

TABLE 4-10.	Summary of experimental studies and field observations regarding ISCO and altered mobility and treatment-induced toxicity.

Study type	Results	Comments
Altered Behavior of Untreated COCs		
Laboratory study of permanganate oxidation of TCE DNAPL (Urynowicz 2000).	Formation of an interfacial coating/film on TCE DNAPL after chemical oxidation with permanganate, which can reduce the oxidative degradation of DNAPLs.	Development of such a film has been shown to occur during laminar flow condition.
Treatment-Induced Toxicity		
Field application of permanganate ISCO by oxidant flushing at a DOE site in Ohio (Lowe et al. 2000).	Microtox assay of groundwater toxicity to *Vibrio fischeri* when exposed to 91% of the total sample concentration demonstrated no measured toxicity in background or post-oxidant flushing samples.	Within 10 days of flushing, TCE concentrations were reduced from 1.5-2.0 mg/L to 0.005 mg/L.
Field monitoring of permanganate-treated groundwater for one month following an oxidant flushing application to evaluate the toxicity of Mn^{2+} (West et al. 1998a,b, 2000).	A *Ceriodaphnia dubia* (water flea) bioassay demonstrated no significant difference in Mn^{2+} toxicity between background well water and water from a well where elevated permanganate had been detected.	Reduced survival and reproduction were noted in the samples with elevated permanganate (as compared to background), but this could be attributed to low alkalinity and high water hardness.
Laboratory and field scale testing of permanganate oxidation of explosives to examine toxicity of resulting solutions (Clayton et al., IT Corp and SM Stollar Corp 2000).	A Microtox bioassy indicated a toxicity effect associated with the permanganate treatment that could be due to: 1. Oxidant effects on bacteria 2. Formation of toxic intermediates or byproducts 3. Reduced system pH 4. Combination of above	The toxicity of the samples from the treatability study with permanganate were equal to or lower than those association with the reference samples of permanganate and clean water, suggesting that the toxicity could be the effect of the oxidant.

TABLE 4-11.	Summary of experimental studies and field observations regarding ISCO and permeability effects.	
Study type	**Results**	**Comments**
Field studies of ISCO application in Ohio examining pre- and post-treatment hydraulic conductivity (Lowe et al. 2000, Siegrist et al. 2000).	Similar ISCO applications at different areas of the same site demonstrated no loss in formation permeability under low TCE (1-2 mg/L) and low oxidant conditions 250 mg/L), yet well pressure build-up occurred in the second study with high TCE (up to 600 mg/L) and oxidant (3000 mg/L) concentrations.	Laboratory experiments were conducted to understand the nature and cause of the permeability problems of the second application. Batch studies indicated greater production of filterable particles under higher TCE concentrations and in the presence of groundwater particles. Column experiments revealed that permeability loss was possible during ISCO, but only under such conditions as high oxidant and reductant concentrations.
Field examinations of hydraulic conductivities resulting from ISCO application of permanganate to treat TCE (West et al. 1998a,b, 2000).	Post-treatment measurements of hydraulic conductivities showed order of magnitude decreases in several wells tested as compared to pre-treatment conditions.	The clogging was attributed to precipitation of Mn-oxide, which could have been aggravated by mobilization of colloidal aquifer material.
Laboratory 1-D column and 2-D test cell studies examining flushing efficiencies resulting from reaction of permanganate with typical aquifer materials (Li and Schwartz 2000).	Column and tank studies demonstrated that the tendency to develop preferential flow paths is promoted by manganese oxides precipitation in zones of higher DNAPL saturation or as carbon dioxide bubbles are trapped in porous media.	The effect is attributed to the rapid rate of TCE oxidation vs. the slow mass transfer rate from DNAPL to aqueous phase, resulting in MnO_2 precipitation around zones of greatest saturation and causing flow to bypass.

continued

TABLE 4-11. (continued)

Study type	Results	Comments
Laboratory 2-D studies to examine permeability effects during ISCO of pooled NAPL at varied reaction rates (Reitsma and Marshall 2000).	Manganese oxides and carbon dioxide production influenced flow patterns. MnO_2 build-up resulted in diversion in flow depending on the location of deposition (which differed for the fast vs. slow reaction rates). Carbon dioxide gas production led to desaturation of DNAPL from the pool and transported organic vapors into the sand above, enhancing the overall mass transfer of the DNAPL.	MnO_2 deposition and carbon dioxide gas transfer differed based on the reaction rate of the DNAPL and permanganate, but proved significant influences on flow patters under both conditions.
Laboratory column permeability studies in support of a California field operation to examine the effect of permanganate oxidation on subsurface permeability (Chambers et al. 2000).	Standard permeability tests demonstrated no reduction in permeability due to treatment of low concentrations of TCE with low concentrations of permanganate.	Hydraulic conductivity actually increased by almost an order of magnitude despite demonstrated production of MnO_2.
Field study of ISCO of DNAPL at a Cape Canaveral, FL site examining impact on permeability in three lithologic units (Mott-Smith et al. 2000).	No significant formation plugging of decreased injection flow rate was noted upon treatment of DNAPL residuals with high concentrations of permanganate.	Significant mass destruction of TCE was achieved.

continued

TABLE 4-11. (continued)		
Study type	**Results**	**Comments**
Field measurements of effects of Fenton's oxidation of DNAPL (primarily TCE) at a test site in Alabama (Levin et al. 2000).	Clean-up goals for TCE, 1-2 DCE, vinyl chloride, methylene chloride, and tetrachloroethene were reached with no noted mobilization of contaminants (including other co-contaminants) outside of test area.	Original site wastes included lead, petroleum hydrocarbons, solvents, and Industrial Waste Water Treatment plant sludge.
Laboratory investigation of MnO_2 concentrations resulting from diffusion of permanganate into a silty clay soil core (Struse 1999).	The MnO_2 detected through chemical analysis of extract from the cores accounted for only 0.082% - 0.086% of the total pore volume.	The investigation was initiated to determine if MnO_2 generation could increase tortuosity in the porous media.
Laboratory study of potential pore volume changes upon treatment of PCE in soil with permanganate (Schnarr and Farquhar 1992).	A 5.6% decrease in pore volume following permanganate injection resulted from treatment of high PCE concentrations.	Results were significant as compared to the 1% volume change attributed to the PCE itself.

TABLE 4-12	Summary of experimental studies and field observations regarding ISCO and geochemical and microbial perturbations.	
Study type	**Results**	**Comments**
Field studies of permanganate ISCO to treat residual DNAPL to examine the resulting form of Mn (West et al. 2000).	High levels of dissolved Mn was attributed to residual permanganate. The majority of permanganate injected, though, was transferred to less soluble Mn-oxide. Very little easily extractable Mn was found in boreholes without detectable permanganate.	Easily extractable Mn was found only in boreholes where residual permanganate was still present. Wherever permanganate had been consumed, the Mn was not easily leachable, although it was extractable under reducing conditions.
Laboratory and field examinations of the impact of permanganate on natural organic matter (NOM) (Struse 1999, Siegrist et al. 1999).	Permanganate reaction with low-permeability clays and silts indicate that only a portion of TOC is oxidized as measured by oxidant demand and pre- and post-treatment TOC measurements.	TOC was decreased through oxidation by 33% or less indicating substantial NOM was apparently resistant to oxidation by permanganate even at high concentrations and long durations.
Field examination of TOC content after ISCO application through hydraulic fractures (Siegrist et al. 1999).	Examination of TOC in soil cores demonstrated appreciable TOC remaining.	TOC remained even after 10 months of exposure to highly oxidizing conditions.
Field examination of TOC content after ISCO application by deep soil mixing (Cline et al. 1997, Gardner et al. 1998).	Pre-treatment TOC content ranged from 4000–9000 mg/kg. Post-treatment TOC content declined to 4000–7000 mg/kg throughout permanganate treated zones.	Permanganate was no longer detected in the subsurface 3-5 days after injection.

continued

TABLE 4-12 (continued)

Study type	Results	Comments
Field measurements of pH effects of permanganate oxidation of TCE (Siegrist et al. 1999, West et al. 2000).	Siegrist et al. found pH depression from 5.5-6.0 down to 4.5 after 10 months of exposure of silty clay soil with low TCE concentrations to permanganate. West et al. observed an increase in groundwater pH from 5.5-6.0 to 6.25-6.75 following flushing of permanganate to treat TCE DNAPL residuals.	Effects of permanganate oxidation on subsurface pH depends on site conditions, reductants present, and oxidant loading.
Pilot-scale TCE treatability study for a contaminated site in CT using 10 g/L NaMnO$_4$ injection into a test cell containing TCE in both DNAPL and aqueous phases (DNAPL directly atop a clay aquitard with aqueous phase in overlying sand) (Huang et al. 2000).	Initially the pH within the test cell dropped from 6.5 to 3.5 within 2 days. The pH increased as the test progressed, above the initial site pH, to the point of alkaline conditions. The increase was sustained throughout the remaining duration of the test and was attributed to permanganate side reactions (e.g., with water and/or with water + Fe).	Permanganate consumption at the site was dictated more by the side reactions than by reactions with dissolved-phase TCE. Heavy MnO$_2$ build-up was noted at the mass transfer paths.

continued

TABLE 4-12 (continued)

Study type	Results	Comments
Field investigations of the impact of ISCO on microbial populations (Cline et al. 1997, Gardner et al. 1998).	Analyses for both anaerobic and aerobic bacteria revealed that neither were greatly influenced by the addition of up to 5 wt.% $KMnO_4$ solution to treat 1500mg/kg or higher of TCE.	Levels of aerobic bacteria ranging from 1×10^7 to 1×10^2 CFU/g remained in the range of 1×10^6 to 1×10^4 CFU/g following treatment.
Field studies of Mn-reducing bacteria in aquifer sediment samples during horizontal well flushing of permanganate for treatment of DNAPL-contaminated groundwater (West et al. 1998a,b, 2000).	The presence of Mn-reducing bacteria were confirmed in the majority of site sediment samples after 1 month of exposure.	The presence of Mn-reducing bacteria is important because of biological facilitation of the reduction of Mn-oxides to release Mn^{2+}, and for the potential of oxidation of residual organic contaminants by the organisms.

TABLE 4-13. Contaminated site conditions and level of concern over ISCO and secondary effects.

ISCO system and subsurface effects	Associated with . . .	Conditions indicating a *lower* level of concern for post-ISCO effect	Conditions indicating a *higher* level of concern for post-ISCO effect
Altered behavior of organic COCs.	Mobilization of COCs outside treatment zone.	Low injection volumes and/or adequate hydraulic control (e.g., through capture wells).	High volumetric injections into groundwater zones without capture wells.
	Reduced degradation of COCs.	Dissolved and sorbed COCs with low oxidant loading.	DNAPLs with high oxidant loading.
Treatment-induced toxicity.	Toxic impurities in permanganate.	Low oxidant loading and/or high groundwater velocities and flushing effects.	High oxidant loading and/or low groundwater velocities and flushing effects.
	Toxic reaction byproducts.	Simple, low molecular weight compounds (e.g., chlorinated ethenes).	High molecular weight, complex organics.
Permeability changes.	MnO_2 genesis pore plugging.	Low oxidant loading and low levels of reductants present.	High oxidant loading and high levels of reductants present (e.g., DNAPLs).
	Gas evolution and entrapment.	Non-calcareous sediments with low oxidant loading and low reductant levels.	Calcareous sediments with high oxidant loading and high reductant levels (e.g., DNAPLs).

continued

CHAPTER 4. Permanganate Effects on Subsurface Conditions   167

TABLE 4-6 (continued)			
ISCO system and subsurface effects	Associated with . . .	Conditions indicating a *lower* level of concern for post-ISCO effect	Conditions indicating a *higher* level of concern for post-ISCO effect
Geochemical and microbial perturbations.	Fate of manganese added in MnO_4^-.	Low oxidant loading and low reductant levels, low groundwater velocities, and no local groundwater uses.	High oxidant loading and high reductant levels, fast groundwater velocities, and local groundwater uses.
	NOM impacts.	Low oxidant loading to low NOM media with no sorbed co-contaminants.	High oxidant loading to high NOM media and sorbed co-contaminants.
	pH effects.	Low oxidant loading and low reductant levels.	High oxidant loading and high reductant levels (e.g., DNAPLs) in a weakly buffered system.
	Changes in microbial biomass and activity.	Lower oxidant loading to finer-grained media with high NOM and short-lived oxidant presence .	High oxidant loading to coarse grained media with limited NOM and sustained oxidant presence.
	Effects on cation exchange.	Coarse grained porous media with low CEC and/or high levels of reductants present.	Fine-grained porous media with high CEC, or high silt and clay content and $NaMnO_4$ oxidant used, and with low levels of reductants present.

4.6. REFERENCES

Bohn, H., B. McNeal, and G. O'Conner (1979). Soil Chemistry. John Wiley & Sons, New York.

Bulich, A.A. (1979). Use of luminescent bacteria for determining toxicity in aquatic environments. In: Aquatic Toxicology, ASTM STP 667, L.L. Marking and R.A. Kimerle (eds.), American Society for Testing and Materials, pp. 98-106.

Case, T.L. (1997). Reactive permanganate grouts for horizontal permeable barriers and in situ treatment of groundwater. M.S. Thesis, Colorado School of Mines, Golden, CO.

Chambers, J., A. Leavitt, C. Walti, C.G. Schreier, and J. Melby (2000). Treatability study – Fate of chromium during oxidation of chlorinated solvents. . In: Wickramanayake, G.B., A.R. Gavaskar, and A.S.C. Chen (ed.). Chemical Oxidation and Reactive Barriers. Battelle Press, Columbus, OH. pp. 57-66.

Cline, S.R., O.R. West, N.E. Korte, F.G. Gardner, R.L. Siegrist, and J.L. Baker (1997). KMnO$_4$ chemical oxidation and deep soil mixing for soil treatment. *Geotechnical News.* December. pp. 25-28.

Gardner, F.G., N.E. Korte, J. Strong-Gunderson, R.L. Siegrist, O.R. West, S.R. Cline, and J. Baker. (1998). Implementation of deep soil mixing at the Kansas City Plant. Final project report by Oak Ridge National Laboratory for the Environmental Restoration Program at the DOE Kansas City Plant. ORNL/TM-13532.

IT Corporation and SM Stollar Corp (2000). Implementation report of remediation technology screening and treatability testing of possible remediation technologies for the Pantex perched aquifer. October, 2000. DOE Pantex Plant, Amarillo, Texas.

Kostka, J., and K. Nealson (1998). Isolation, cultivation and characterization of iron- and manganese reducing bacteria. In: Burlage, R., R. Atlas, D. Stahl, G. Geesey, and G. Sayler, (eds.), *Techniques in Microbial Ecology*, Oxford University Press, New York (1998).

Li, X.D. and F.W. Schwartz (2000). Efficiency problems related to permanganate oxidation schemes. . In: Wickramanayake, G.B., A.R. Gavaskar, and A.S.C. Chen (ed.). Chemical Oxidation and Reactive Barriers. Battelle Press, Columbus, OH. pp. 41-48.

Lovley, D. and E. Phillips (1988). Novel mode of microbial energy metabolism: Organic carbon oxidation coupled to dissimilatory reduction of iron or manganese. *Applied and Environmental Microbiology*, 54:1472-1480.

Lowe, K.S., F.G. Gardner, R.L. Siegrist, and T.C. Houk (2000). Field pilot test of in situ chemical oxidation through recirculation using vertical wells at the Portsmouth Gaseous Diffusion Plant. EPA/625/R-99/012. U.S. EPA Office of Research and Development, Washington, D.C. 20460. pp. 42-49.

Mott-Smith, E., W.C. Leonard, R. Lewis, W.S. Clayton, J. Ramirez, and R. Brown (2000). In situ oxidation of DNAPL using permanganate: IDC Cape Canaveral demonstration. . In: Wickramanayake, G.B., A.R. Gavaskar, and A.S.C. Chen (ed.). Chemical Oxidation and Reactive Barriers. Battelle Press, Columbus, OH. pp.125-134.

Persoone, G., C. Janssen and W. De Coen (1994). Cyst-based toxicity tests X: Comparison of the sensitivity of the acute daphnia magna test to two crustacean microbiotests for chemicals and wastes. Chemosphere 29 (12): 2071.

Reitsma, S. and M. Marshall (2000). Experimental study of oxidation of pooled NAPL. . In: Wickramanayake, G.B., A.R. Gavaskar, and A.S.C. Chen (ed.). Chemical Oxidation and Reactive Barriers. Battelle Press, Columbus, OH. pp.25-32.

Ross, P.E. (1993). The use of bacterial luminescence systems in aquatic Toxicity testing. Chapter 13 in: Richardson, M. (ed.) *Ecotoxicology Monitoring*. VCH Publishers, New York, NY USA 384p.

Schnarr, M.J. and G.J. Farquhar (1992). An in situ oxidation technique to destroy residual DNAPL from soil. Subsurface Restoration Conference, Third International Conference on Ground Water Quality, Dallas, Texas, Jun 21-24, 1992.

Siegrist, R.L., K.S. Lowe, L.C. Murdoch, T.L. Case, and D.A. Pickering. 1999. In situ oxidation by fracture emplaced reactive solids. *J. Environmental Engineering*. Vol.125, No.5, pp.429-440.

Siegrist, R.L., M.A. Urynowicz, M. Crimi, and A.M. Struse (2000). Experimental studies of particle genesis and effects during in situ chemical oxidation of chlorinated solvents in groundwater using permanganate. CSM project report submitted to Oak Ridge National Laboratory, Grand Junction, CO. January, 2000.

Struse, A.M. 1999. Mass transport of potassium permanganate in low permeability media and matrix interactions. M.S. Thesis, Colorado School of Mines, Golden, CO.

Struse, A.M. and R.L. Siegrist (2000). Permanganate transport and matrix interactions in silty clay soils. In: Wickramanayake, G.B., A.R. Gavaskar, and A.S.C. Chen (ed.). Chemical Oxidation and Reactive Barriers. Battelle Press, Columbus, OH. pp. 67-74.

Stumm, W. (1997). Reactivity at the mineral-water interface: dissolution and inhibition. Colloids and Surfaces A: Physicochemical and Engineering Aspects, 120:143-166.

Stumm, W. and R. Wollast (1990). Coordination chemistry of weathering: kinetics of the surface-controlled dissolution of oxide minerals. *Reviews of Geophysics*, 28(1):53-69.

Urynowicz, M.A. (2000). Reaction kinetics and mass transfer during in situ oxidation of dissolved and DNAPL trichloroethene with permanganate. Ph.D. dissertation, Environmental Science & Engineering Division, Colorado School of Mines. May.

USEPA (1985). Methods for measuring the acute toxicity of effluents to freshwater and marine organisms. EPA 600/4-85-013. Environmental Monitoring and Support Laboratory, Cincinnati, OH.

USEPA (1994). Short-term methods for estimating the chronic toxicity of effluents and receiving waters to freshwater organisms. EPA/600/4-91/002. U.S. Environmental Protection Agency, Cincinnati, OH. (1994).

West, O.R., S.R. Cline, W.L. Holden, F.G. Gardner, B.M. Schlosser, J.E. Thate, D.A. Pickering, and T.C. Houk (1998a). A Full-scale field demonstration of in situ chemical oxidation through recirculation at the X-701B Site. Oak Ridge National Laboratory Report, ORNL/TM-13556.

West, O.R., Cline, S.R., Siegrist, R.L., Houk, T.C., Holden, W.L., Gardner, F.G. and Schlosser, R.M. (1998b). A Field-scale test of in situ chemical oxidation through recirculation. Proc. Spectrum '98 International Conference on Nuclear and Hazardous Waste Management. Denver, Colorado, Sept. 13-18, pp. 1051-1057.

West, O.R., R.L. Siegrist, S.R. Cline, and F.G. Gardner (2000). The effects of in situ chemical oxidation through recirculation (ISCOR) on aquifer contamination, hydrogeology and geochemistry. Oak Ridge National Laboratory, Internal report submitted to the Department of Energy, Office of Environmental Management, Subsurface Contaminants Focus Area.

Wiesner, M.R., M.C. Grant, and S.R. Hutchins (1996). Reduced permeability in groundwater remediation systems: Role of mobilized colloids and injected chemicals. *Environ. Sci. & Technol.* (30):3184-3191.

CHAPTER 5

Oxidant Delivery and Subsurface Distribution

Successful in situ oxidation with permanganate relies on an oxidant handling and delivery method that enables dispersal of oxidant throughout a region of the subsurface within which treatment is desired. Permanganate oxidants in liquid or solid form at low mg/L to high wt.% dose concentrations can be delivered in a continuous or cyclic fashion using injection probes, soil fracturing, soil mixing, vertical and horizontal well flushing, and treatment fences. This chapter contains a discussion of the factors affecting handling and delivery and a description of the optional methods that have been employed. There are important worker health and safety concerns associated with handling of reactive chemicals such as sodium and potassium permanganate oxidants. Site-specific health and safety plans must be carefully developed and stringently adhered to as discussed in Chapter 6.

5.1. ABOVEGROUND HANDLING AND SUBSURFACE DELIVERY

Aboveground Handling

Permanganate oxidants can be obtained in different forms, and grades, and in various quantities and container types (Appendix A). The oxidants are normally shipped from a local supplier to the site by ground transportation. Solid potassium permanganate ($KMnO_4$) is used most commonly and is widely available in small (e.g., 40-kg pails) to large quantities (e.g., bulk hoppers). Sodium permanganate, which is a liquid

(40 wt.% $NaMnO_4$), is not widely available and most often is purchased directly through the manufacturer. Potassium permanganate has the benefit of delivery as a liquid or a solid and is more widely available, less expensive, and less hazardous to handle. Sodium permanganate has the benefits of lower levels of metal impurities, enabling higher oxidant dose concentrations, and easier in-line mixing, but it is more expensive and more hazardous to handle.

There are several methods of handling permanganate oxidant above ground and thereby enabling its delivery into the subsurface in either a liquid or solid form (Table 5-1). Preparation of a solution of $KMnO_4$ or $NaMnO_4$ is most commonly done aboveground as illustrated by one of the methods shown in Figures 5-1 to 5-5. The most appropriate method of preparation will depend on site specific factors, including the source of water to be used (e.g., clean water vs. extracted groundwater), the size of the region to be treated (e.g., small vs. large), and the method of delivery of the oxidant solution to the subsurface (e.g., intermittent delivery through injection points vs. continuous delivery via wells).

TABLE 5-1.	Methods of handling oxidant and enabling delivery to the subsurface.			
Type	Oxidant form	Oxidant concentration	Description	Comments
S-1	$KMnO_4$ solution	Low ppm to solubility limit (4 to 6 wt.% depending on temperature)	Dissolving $KMnO_4$ crystals in clean water from a hydrant or a hydrant or a potable source	Need to provide vigorous mixing to ensure dissolution Avoids above-ground particle generation by oxidation reactions such as in S-2 Injection of clean water may cause hydraulic gradients and plume movement unless groundwater extraction wells are provided
S-2	$KMnO_4$ solution	Low ppm to solubility limit (4 to 6 wt.% depending on temperature)	Dissolving $KMnO_4$ crystals in extracted contaminated groundwater	Need to provide vigorous mixing to ensure dissolution Precautions are needed for above-ground particle generation by oxidation reactions with reductants in extracted groundwater Re-injection of extracted groundwater may enable hydraulic control over plume movement

continued

TABLE 5-1.	(continued)			
Type	**Oxidant form**	**Oxidant concentration**	**Description**	**Comments**
S-3	$NaMnO_4$ solution	Low ppm to 40 wt.%	Dilution of $NaMnO_4$ liquid in clean water from a hydrant or a potable source	Combining two liquids can enable more accurate and uniform dose concentrations In some formations Na+ may cause problems with clay dispersion Avoids above-ground particle generation by oxidation reactions such as in S-2 Injection of clean water may cause hydraulic gradients and plume movement unless ground
S-4	$NaMnO_4$ solution	Low ppm to 40 wt.%	Dilution of $NaMnO_4$ liquid in contaminated groundwater pumped from the site	Need to provide vigorous mixing to ensure dissolution Precautions are needed for above-ground particle generation by oxidation reactions with reductants in extracted groundwater Re-injection of extracted groundwater may enable hydraulic control over plume movement
S-5	$KMnO_4$ OPM	1 to 50 wt.%	Mixture of $KMnO_4$ crystals in a mineral grout formulation	Provides a pumpable solid containing MnO_4^- ions which can be used in delivery systems where solutions are not as compatible Waste management is a concern with OPM that is not used soon after formulation

A typical method of creating a permanganate solution is to dissolve $KMnO_4$ crystals into relatively clean water from a nearby hydrant or a potable source (Figures 5-1 and 5-3). An alternative is to use $NaMnO_4$ liquid and an in-line mixer unit (Figure 5-2). The solutions can be created in batch mixing tanks (Figures 5-1 and 5-3) or by in-line liquid (Figure 5-2) or dry solids feeding equipment (Figure 5-4). In either case, vigorous

FIGURE 5-1. Schematic of a simple permanganate delivery system using a pressurized clean water source (e.g., fire hydrant) and manual batch preparation of KMnO₄ solutions.

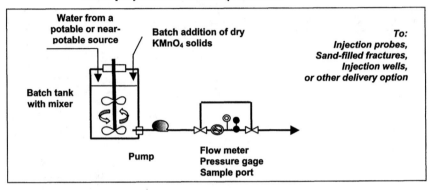

FIGURE 5-2. Schematic of a simple permanganate delivery system using a pressurized clean water source (e.g., fire hydrant) and in-line injection of NaMnO₄ solution.

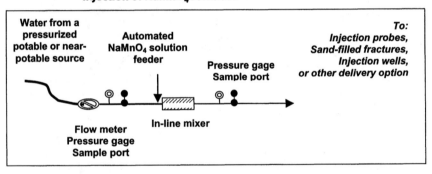

FIGURE 5-3. Schematic of a permanganate delivery system using a low pressure clean water source and batch tanks for manual addition of KMnO₄ solids.

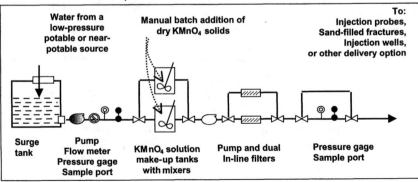

FIGURE 5-4. Schematic of a larger scale permanganate delivery system using a clean water source and automated feeding of KMnO₄ solids.

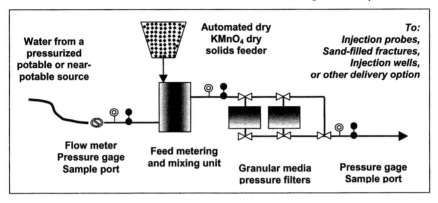

FIGURE 5-5. Schematic of a permanganate delivery system using manual batch addition of dry KMnO₄ solids with aboveground particle removal operations.

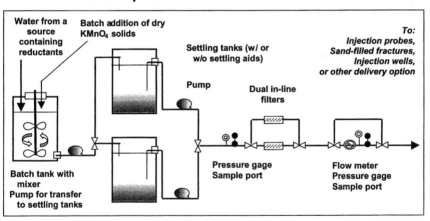

mixing is very important to achieving dissolution of the solid permanganate crystals before subsurface delivery. With $NaMnO_4$, mixing is also important due to the potential for density effects yielding depth-varying concentrations within a drum or container. The resulting MnO_4^- solutions can be created to have aqueous permanganate concentrations ranging from low mg/L to wt.% levels. With $KMnO_4$, it is important to keep the concentration below about 2 to 3 wt.% to avoid problems with undissolved crystals. The $KMnO_4$ solutions created aboveground can then be injected into the subsurface as described later in this chapter. Creating $KMnO_4$ solutions with clean water avoids the generation of MnO_2 particles in the feed that result when $KMnO_4$ is dissolved in water containing reductants (e.g., organic contaminants, NOM, Fe^{+2}).

Groundwater from the contaminated site can be used to prepare MnO_4^- solutions by dissolution of $KMnO_4$ crystals or dilution of $NaMnO_4$ liquid. However, when MnO_4^- (either as $KMnO_4$ or $NaMnO_4$) is added to extracted groundwater that contains chlorinated ethenes and other reductants (e.g., NOM, Fe^{+2}), precautions need to be taken to manage MnO_2 particles that are generated in batch mixing tanks or in-line feed equipment. If not removed, these particles can be delivered to the subsurface and create problems with fouling of well screens or filter packs. Process operational considerations include: (1) minimizing oxidant dose to minimize MnO_2 solids generated, (2) maximizing quiescent settling to facilitate particle growth and filterability, (3) adding a coagulant aid and providing for concomitant secondary waste handling, (4) using enhanced particle-removal methods such as cross-flow filtration, rapid sand filtration, or other media filtration as opposed to simple cartridge filtration, and (5) within the constraints of maintaining hydraulic control within the treated region, limiting down-hole injection rates and periods of delivery in a single well so as to limit mass flux loadings of particulates across a given length of well-screen and borehole interface.

Well screens are very susceptible to clogging when water that contains particulates is injected. Particulate concentrations as low as a few mg/L in the injected water can result in substantial losses in well screen permeability (e.g., 30% or more). Driscoll (1986) notes concentrations as low as 3.3 mg/L clogged a 6-in well during 9 days of injection at 32 gpm. A rule of thumb for preventing well clogging during continuous injection recommends that the quantities of total suspended solids be kept very low (e.g., 1 mg/L) (Driscoll 1986).

Permanganate can also be delivered as a solid rather than through creation of an aqueous solution as described above. For example, solid $KMnO_4$ crystals can be dispensed and mechanically mixed (e.g., by use of a backhoe) to treat small hotspot areas of contaminated soils. Permanganate can also be mixed with mineral fluids to create an oxidative particle mixture (OPM) that is pumpable (Siegrist et al. 1999). This OPM can be made with from 1 to 50 wt.% $KMnO_4$ crystals and can be used as a proppant in hydraulic fracturing or as a filling agent in vertical boreholes.

The MnO_4^- dissolves from the OPM into soil solution or groundwater and migrates by diffusion and/or advection depending on site conditions.

Construction materials in the above- and below-ground oxidant delivery system must be compatible with the oxidant form and concentrations and with the anticipated duration of operation. Appendix A provides information regarding compatibility with different materials. Care must be exercised to avoid corrosion problems as well as energetic reactions and combustion.

Subsurface Delivery and Distribution

Delivery of oxidant into the subsurface must be done such that the MnO_4^- can be transported from the point of release into the subsurface throughout the region to be treated. Also, even though permanganate does not tend to autodecompose, it does react rapidly with reductants, including target organics as well as the NOM and reduced minerals in the subsurface. Although increasing the dose concentration can increase the mass delivered per unit volume of solution, enhance concentration gradients, and accelerate transport by diffusion, it also tends to result in increased oxidant consumption under otherwise comparable conditions (Chapter 2). In general, the rate of reaction is much greater than the rate of transport, even if forced advection in groundwater is employed (see below). Thus, the shorter the distance the oxidant has to travel from the point of release into the subsurface, the lower the dose concentration can be and the more likely the target organics will be degraded at minimum oxidant loading.

Subsurface Transport Processes

The movement of MnO_4^- through the subsurface occurs mainly through advection, mechanical dispersion, and molecular diffusion. *Advection* is an important transport process for moving bulk fluids away from injection wells and probes in both the vadose and saturated zones. In the vadose zone, the motive force is created by pressurized injection, while gravity and matric potentials within the formation cause further transport. In the saturated zone, advection can occur with the ambient groundwater flow or under a forced hydraulic gradient created by the injection and/or extraction of fluid. *Mechanical dispersion* results from microscopic velocity variations in a porous medium, which causes the spreading of an otherwise sharp MnO_4^- front. In most cases, mass fluxes due to mechanical dispersion are likely to be low compared to advective flux (Bear and Verruijt 1987), particularly when groundwater flow rates are relatively fast, such as those anticipated in forced-gradient in situ chemical oxidation (ISCO) operations. *Molecular diffusion* can be an important transport process for distributing oxidant into low permeability vadose zones or into fine-grained, low permeability zones within otherwise permeable aquifer formations. Such oxidant transport by diffusion can occur during ISCO where hydraulic fractures are filled with permanganate OPM, where injection probes are used for delivering MnO_4^- solutions, or where wells are used for groundwater flushing operations. Molecular diffusion is a slow process and generally cannot be relied on for distribution of oxidant over more than a few decimeters distance from the point of release even over months of time.

The movement of permanganate through soil and groundwater also can be caused by density gradients. For example, if a high concentration of permanganate (e.g., 10,000 mg-MnO_4^-/L) is delivered into groundwater via an injection well and transported horizontally under a hydraulic gradient, it would also have a tendency to move downward if there were any underlying zone of groundwater with little or no MnO_4^- in it. The relative rates of horizontal and vertical movement would be highly dependent on injection system operation and site conditions.

The one-dimensional (1-D) mass transport equation for MnO_4^- under saturated conditions is as follows (de Marsily, 1986):

$$(\theta D^* + D)\frac{\partial^2 C_P}{\partial x^2} - U\frac{\partial C_P}{\partial x} + r = \theta\frac{\partial C_P}{\partial t} \qquad (5.1)$$

where C_P is the concentration of MnO_4^-, D^* is the effective molecular diffusion coefficient, D is the dispersion coefficient, U is the Darcy groundwater velocity, θ is the porosity of the porous medium, and r represents the rate of degradation or consumption of MnO_4^- by oxidizable materials in the flowpath (e.g., organic contaminants, natural organic matter, Fe^{+2}) Because the subsurface can be replete with oxidizable matter and the reaction rates for MnO_4^- are relatively rapid (see Chapter 2), the degradation term can significantly affect the overall transport rate for MnO_4^-. The impact of MnO_4^- consumption on its transport can be demonstrated by solving the following simplified mass transfer equation:

$$-U\frac{\partial C_P}{\partial x} - \theta k_P C_P = \theta\frac{\partial C_P}{\partial t} \qquad (5.2)$$

where the effects of molecular diffusion and mechanical dispersion are ignored, and the degradation of MnO_4^- is represented by a first-order kinetic model (Yan and Schwartz 1999). For a sustained injection of oxidant at a concentration C_{Po}, a steady-state $\dfrac{\partial C_P}{\partial t} = 0$ condition will be attained eventually and the solution for eqn. 5.2 is as follows:

$$C_P = C_{Po} \exp\left(\frac{-\theta k_P}{U} x\right) \qquad (5.3)$$

In Figure 5-6, eqn. 5.3 is plotted for probable groundwater velocities in ISCO applications assuming that $k_P = 2 \times 10^{-3}$ s⁻¹, the minimum degradation constant measured by Yan and Schwartz (1999) in solutions of $KMnO_4$ and natural organic matter obtained from a landfill leachate. This

FIGURE 5-6. **Concentration profile as a function of groundwater velocity during sustained oxidant injection under steady state conditions.**

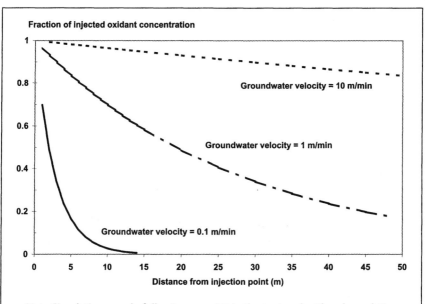

Note: Simulations made following eqn. 5.3 in the text and with a degradation constant (kp) of 2×10^{-3} s⁻¹.

plot shows that the penetration distance for MnO_4^- can be severely limited (e.g., < 15 m) if the groundwater velocity is on the order of 0.1 m/min (144 m/d). Note that the first-order kinetic model for MnO_4^- degradation assumes an infinite supply of oxidizable material. In reality, the mass of material with which MnO_4^- can react is finite, and there is a lower limit to the consumption of MnO_4^-. That limit potentially can be modeled by use of the same mathematical representation for irreversible sorption (Ruthven 1984), where the maximum MnO_4^- consumed per unit mass of soil is analogous to the mass of irreversibly sorbed material per unit mass of sorbent. Nevertheless, the analysis in Figure 5-6 shows that the consumption of MnO_4^- is a key factor in its transport; that consumption must be considered when delivery systems, particularly those in subsurface media with high natural organic content, are being designed.

In systems where diffusion is relied upon for the transport of MnO_4^-, say through a fine-grained material, the mass transport equation becomes:

$$D * \frac{\partial^2 C_P}{\partial x^2} - k_P C_P = \frac{\partial C_P}{\partial t} \tag{5.4}$$

Under steady state conditions and for a constant oxidant concentration C_{Po} at x=0, the solution for eqn. 5.4 is:

$$C_P = C_{Po} \exp\left(-\sqrt{\frac{k_P}{D*}} x\right) \tag{5.5}$$

In practice, the ratio of the effective diffusion coefficient to the molecular diffusion coefficient varies from 0.1 (clays) to 0.7 (sands) (de Marsily 1986), and molecular diffusion coefficients in water ranges from 1×10^{-5} to 2×10^{-5} cm^2/s at 20C for most common ions (de Marsily 1986, Domenico and Schwartz 1990). Using an effective diffusion coefficient of 1×10^{-5} cm^2/s and a degradation constant of 2×10^{-3} s^{-1}, the resulting concentration profile from eqn. 5.5 shows a penetration depth of less than 0.5 cm (see Figure 5-7, curve B). Reducing the degradation constant to 2×10^{-4} s^{-1} still results in a penetration depth that is less than 1 cm. As in the analysis for penetration depths by advection (i.e., eqn. 5.3), it is assumed that there is an excess of oxidizable material that can react with MnO_4^- along the diffusion path. If a lower limit to oxidant consumption is applied, the penetration depth will probably increase, consistent with

| FIGURE 5-7. | Concentration profile as a function of degradation constant and effective diffusion coefficient during oxidant diffusion under steady state conditions. |

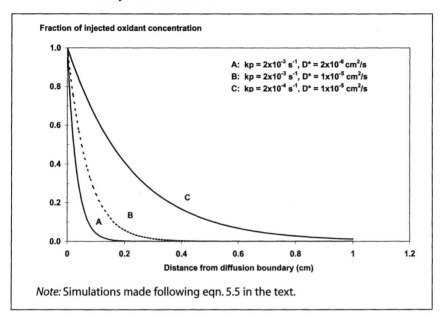

Note: Simulations made following eqn. 5.5 in the text.

the observations of Struse (1999) and Siegrist et al. (1999). Struse (1999) observed a diffusion rate of the MnO_4^- from a source of 5000 mg/L that was approximately 0.1 cm/d through intact cores of silty clay soil (Figure 5-8). This rate is equivalent to a transport distance of 30 cm over 300 days (about 10 months). This rate is consistent with field results reported by Siegrist et al. (1999) wherein a MnO_4^- impacted zone extending 30 cm above and below hydraulic fractures filled with permanganate OPM was observed within 10 months after initial emplacement. However, as in advection, the above analysis shows that migration of MnO_4^- by diffusion can be severely hindered by its reaction with oxidizable materials.

5.2. DESCRIPTION OF SUBSURFACE DELIVERY METHODS

A variety of methods for recovery and delivery of fluids into the subsurface are available, and these can be used for delivery of permanganate oxidant into the subsurface vadose and saturated zones at a contaminated site. Several methods are summarized in Table 5-2 and 5-3 and illustrated in Figures 5-9 to 5-14. They are described further in this section.

FIGURE 5-8.   **Permanganate retardation during diffusive transport in silty clay media due to matrix interactions and TCE DNAPL (after Struse 1999).**

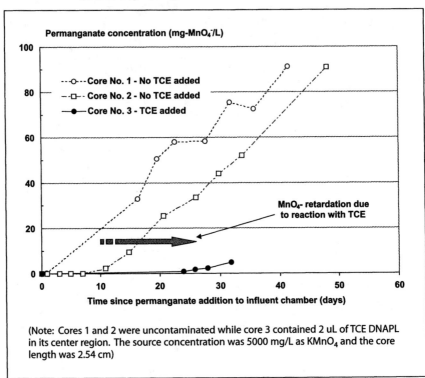

(Note: Cores 1 and 2 were uncontaminated while core 3 contained 2 uL of TCE DNAPL in its center region. The source concentration was 5000 mg/L as $KMnO_4$ and the core length was 2.54 cm)

Injection Probes

In many subsurface settings where in situ treatment is hindered by very low permeability and/or heterogeneity of the formation, coupling a high-density delivery system with a selected treatment process such as ISCO can enhance remediation. Injection probe technologies have been developed for this purpose (Table 5-2, Figure 5-9). They employ a vertical or angled small-diameter (2 to 4 cm diam.) probe that is rotated or pushed into a subsurface region while one or more reagents are injected under low pressure. The reagents are delivered down the hollow core of the probe and exit via small perforations or slots in the leading 30 to 60 cm length of the probe (Figure 5-9). The reagents migrate away from the probe by entering existing pore and fracture networks and creating a halo of reactivity. Slower diffusion processes can supplement this advective

TABLE 5-2.	Optional delivery systems for in situ chemical oxidation using permanganates.		
System type	Description	Application depth *Suitable media*	Comments
Surface application	Application of per- manganate solids or solutions to the ground surface at shallow contamin- ated sites or in above-ground piles	Ground surface *Permeable vadose zone* *formations*	High NOD in surface soils will limit pene- tration Could be appropriate for treatment of con- taminated soil in emptied impound- ments
Probe permeation	Direct-push or augured probes with discrete inter- vals of perforations for pressurized deliv- ery of oxidant	Ground surface to 50 ft. or more bgs *Permeable to low per-* *meable vadose and* *saturated zones*	Can deliver oxidant on close spacings to reduce path length and minimize NOD exerted Can target depths of interest
Soil mixing	Construction equip- ment or single or multiple blade augers with hollow stems for mixing and oxidant delivery	Ground surface to 50 ft. bgs *Moderate to low per-* *meability media in* *both vadose and sat-* *urated zones*	Can achieve high degree of mixing and contact of oxidant with contaminants Very disruptive to site and not possible at obstructed sites
Soil fracturing	Hydraulic fracturing with sand or oxida- tive particle mixtures	Ground surface to 50 ft. bgs *Low permeability* *media with proper* *state-of-stress in* *vadose and saturated* *zones*	Can provide a horizontal reactive barrier
Vertical well flushing	Ground water delivery via a vertical injec- tion well with flush- ing to vertical or horizontal extraction wells (possibly including recirculation)	Ground surface to 100 ft. depth or more bgs *Permeable formations;* *heterogeneities* *impact spacing* *and/or treatment* *time*	Well networks can be configured to site conditions and yield effective mixing Above-ground oxidant amendment of extracted ground- water and reinjec- tion can enable hydraulic control with minimum oxidant

continued

TABLE 5-2.	(continued)		
System type	Description	Application *Suitable media*	Comments
Horizontal well flushing	Ground water delivery via a horizontal well with flushing to horizontal or vertical extraction wells (possibly including recirculation)	Ground surface to 100 ft. depth or more bgs *Permeable formations without significant heterogeneities*	Can be used in sites with surface obstructions (e.g., under buildings) Can be effective in thin aquifers
Treatment walls	Dispersal into ground water to intercept and treat a plume using vertical or horizontal dispersal tubes or trenches	Ground surface to 100 ft. depth or more bgs *Permeable formations without significant heterogeneities*	Can create passive or semi-passive permeable barriers to horizontal groundwater flow

FIGURE 5-9.   Schematic of injection probe delivery of permanganate solution to soil or groundwater.

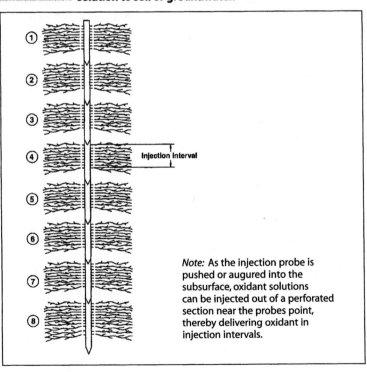

Injection Interval

Note: As the injection probe is pushed or augured into the subsurface, oxidant solutions can be injected out of a perforated section near the probes point, thereby delivering oxidant in injection intervals.

movement. To ensure complete coverage in a region of interest, the probe injections are made at relatively close spacings, typically within 0.6 to 1.2 m of each other. Equipment that achieves penetration depths of up to 10-m has been developed with up to four probes rack-mounted on a tractor. Other equipment, such as cone penetrometers, conceivably enable vertical penetration up to 50 m in some settings. The injected reagents are selected to match the contaminants of concern and a desired treatment effect (e.g., destruction or immobilization). In the vadose zone, permanganate oxidants can be injected up to the air-filled porosity of the media but might be constrained in the range of 5% v/v. Lower pressures of 30 to 75 psig are employed to control fracturing of the formation.

Injection probes have been used for subsurface delivery of permanganate oxidant solutions (Siegrist et al. 2000, Moes et al. 2000, Mott-Smith et al. 2000) (Table 5-3). There are many factors affecting injection

TABLE 5-3.	Methods of handling oxidant and enabling delivery to the subsurface.	
Probe permeation	Multipoint injection to 10-ft. depth with $KMnO_4$ in silty clay soils at a clean test site in Ohio	Siegrist et al. 2000
	Injection of $KMnO_4$ to 30-ft. depth in massive clay sediments contaminated with TCE in California	Moes et al. 2000
	Injection of $KMnO_4$ to 40 ft. depth in sands contaminated with DNAPL TCE in Florida	Clayton et al. 2000, Mott-Smith et al. 2000
Soil mixing	Delivery of $KMnO_4$ to 47-ft depth in silty clay soils contaminated with TCE and chlorinated ethenes at the DOE Kansas City Plant	Cline et al. 1997, Gardner et al. 1998
Soil fracturing	Fracturing with a $KMnO_4$ oxidative particle mixture to 16-ft. depth in silty clay soils contaminated with TCE at a land treatment site in Ohio	Siegrist et al. 1999
Vertical well flushing	Flushing within a 7.5 m³ test cell artificially contaminated with TCE and PCE at the Borden field site in Canada	Schnarr et al. 1998
	Delivery of $KMnO_4$ or $NaMnO_4$ in 5-spot recirculation patterns to 30-ft. depth in 100-ft. x 100-ft areas at the DOE Portsmouth site	Lowe et al. 2000
Horizontal well flushing	Delivery of $KMnO_4$ in a parallel horizontal well system to treat TCE DNAPL contaminated groundwater at the DOE Portsmouth site	West et al. 1998a,b

probe performance, including those associated with the subsurface conditions, the contaminants of concern, and the interaction between those conditions and contaminants. The formation must be sufficiently unconsolidated and free of boulders or buried debris that would inhibit penetration to the depth desired. Subsurface obstructions such as buried utilities and conduits can create similar problems. Surface obstructions such as parking areas and buildings may preclude use of injection probes or greatly increase its cost. The permeability of the formation and its porosity are important considerations; if the formation is massive in character with near zero permeability to fluids, then permeation will not be effective. Some degree of advection away from the lance is important to yield a rapidly evolving halo of effect around the probe, allowing diffusive transport to expand the treated region over the subsequent period. However, if the rate of reaction of the delivered reagent is too rapid, the transport will be limited and there may be gaps in treatment effectiveness. If contaminant concentrations are too high, the mass of treatment agent that can be delivered in the volume of fluid may be insufficient. In that case, multiple treatment cycles may be required. This is a positive attribute of injection probe delivery, since re-treatment or increased delivery in hot spot areas is possible. The presence of an underlying groundwater zone and its susceptibility to contamination by leaching of contaminants or treatment agents during permeation must also be carefully considered.

Hydraulic Fracturing

Application of hydraulic fracturing to environmental remediation is adapted from petroleum recovery techniques used for more than 50 years. In conventional hydraulic fracturing, horizontal fractures are created and filled with coarse sand proppant that can enable the delivery of fluids, including oxidants, into the subsurface (Murdock et al. 1994, 1995). Alternatively, with advanced fracturing methods, fractures can be filled with chemically reactive media such as potassium permanganate solids (Murdock et al. 1997a,b). A permanganate OPM was created and used for this purpose (Siegrist et al. 1999). In conventional hydraulic fracturing, the fractures typically are created by use of a 2-in. steel casing and PVC drive point driven into the subsurface by a pneumatic hammer (Figure 5-10). After the casing has been driven to the desired depth, the drive point is dislodged downward an additional 1 to 4 inches, exposing an open hole in the subsurface from which fractures are nucleated by the

FIGURE 5-10. Illustration of hydraulic fracturing for delivery of permanganate solids or solutions in low permeability zones.

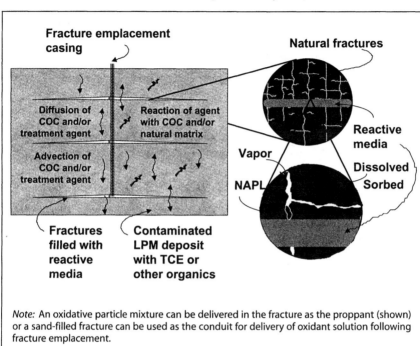

Note: An oxidative particle mixture can be delivered in the fracture as the proppant (shown) or a sand-filled fracture can be used as the conduit for delivery of oxidant solution following fracture emplacement.

following process: a high-pressure (2,500 psi) water jet cuts a horizontal notch into the soil; a viscous fluid is injected at a constant rate; the pressure exceeds a critical value (30–60 psi) and the fracture is nucleated. Coarse-grained sand is then injected as slurry while the fracture grows away from the borehole. Guar gum gel, a viscous fluid, typically is used to suspend the sand and facilitate transport of the sand grains into the fracture. Guar gum is mixed with water to form a short-chain polymer and a crosslinker is added to form a thick gel capable of suspending high concentrations of coarse-grained sand. The guar gum gel contains a time-activated enzyme that degrades the gel matrix into a liquid with a water-like viscosity approximately 8 to 36 hours after the initial mixing of the guar gum gel solution. After pumping, the fracture remains propped open by the sand, and the guar gum gel is decomposed by the enzyme added during injection. The gel is subsequently recovered. The same procedure is used for the placement of reactive fractures, but the reactive media (e.g., $KMnO_4$ OPM) is injected into the fractures.

The fracturing equipment normally includes a continuous mixer consisting of a hopper with an auger feeder to store and meter the proppant (e.g., sand); metering pumps for gel, crosslinker, and breaker; and an inclined auger mix tube to blend the ingredients. The mixer feeds a progressive cavity pump. Both the pump and mixer are mounted on an equipment trailer, along with a suitable prime-mover, approximately 8 ft wide, 18 ft long, and 8 ft high.

The pressure required to initiate a fracture in a borehole depends on the confining stress, the toughness of the enveloping formation, the initial rate of injection, the size of incipient fractures, pores, or defects in the borehole wall, and other factors. In general, the injection pressure will increase as depth, injection rate, and fluid viscosity increase. The pressure required to propagate a fracture created by injecting liquid into soil at 75 L/min at a 6-ft depth is on the order of 50 to 75 kPa (7 to 11 psig) and increases approximately 20 kPa (3 psig) every 3 ft. The pressure during propagation decreases during most operations, but may vary based on several factors; e.g., slight increases in pressure may occur when the sand concentration in the slurry increases.

Hydraulic fracturing generally produces a single seam (multiple fractures require repeated operations) that assumes one of two typical forms. One form consists of a steeply dipping fracture that has a greater vertical than lateral dimension. This type of fracture climbs rapidly and reaches the ground surface in the vicinity of the borehole after modest volumes have been injected. Significant propagation ceases after this has occurred. The second fracture form is elliptical in plan and dips gently toward the nucleation borehole. In some cases, the fracture essentially is flat lying; in others it is nearly flat lying in the vicinity of the borehole but then dips increasingly to approximately 20° at some distance away. In other cases, hydraulic fractures maintain a roughly uniform dip from the borehole to the termination. Fractures commonly have a preferred direction of propagation so that the borehole is off the center of the fracture. However, the area of the fracture containing the thickest sand nearly always occurs in the vicinity of the center, so that the thickest point rarely coincides with the borehole. The preferred direction of propagation is commonly related to distribution of vertical load at the ground surface, with the fractures propagating toward regions of diminished vertical load. In general, fracture radius and aperture increase as the volume of injected slurry increases at a given depth and formation. In most cases, hydraulic fractures at shallow depths in an overconsolidated formation have a maximum dimension of 25 to 60 ft (7 to 18 m) and an average fracture thickness of 1/10 to 1/4 in. (5 to 10 mm) (Murdoch et al. 1997a,b). Radius and aperture are functions of depth, however, with fractures of a given volume becoming broader and thinner as depth increases. Accordingly, the final thickness of the propped fracture increases with the concentration of sand in the injected slurry.

The feasibility of using hydraulic fracturing as a delivery method for ISCO requires consideration of the horizontal continuity of the fractures emplaced. If permanganate OPM is used as the proppant, then consideration must also be given to the OPM's degradation capacity and longevity. If permanganate solutions are to be delivered, consideration must be given to the feasibility of delivering reactive solutions into a sand-propped fracture from which they can infiltrate into undisturbed surrounding formation. The results obtained from work by Siegrist et al. (1999) demonstrated the use of reactive fractures created with reactive media that dissolves and permeates into the surrounding soil to produce a wide reactive zone (i.e., $KMnO_4$ particles in a mineral gel). During a field trial in Ohio in which a permanganate OPM was handled and emplaced by conventional hydraulic fracturing equipment and methods, handling of the permanganate was problematic in some respects, but modifications to fracturing equipment or development of encapsulation methods should resolve this issue. In general, the geometry of the reactive media fractures created was similar to that of conventional sand-filled fractures emplaced at the same site. Thus, there was no unusual behavior associated with the different fracturing fluids (i.e., iron particles in guar gum gel; permanganate particles in mineral-based gel; sand in guar gum gel).

Hydraulic fractures may bifurcate to form offset segments, a circumstance under which local areas avoided by the injected material are produced (Murdoch 1995). This challenges the fracture emplacement to be continuous and uniform horizontally with limited breaches, an imperative that may require overlapping fractures created at several depths. Fractures filled with a $KMnO_4$ OPM yield MnO_4^- ions that migrate away from their original location, dominantly by diffusion in a low permeability deposit but possibly aided by advection. This behavior will produce a zone at least several dm wide where resident TCE will be degraded rapidly; it could provide a barrier that would degrade mobile TCE as well. As a result, the gaps between offset fracture lobes or discontinuities between neighboring fractures might be "healed" by the migration of permanganate ions. A 5-mm-thick permanganate-filled fracture contains about 0.4 g $KMnO_4$ per cm^2 of fracture horizontal area. Based on complete oxidation and a stoichiometric TCE demand of 2.5 wt./wt., each cm^2 of fracture can treat about 0.16 g of TCE. This oxidant loading is sufficient to degrade an initial TCE concentration of 1000 mg/kg within a zone of LPM that is 90 cm thick. Alternatively, it is sufficient to treat 16 L of percolate with a concentration of 10 mg/L of TCE that is equivalent to a 50-yr life at a deep percolation flux of 1 cm/d. Realistically though, it is anticipated that the oxidant demand of NOM or the advective loss of oxidant out of the treatment region could markedly diminish this life.

Soil Mixing

Mixing approaches can be used to deliver solid or liquid permanganate into contaminated soils and sediments. For near surface applications (e.g., to treat residual organic COCs in shallow excavations made during cleanup of surface spills, impoundments, or leakage from underground tanks), mixing of permanganate solids (e.g., dry $KMnO_4$ crystals) by use of standard construction equipment (e.g., backhoe) represents a simple approach. For in situ applications at greater depths (e.g., contaminated vadose and saturated zones), deep soil mixing techniques can be employed.

Deep soil mixing is an aggressive subsurface manipulation technique for source areas and is suitable for delivering reagents to low permeability soils, such as shallow deposits of massive silts and clays (Figure 5-11). Oxidant delivery during deep soil mixing is accomplished by pumping an oxidant solution down a hollow kelly bar and into the mixed region either out of the bottom of the kelly or through nozzles on the mixing blades. The oxidant solution is beneficial to mixing in that the liquid can lubricate the cutting of the dense soils and sediments.

Deep soil mixing has been used for in situ remediation of contaminated sites, but mostly with vapor stripping of volatile COCs (Siegrist et al. 1995a) or for solidification and stabilization of metals (Siegrist et al. 1995b). Deep soil mixing was demonstrated in July 1996 at the Kansas City Plant for delivery of $KMnO_4$ to oxidize TCE in the subsurface (Cline et al. 1997, Gardner et al. 1998). Approximately 500 tons (456,710 kg) of soil were treated with $KMnO_4$ during the mixing of six 8 ft diameter columns to a depth of 25 ft to 47 ft. Approximately 67% of the TCE mass in the test area was removed, close to the 70% VOC removal objective. Higher TCE removal rates (90 wt %) are suggested by bench-scale studies

---

FIGURE 5-11. Schematic of deep soil mixing for delivery of permanganate solutions.

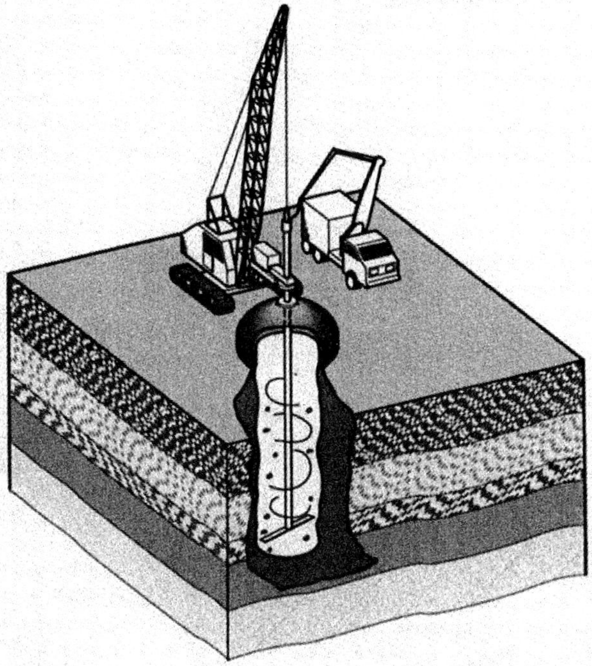

Note: Oxidant can be delivered in water pumped through a hollow auger stem and out of nozzles on the mixing blades or by injection probes inserted into the disrupted region after mixing is completed.

---

at higher $KMnO_4$ injected solutions (4 wt %). A lower oxidant loading was chosen for the field demonstration because the volume of oxidant that could be added to the low permeable soils was limited. Comparison of the distribution of pre-and post-treatment soil properties with depth suggest that the deep soil mixing process effectively mixed and homogenized the treatment region. Additionally, post-treatment soil properties measured suggest that the soil was not adversely affected by the oxidant.

There are several critical factors that affect the performance of soil mixing for oxidant delivery. Surface topography must be generally level (or must be made so by grading) to provide a stable base for the mixing equipment. Surface obstructions such as parking lots, buildings, and overhead power or steam lines can limit access totally or make it exceedingly expensive. Subsurface obstructions such as buried utilities, boulders/rock, or construction debris can similarly limit application. Depth and breadth of contamination are important. To date, most field applications have been accomplished at depths <16 m (50 ft) and in areas smaller than 1 hectare. Planned land use following mixed region treatment is very important. Mixing of soil yields an absolute volume expansion (i.e., soil will be lifted above the original ground surface) on the order of 15% v/v (i.e., 1.5 m^3 within an above-ground berm per 10 m^3 of media treated in situ) (Siegrist et al. 1995a).

Vertical Wells

Vertical wells can be used for oxidant delivery into subsurface regions of contaminated groundwater (Figure 5-12). The wells are similar to conventional groundwater wells except that the materials of construction need to be compatible with permanganate oxidants (see Appendix A). A variety of well network designs can be employed depending on the site conditions and the mode of aboveground oxidant handling. For example, individual vertical wells can be used strategically to deliver oxidant to hotspots near source areas in the subsurface (West et al. 1998a,b, IT Corporation and SM Stoller Corporation 2000). Alternatively, networks of multiple wells can be used to treat a broader region of a groundwater plume. The use of well networks with groundwater extraction and injection, such as a 5-spot pattern with center well injection and perimeter well extraction, can be effective for maintaining hydraulic control (Lowe et al. 2000). Treatment and disposal of extracted groundwater can be avoided if extracted groundwater is used for oxidant amendment followed by reinjection. Lowe et al. (2000) reported that the advantages of using liquid NaMnO$_4$ and a vertical well recirculation approach included: (1) better control of oxidant and contaminant migration within the treatment zone when compared to single well injections, (2) the introduction of higher volumes of oxidant solutions in a given time because existing soil pore water is extracted concurrent with oxidant injection, (3) easier oxidant handling, and (4) potentially lower overall cost for treating larger volumes. However, for application to highly contaminated sites, particles generated aboveground need to be carefully managed.

FIGURE 5-12. Schematic of vertical wells for delivery of permanganate solutions to a region of contaminated groundwater.

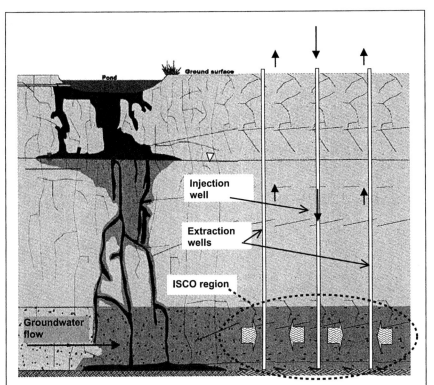

Note: Injection and recovery wells can be placed in various configurations depending on the region to be treated. For example, a 5-spot pattern can be used with a center injection well and four corner extraction wells. Oxidant can be delivered by amending clean water in which case the extracted groundwater may need to be treated before disposal.

The factors affecting vertical well application and performance are basically the same as those that affect groundwater extraction and injection. However, depending on the formation properties, well spacing may need to be far closer than simple well hydraulics would dictate. Often, this is needed to minimize transport distances where reaction is occurring and thereby enable effective delivery of oxidant throughout a region to be treated.

Horizontal Wells

Horizontal wells can be used for delivery of oxidants in thin zones of saturation or to regions where surface obstructions prohibit the use of vertical wells (Murdoch et al. 1994) (Figure 5-13). Horizontal wells for oxidant delivery are similar to conventional horizontal groundwater wells except that the materials of construction need to be compatible with permanganate oxidants (see Appendix A). A variety of well network designs can be employed depending on the site conditions and the mode of aboveground oxidant handling. For example, individual horizontal wells can be used strategically to deliver oxidant to areas in the subsurface (West et al.

FIGURE 5-13. Schematic of a pair of horizontal wells for permanganate flushing of contaminated aquifers.

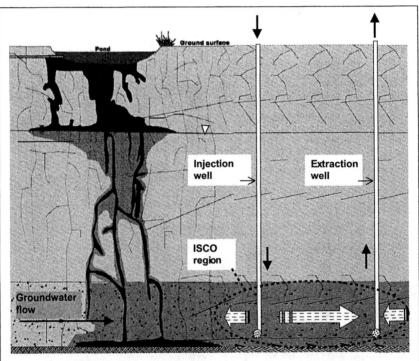

Note: Horizontal injection and recovery wells can be placed in various configurations depending on the region to be treated. For example, a pair of wells can be used with an upgradient injection well and a downgradient extraction well (shown). Oxidant can be delivered in (1) extracted groundwater that is amended with oxidant and reinjected or (2) by amending clean water in which case the extracted groundwater may need to be treated before disposal.

1998a,b). Alternatively, one or more wells can be used to treat a broader region of a groundwater plume. The use of well networks with groundwater extraction and injection, such as a pair of horizontal wells with one well for injection and the other for extraction, can be used for maintaining hydraulic control. Treatment and disposal of extracted groundwater can be avoided if extracted groundwater is used for oxidant amendment followed by reinjection. However, for application to highly contaminated sites, particles generated aboveground need to be carefully managed.

The factors affecting horizontal well application and performance are similar to those that affect groundwater extraction and injection. However, well spacing may need to be closer than simple well hydraulics would require to minimize transport distances where reaction is occurring.

Treatment Walls

Treatment-wall approaches can be used to deliver oxidant to a distal end of a groundwater plume and prevent the oxidant's migration (Figure 5-14). The walls can consist of continuous trenches filled with aggregate and a delivery-piping network into which permanganate solutions can be fed continuously or intermittently. Alternatively, disconnected vertical wells can be used for pressurized injection of permanganate solutions. Widely available standard construction equipment is used to install trenches and wells. Trenches provide a contiguous horizontal zone for contaminant collection, typically resulting in the greatest vertical access compared to other enabling methods. Trenches significantly increase the access (or exposure) to contaminants compared to vertical wells and, consequently, may be preferred for applications requiring shallow surface access.

The factors that affect treatment-wall application and performance are basically the same as those that affect the application of permeable reactive barriers (Gavaskar et al. 1997, 1998). However, depending on the plume velocity and the concentrations of reductants, well spacing may need to be quite close. Given the rapid rate of reaction, trenches of very narrow thickness can be used if oxidant can be delivered and mixed with the groundwater as it flows through the trench. Reliance on diffusion for passive mixing of permanganate with a groundwater plume is not advised; it is not reliable. If concentrations of reductants in the groundwater are very high, an aggregate-filled trench or the aquifer formation may be subject to plugging and hydraulic conductivity loss due to MnO_2 solids that are generated in situ.

FIGURE 5-14. **Illustration of a treatment wall approaches for delivery of permanganate to intercept a plume.**

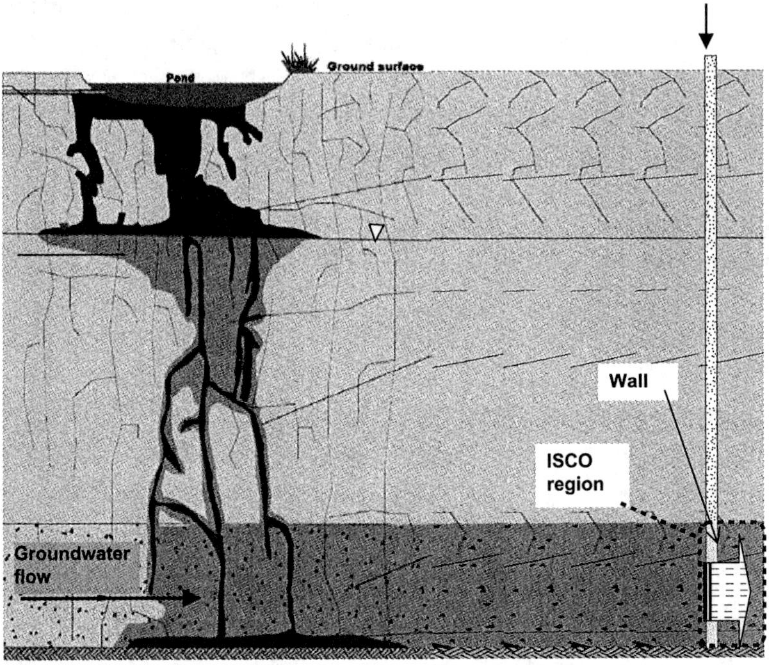

Note: The wall can be a continuous trench with permanganate solution delivery via a piping network within the aggregate filled trench. Alternatively the wall can be disconnected vertical wells into which permanganate solution is actively fed.

5.3. FACTORS AFFECTING OXIDANT DISTRIBUTION

Environmental and design factors can impact the uniformity of oxidant distribution in the subsurface and the degree of contact between the oxidant and the target organics. They include:

(1) Method of subsurface delivery and appropriateness for site conditions

(2) Degree of site heterogeneity impacting transport processes

(3) Extent of NOD and depletion of the oxidant dose

(4) Permeability loss due to ISCO-induced particle generation and pore filling

(5) Presence of DNAPLs and interphase mass transfer film formation

Several of the factors listed above will impact the effectiveness of any in situ remediation method (1, 2 and 5), while some are unique to ISCO (3 and 4). It is difficult to generalize regarding which of the factors will exert the greatest influence on oxidant distribution. Rather, the relative importance of the above five factors with respect to ISCO performance is dependent on contaminant properties and site conditions.

In situ delivery normally will also require hydraulic control to ensure that the oxidant is contained within the subsurface region to be treated and that contaminants do not exit the treatment zone because of unforeseen hydraulic gradients imposed by groundwater injection. For this purpose, well networks (either vertical or horizontal) can be used with groundwater extraction and reinjection (e.g., Figures 5-12 and 5-13). Examples include a 5-spot vertical well system with extraction from the perimeter wells and reinjection into the center well (Lowe et al. 2000) or a horizontal well pair with extraction from one horizontal well and reinjection into the other (West et al. 1998a,b). As noted above, precautions need to be taken to manage particles generated aboveground to avoid plugging of well screens and filter packs.

5.4. SUMMARY

There are advantages and disadvantages associated with each method of oxidant delivery; one method may be better suited for the remediation of a particular site than another. In addition to the method of delivery, the location and number of delivery points, the rate and duration of delivery, and the concentration of permanganate delivered must be determined. Also, in addition to introducing the oxidant into the subsurface, the delivery method often includes an aboveground fluid and material handling. For example with a 5-spot vertical well recirculation system for oxidant delivery, groundwater is extracted and then amended with oxidant before delivery into the subsurface via another injection well.

Particulates inevitably will form during permanganate oxidation, and the possibility that they will cause problems in fluid-handling piping, well systems, and in the formation itself must be carefully considered. In recirculation systems, canister cartridge filtration devices with 5 to 1 μm nominal pore sizes can be used for smaller applications. For larger applications it may be cost effective to consider other separation methods (e.g., sand filtration, coagulation/flocculation and cross-flow filtration, hydrocyclones).

5.5. REFERENCES

Bear, J. and A. Verruijt (1987). Modeling Groundwater Flow and Pollution. D. Reidel Publishing Company. Dordrecht, Holland.

Clayton, W.S., B.K. Marvin, T. Pac, and E. Mott-Smith (2000). A multisite field performance evaluation of in situ chemical oxidation using permanganate. In: G.B. Wickramanayake, A.R. Gavaskar, A.S.C. Chen (ed.). Chemical Oxidation and Reactive Barriers: Remediation of Chlorinated and Recalcitrant Compounds. Battelle Press. Columbus, OH. pp. 101-108.

Cline, S.R., O.R. West, N.E. Korte, F.G. Gardner, R.L. Siegrist, and J.L. Baker (1997). KMnO$_4$ chemical oxidation and deep soil mixing for soil treatment. *Geotechnical News*. December. pp. 25-28.

Cussler, E.L. (1997). Diffusion. Cambridge University Press.

de Marsily, G. (1986). Quantitative Hydrogeology: Groundwater Hydrology for Engineers. Academic Press, Inc., San Diego, CA.

Domenico, P. and F. Schwartz (1990). Physical and Chemical Hydrogeology. John Wiley and Sons, Inc., New York, NY.

Driscoll, F.G. (1986). Groundwater and Wells. Johnson Division, St. Paul, MN.

Gardner, F.G., N.E. Korte, J. Strong-Gunderson, R.L. Siegrist, O.R. West, S.R. Cline, and J. Baker (1998). Implementation of deep soil mixing at the Kansas City Plant. Final project report by Oak Ridge National Laboratory for the Environmental Restoration Program at the DOE Kansas City Plant. ORNL/TM-13532.

Gavaskar, A., N. Gupta, B. Sass, T. Fox, R. Janosy, K. Cantrell, and R. Olfenbuttel (1997). Design guidance for application of permeable barriers to remediate dissolved chlorinated solvents. U.S. Air Force Armstrong Laboratory. AL/EQ-TR-1997-0014.

Gavaskar, A.R., N. Gupta, B.M. Sass, R.J. Janosy and D. O'Sullivan (1998). Permeable barriers for groundwater remediation: Design, construction, and monitoring. Battelle Press, Columbus, OH.

IT Corporation and SM Stollar Corporation (2000). Implementation report of remediation technology screening and treatability testing of possible remediation technologies for the Pantex perched aquifer. October, 2000. DOE Pantex Plant, Amarillo, Texas.

Lowe, K.S., F.G. Gardner, R.L. Siegrist, and T.C. Houk (2000). Field pilot test of in situ chemical oxidation through recirculation using vertical wells at the Portsmouth Gaseous Diffusion Plant. EPA/625/R-99/012. U.S. EPA Office of Research and Development, Washington, D.C. 20460. pp. 42-49.

Moes, M. C. Peabody, R. Siegrist, and M. Urynowicz (2000). Permanganate Injection for Source Zone Treatment of TCE DNAPL. In: Wickramanayake, G.B., A.R. Gavaskar, and A.S.C. Chen (ed.). Chemical Oxidation and Reactive Barriers. Battelle Press, Columbus, OH. pp. 117-124.

Mott-Smith, E., W.C. Leonard, R. Lewis, W.S. Clayton, J. Ramirez, and R. Brown (2000). In situ oxidation of DNAPL using permanganate: IDC Cape Canaveral demonstration. In: Wickramanayake, G.B., A.R. Gavaskar, and A.S.C. Chen (ed.). Chemical Oxidation and Reactive Barriers. Battelle Press, Columbus, OH. pp.125-134.

Murdoch, L.C., D. Wilson, K. Savage, W. Slack, and J. Uber (1994). Alternative methods for fluid delivery and recovery. EPA/625/R-94/003. U.S. Environmental Protection Agency, Cincinnati, OH.

Murdoch, L.C. (1995). Forms of hydraulic fractures created during a field test in fine-grained glacial drift. *Quarterly Journal of Engineering Geology*, 28:23-35.

Murdoch, L., W. Slack, R. Siegrist, S. Vesper, and T. Meiggs (1997a). Advanced hydraulic fracturing methods to create in situ reactive barriers. Proc. International Containment Technology Conference and Exhibition. February 9-12, 1997, St. Petersburg, FL.

Murdoch, L., B. Slack, B. Siegrist, S. Vesper, and T. Meiggs (1997b). Hydraulic fracturing advances. *Civil Engineering*, May 1997. pp. 10A-12A.

Ruthven, D. (1984). Principles of Adsorption and Adsorption Processes. John Wiley and Sons, Inc., New York, NY.

Schnarr, M., C. Truax, G. Farquhar, E. Hood, T. Gonully, and B. Stickney (1998). Laboratory and controlled field experimentation using potassium permanganate to remediate trichloroethene and perchloroethene DNAPLs in porous media. *Journal of Contaminant Hydrology*, 29:205-224.

Siegrist, R.L., O.R. West, M.I. Morris, D.A. Pickering, D.W. Greene, C.A. Muhr, D.D. Davenport, and J.S. Gierke (1995a) In Situ Mixed Region Vapor Stripping of Low Permeability Media. 2. Full Scale Field Experiments. *Environ. Sci. & Technol.*, 29(9): 2198-2207.

Siegrist, R.L., S.R. Cline, T.M. Gilliam, and J.R. Conner (1995b). In situ stabilization of mixed waste contaminated soil. In: Gilliam, T.M. and C.C. Wiles (ed.). Stabilization and Solidification of Hazardous, Radioactive, and Mixed Wastes. STP 1240. ASTM, West Conshohocken, PA. pp. 667-684.

Siegrist, R.L., K.S. Lowe, L.C. Murdoch, T.L. Case, and D.A. Pickering (1999). In situ oxidation by fracture emplaced reactive solids. *J. Environmental Engineering*, 125(5):429-440.

Siegrist, R.L., D.R. Smuin, N.E. Korte, D.W. Greene, D.A. Pickering, K.S. Lowe, and J. Strong-Gunderson (2000). Permeation dispersal of treatment agents for in situ remediation in low permeability media: 1. Field studies in unconfined test cells. Oak Ridge National Laboratory Report, ORNL/TM-13596.

Struse, A.M. (1999). Mass transport of potassium permanganate in low permeability media and matrix interactions. M.S. Thesis, Colorado School of Mines, Golden, CO.

West, O.R., S.R. Cline, W.L. Holden, F.G. Gardner, B.M. Schlosser, J.E. Thate, D.A. Pickering, and T.C. Houk (1998a). A Full-scale field demonstration of in situ chemical oxidation through recirculation at the X-701B site. Oak Ridge National Laboratory Report, ORNL/TM-13556.

West, O.R., Cline, S.R., Siegrist, R.L., Houk, T.C., Holden, W.L., Gardner, F.G. and Schlosser, R.M. (1998b). A Field-scale test of in situ chemical oxidation through recirculation. Proc. Spectrum '98 International Conference on Nuclear and Hazardous Waste Management. Denver, Colorado, Sept. 13-18, pp. 1051-1057.

Yan, Y.E. and F.W. Schwartz (1999). Oxidative degradation and kinetics of chlorinated ethenes by potassium permanganate. *Journal of Contaminant Hydrology*, 37:343-365.

CHAPTER 6

Design of Permanganate Systems for ISCO

The effective design and implementation of in situ chemical oxidation (ISCO) using permanganate must couple the technology capabilities with site-specific contaminant properties and subsurface conditions to achieve performance goals (Figure 6-1). This chapter describes the site-specific application of in situ chemical oxidation using permanganate and provides guidance on initial screening, treatability studies, design, and implementation. The reader is referred to the previous chapters for details on permanganate as a chemical oxidant for in situ remediation. Chapter 7 gives detailed information on five field applications of ISCO using potassium or sodium permanganate that illustrate many of the activities described in this Chapter. Additional information on pilot-scale studies and full-scale applications of permanganate as well as other oxidants may be found in recent status reports (e.g., USEPA 1998a, ESTCP 1999).

6.1. INITIAL SCREENING

The initial screening is intended to rule out sites that are clearly not suited to chemical oxidation using permanganates; type or level of COCs, site conditions, and/or performance goals are the usual items to be considered. Table 6-1 lists some common COCs at hazardous waste sites and provides a generalized assessment of their amenability to oxidation by permanganates. Tables 6-2 and 6-3 summarize conditions that are favorable or challenging to application of ISCO using permanganate. In the

FIGURE 6-1.   Sequence of activities during evaluation of in situ chemical oxidation using permanganate.

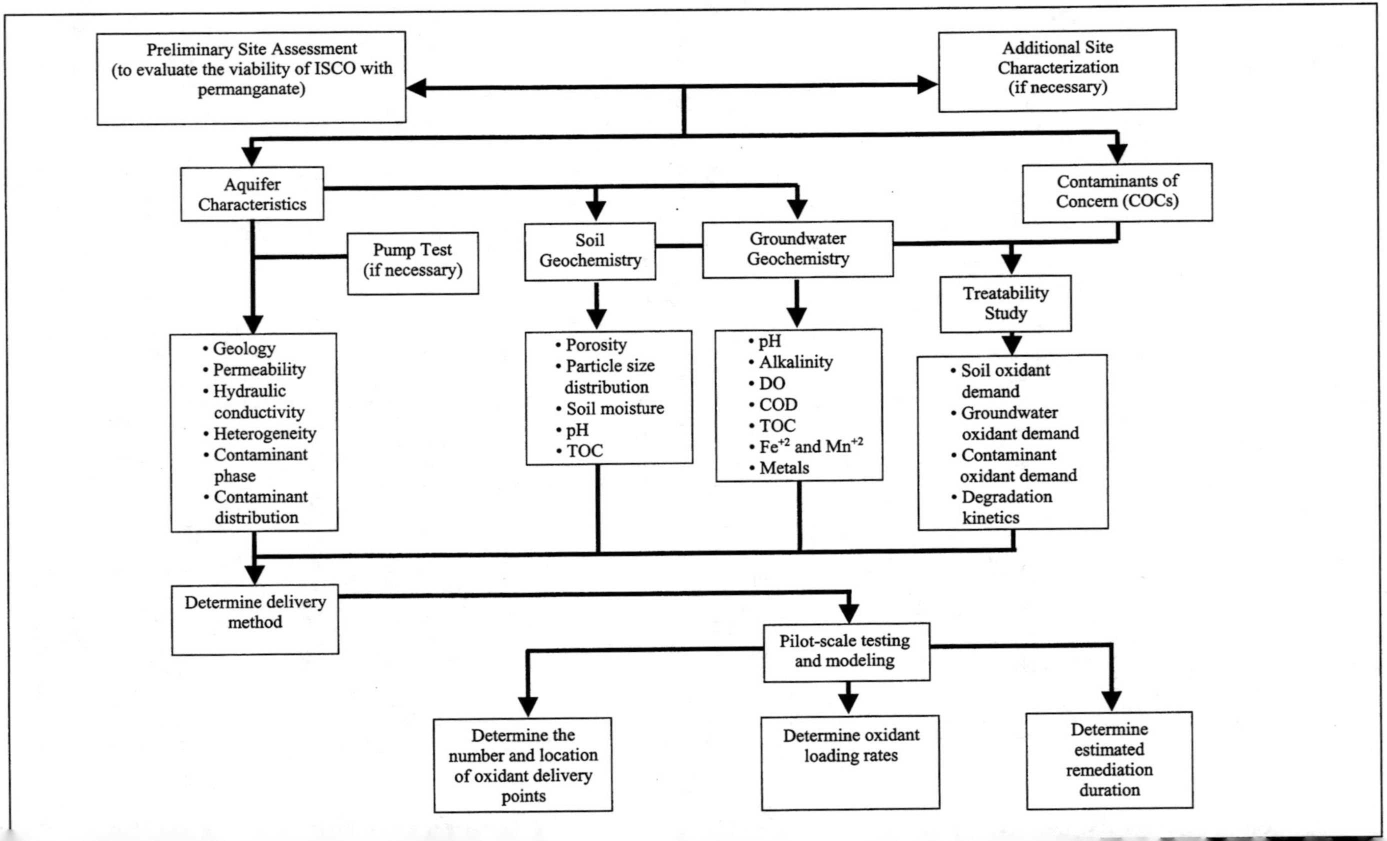

TABLE 6-1.	Amenability of selected organic contaminants to oxidation by permanganate.[1]		

| | COC phase | | |
Target organic chemical	Dissolved	Sorbed	NAPL[2]
Aliphatic compounds			
Alkanes			
Carbon tetrachloride	No	No	No
Trichloroethane	Not likely	Not likely	Not likely
Alkenes			
Tetrachloroethene	Yes	Yes	Yes
Trichloroethene	Yes	Yes	Yes
Dichloroethene	Yes	Yes	Yes
Vinyl chloride	Yes	Yes	Yes
Aromatic compounds			
Single ring			
Benzene	Uncertain	Uncertain	Uncertain
Ethylbenzene, toluene, xylenes	Yes	Yes	Yes
Phenols	Yes	Yes	Yes
Polyaromatic			
Naphthalene, phenanthrene,	Yes	Yes	Yes
Pyrene	Yes	Yes	Yes
Explosives			
TNT	Yes	Yes	NA
RDX	Yes	Yes	NA
HMX	Yes	Yes	NA
2,4-DNT	Yes	Yes	NA

[1]Information presented is for general insight only and does not construe a particular rate or extent of permanganate oxidation under a particular set of conditions.
[2]NAPL degradation requires dissolution into the aqueous phase where oxidation reactions occur. The rate and extent of NAPL degradation can be impacted by interfacial films that develop at the NAPL-water interface.

absence of data in the available technical literature, initial laboratory screening experiments can be used to determine the validity of chemical oxidation with permanganate for a particular contaminant or host of contaminants. If the results from the initial laboratory screening experiments are favorable, further laboratory and pilot scale tests may be performed.

Contaminant Type, Phase, and Distribution

The first question that a site manager assessing the applicability of ISCO must ask is, "Are the contaminants of concern amenable to

TABLE 6-2.	Conditions favorable to effective in situ oxidation by potassium or sodium permanganate.

Condition	Comments
Contaminants are alkenes (PCE, TCE,…)	Demonstrated to be rapidly oxidized by MnO_4^-
Contaminants are present in dissolved and sorbed phases (i.e., no NAPLs)	Demonstrated ability to oxidize these phases with comparably low and deliverable oxidant loadings
Co-contaminants are not redox sensitive or otherwise affected by permanganates	Mitigates concern over co-contaminant mobility
Treatment goal = mass reduction (e.g., 90%)	Demonstrated ability to accomplish this goal
Small subsurface region to be treated	Ability to complete the field activities quickly
High permeability and low heterogeneity	Simplifies delivery and distribution
Low natural oxidant demand (e.g., GW TOC <50 mg/L and soil f_{oc} <0.005)	Reduces oxidant consumption and cost
Subsurface temperature	Rate of reaction will not be limiting
No nearby withdrawals of GW for potable use	Mitigates concern over trace impurities and possible byproducts

degradation with permanganate?" As a general rule, permanganate will degrade unsaturated organic compounds. For example, chlorinated alkenes such as trichloroethene and tetrachloroethene are rapidly oxidized by permanganate (Vella and Veronda 1994, Yan and Schwartz 1999, Huang et al. 1999, 2000, Urynowicz 2000). Aromatic compounds such as phenols (Vella et al. 1990) and PAHs such as naphthalene and phenanthrene also have been shown to be readily oxidized by permanganate (Gates-Anderson et al. 2001). However, permanganate has little or no effect on chlorinated alkanes (saturated compounds) such as carbon tetrachloride and 1,1,1-trichloroethane (Tratnyek et al. 1999, Gates-Anderson et al. 2001).

Contaminant phase is an important additional consideration when the viability of chemical oxidation is being determined. Challenging factors include the presence of DNAPLs at high saturation and a high degree of

TABLE 6-3.	Conditions challenging effective in situ oxidation by potassium or sodium permanganate.

Condition	Comments
Contaminants are alkanes (e.g., CT, TCA)	Demonstrated to be resistant to oxidation by MnO_4^-
Contaminants are present in nonaqueous phases (i.e., as DNAPLs)	Comparably high oxidant loadings required and increased potential for secondary effects
Co-contaminants are redox sensitive or otherwise affected by permanganates	Potential concern over co-contaminant mobility (e.g., Cr, As)
Treatment goal = MCL	Ability to destroy high concentrations to sustained MCL levels is uncertain
Larger subsurface region to be treated	Implementation time can be long
Low permeability and/or high heterogeneity	Complicates delivery and distribution of oxidant
High natural oxidant demand (e.g., GW TOC >50 mg/L and soil f_{oc} >0.005)	High oxidant consumption and cost
Temperature is low (e.g., <10C)	Rate of reaction may be limiting (e.g., in ex situ batch tanks)
Nearby withdrawals of GW for potable use	Concern over trace impurities and possible byproducts and public health effects

contaminant sorption onto formation media. In addition to providing a long-term source of contamination, the presence of DNAPL residuals and pools also may reduce the migration of contaminants or reagents by displacing the available pore volume, thereby lowering the effective permeability available for fluid migration. Furthermore, DNAPL pools generally will take longer to treat than DNAPL residual or dissolved/sorbed phase COCs because of the relatively small surface area to volume ratios associated with the former. Although permanganate has been shown to accelerate DNAPL dissolution and degradation (Schnarr and Farqhuar 1992, Schnarr et al. 1998, Urynowicz 2000, Siegrist and Urynowicz 2000), the presence of DNAPL will require that a greater mass of oxidant is delivered over a longer period of time. In addition, the formation of an interfacial film at the groundwater-DNAPL interface can reduce the rate of DNAPL degradation substantially (Urynowicz and Siegrist 2000, Li and Schwartz 2000, Reitsma and Marshall 2000). Contaminant sorption onto organic matter within the fine matrix of a formation may directly retard the movement of COCs to the pore channels containing oxidant and/or that of the oxidant into the fine pores with the COCs.

A relatively accurate estimate of the total contaminant mass in both the aqueous, sorbed, and non-aqueous phase is essential for the successful design and implementation of ISCO. Site-characterization data can provide the necessary parameter values (e.g., bulk density, porosity, fraction organic carbon, water content) to estimate the total mass and phase distribution of organic chemicals by use of simple equilibrium partitioning models (e.g., Dawson 1997). However, the delineation of organic contaminants in the subsurface remains very challenging and imprecise, and in most cases the estimated total mass turns out to be lower than the mass actually present (Siegrist and Van Ee 1994, West et al. 1995).

Plume size and the spatial distribution of contamination within the subsurface will also have a direct bearing on the amount of permanganate required. Natural oxidant demand of the native soil and groundwater is expressed on a unit mass or unit volume basis; therefore, more permanganate will be used to oxidize "non-target" compounds like NOM when contaminants are spread over larger volumes. All else being equal, lower contaminant concentrations spread over larger volumes will require much more oxidant and take longer to remediate than higher contaminant concentrations spread over smaller volumes. However, in comparison with the costs of conventional technologies like pump-and-treat, ISCO with permanganate may still be the most cost-effective approach.

Geologic and Hydrologic Site Characteristics

The lithology and hydrogeology characteristics of a site are important because they affect the transport of oxidants as well as their fate in the subsurface. The depth and thickness of the contaminated region and its permeability must be known from prior investigations or be determined through subsurface characterization methods (e.g., USEPA 1993). Heterogeneities can dramatically influence oxidant delivery, and they must be defined to enable effective delivery system design and implementation. Geologic features, such as low permeability bulk deposits and clay lenses or fractured bedrock, may make oxidant delivery more difficult but do not necessarily preclude the use of ISCO with permanganate (Tables 6-2 and 6-3). Unlike conventional advanced oxidation processes (AOPs) that rely on the generation of short-lived hydroxyl radicals, permanganate is not consumed by carbonate and bicarbonate and may persist in the subsurface for months or even years after the initial application. Consequently, permanganate has some ability to diffuse into low permeability zones and destroy contaminants that would be difficult to reach advectively. However, remediation time frames of months are required for appreciable matrix diffusion to occur. Furthermore, the penetration distance of the MnO_4^- front by diffusive transport may be limited if the porous medium has a high NOD (see Chapter 5).

Potential surface water receptors such as lakes and streams need to be considered when permanganate is to be delivered into the subsurface. If high concentrations are released, permanganate may have a detrimental effect on aquatic life. The probability and risk associated with a surface water release of permanganate must be evaluated during the initial screening activities. If the risk is considered manageable, a release contingency plan should be developed and included in the site health and safety plan.

6.2. SITE CHARACTERIZATION FOR SYSTEM DESIGN

Although the initial screening may show that the site is suitable for ISCO with permanganate, available site characterization data may not be sufficient to support a preliminary design. Contaminant type, phase, and distribution should be quantified across the entire site in three dimensions. Accurate delineation of the contaminant source(s) is of even greater importance. If DNAPLs are suspected, partitioning tracer tests (PTT) may be used to provide insight into the pre-ISCO mass and distribution. However, the effective application of PTTs to assess subsurface regions following ISCO has yet to be demonstrated. The following geologic and aquifer characteristics should be known: permeability, vertical and lateral hydraulic conductivity, and heterogeneity. Depending on the complexity of the site and type of delivery method proposed, groundwater pump tests may also be required to determine aquifer yield and to establish well extraction and re-injection rates. General properties of the media, including grain size distribution, soil pH, soil Eh, and total organic carbon content can be measured using established methods (e.g., Klute et al. 1986, Carter 1993, Sparks et al. 1996, Tan 1996, USEPA 1986, 1990, 1993). In addition to the groundwater contaminants and their concentrations (USEPA 1986, 1990), groundwater pH, alkalinity, dissolved oxygen, chemical oxidant demand (COD), reduced metal species (e.g., Fe^{+2} and Mn^{+2}), sulfides, and co-contaminant metals concentrations should also be determined (e.g., APHA 1998). If a natural or active bioremediation process is planned as a post-oxidation treatment method, then appropriate microbial measurements regarding the effects of MnO_4^- treatment on indigenous microbial community biomass and activities can be evaluated in the laboratory (e.g., biomass, activities, yield, degradation efficiency, nutrient and electron acceptor levels). (Phelps et al. 1989, NATO 1999, Weaver et al. 1996, Burlage et al. 1998, USEPA 1998b).

6.3. PRELIMINARY DESIGN

Preliminary design of an ISCO system requires the definition of several key components as outlined in Table 6-4. The goal of the preliminary design phase is to select the appropriate oxidant concentration/dose and method of delivery to achieve a treatment efficiency throughout a region of interest and thereby reach a target performance goal (e.g., reduce original COC mass by 90%).

TABLE 6-4. Preliminary design components for an ISCO system using permanganate.

Components

✓ Performance goals and constraints

✓ Oxidant type (e.g., $KMnO_4$ or $NaMnO_4$), delivery concentration, and mass loading

✓ Oxidant delivery method (e.g., vertical wells, permeation probes, soil mixing)

✓ Above-ground fluid and material handling (e.g., oxidant amendment to water)

✓ Operational strategy and time (e.g., continuous one-time delivery, repeated delivery)

✓ Process control monitoring (e.g., flow rates, pressures, chemistry)

✓ Performance monitoring and assessment

✓ Worker and environmental safety

✓ Regulatory considerations and permitting

✓ Waste management

✓ Capital and operational costs

✓ Post-ISCO land use

Performance Goals and Constraints

If realistic performance goals are to be developed, a thorough understanding of site-specific characteristics and the risk to be mitigated is required. Performance goals should ultimately be linked to a risk reduction that is deemed necessary and technologically feasible for soil and/or groundwater media, and the goals should be applied within the zone to be treated or downgradient of the treatment zone. Example goals include: (1) a percent decrease in the COC mass based on pre- and post-treatment soil sampling and analysis, (2) reaching a post-treatment residual soil concentration based on soil sampling and analysis, (3) reducing groundwater concentrations to a risk-based level based on monitoring wells, or (4) a combination of the above. Certain contaminants (e.g., COCs that are not readily oxidized or DNAPLs) and site conditions (e.g., subsurface heterogeneity and low permeability media) do not necessarily preclude setting high performance expectations, but they will certainly complicate remediation efforts and may require greater expenditures over longer periods of time.

Oxidant Form, Delivery Concentration, and Mass Dose

Since the permanganate ion is provided in $KMnO_4$ and $NaMnO_4$, both compounds are equally effective oxidants on a mass permanganate basis. However there are considerations in choosing one form over the other. Potassium permanganate is less costly than sodium permanganate. However, $KMnO_4$ is generally provided as a solid whereas $NaMnO_4$ is provided in a concentrated liquid form (40% by wt. or more). Thus, there is a need for on-site dissolution of $KMnO_4$, which adds a unit process and presents potential dust hazards. Concentrated $NaMnO_4$ solutions can be fed via commercially available chemical feed pumps and systems, but they require more care in handling than solid and aqueous $KMnO_4$ solutions, the latter being limited in concentration due to the solubility of $KMnO_4$ in water. Potassium permanganate presents difficulties at some facilities due to its ^{40}K content and impurities in the potassium ores. The levels of ^{40}K from $KMnO_4$ are not likely to be of concern, but, if radioactivity is detected at the site (e.g., DOE facilities), it can be difficult to distinguish between harmless levels of ^{40}K from $KMnO_4$ as the source versus harmful radioactivity from contaminant sources (e.g., Tc^{99}). Sodium permanganate presents issues related to Na^+ and its water quality impacts and potential permeability effects (see Chapter 4).

Once the oxidant type is selected, it is necessary to determine the appropriate concentration and the requisite mass dose. In general, it is advantageous to keep the dose concentration as low as feasible given the target chemical demand and that of the natural matrix. However, the dose concentration must be set high enough to deliver sufficient oxidant to satisfy the target chemical demand and any NOD. Consequently, the dose is controlled in large part by the contaminant concentration/distribution, as well as the method of delivery. In permeable systems, it is recommended that the oxidant be advected into the region of interest quickly so as to degrade dissolved phase COCs and allow sorbed and entrapped contaminants to migrate into the pore channels filled with oxidant. At many sites containing fine-grained sediments, there are situations where diffusive transport pathways may be important. However, in systems with high NOD, there will be a need for handling and delivery of high concentrations of permanganate—a problematic and/or costly process. A recently developed oxidative particle mixture can enable delivery of high masses of $KMnO_4$ in the form of a slurry, providing a long-term oxidant source (Siegrist et al. 1999).

Delivery Method and Above-Ground Fluid and Material Handling

Successful ISCO with permanganate relies on careful selection of the delivery method as well as proper design and construction. There are optional methods of delivery for soil and groundwater applications as highlighted in Table 6-5 and described in Chapter 5. As shown, several methods of delivery are available including: injection probes, soil fracturing, soil mixing, vertical and horizontal well flushing, and treatment fences. Depending on the type of delivery method, changes or modifications usually can be made as process monitoring continues and the understanding of the site grows.

There are advantages and disadvantages associated with each method of oxidant delivery; one method may be better suited for the remediation of a particular site than another. In addition to the method of delivery, the location and number of delivery points, the rate and duration of delivery, and the concentration of permanganate delivered must be determined. In addition to introducing the oxidant into the subsurface, the delivery method often includes above-ground fluid and material handling. For example with a 5-spot vertical well recirculation system for oxidant delivery, groundwater is extracted and then amended with oxidant before delivery into the subsurface via another injection well (Lowe et al. 2000).

TABLE 6-5.	Optional delivery systems for in situ chemical oxidation using permanganate.	

Condition	Description	Example application
Surface infiltration	Application of permanganate solutions to the ground surface at shallow contaminated sites or in above-ground piles	Conceivable method of delivery but no known applications
Injection probes	Direct-push or augured probes with discrete intervals of perforations for pressurized delivery of oxidant	Delivery of $KMnO_4$ to 40 ft. depth in sands at Cape Kennedy (Mott-Smith et al. 2000)
Soil mixing	Construction equipment or specific augers for mixing solids or liquids	Delivery of $KMnO_4$ to 47-ft depth in silty clay soils at the DOE Kansas City Plant (Cline et al. 1997)
Soil fracturing	Hydraulic fracturing with sand or oxidative particle mixtures	Application for delivery of $KMnO_4$ OPM to 16-ft. depth in silty clay soils at the DOE Portsmouth site (Siegrist et al. 1999)
Vertical well flushing	Groundwater delivery via a vertical injection well with flushing to vertical or horizontal extraction wells (possibly including recirculation)	Delivery of $KMnO_4$ and $NaMnO_4$ in 5-spot recirculaton patterns to 30-ft. depth in 100-ft. x 100-ft areas at the DOE Portsmouth site (Lowe et al. 2000)
In-well recirculation	Vertical well recirculation by pump or air-lift with oxidant injection and dispersal around the well	Conceivable method of delivery but no known applications
Horizontal well flushing	Groundwater delivery via a horizontal well with flushing to horizontal or vertical extraction wells (possibly including recirculation)	Delivery of $KMnO_4$ in a parallel horizontal well recirculation system at the DOE Portsmouth site (West et al. 1998a,b)
Treatment walls	Dispersal into groundwater to intercept and treat a plume using vertical or horizontal dispersal tubes or trenches	Conceivable method of delivery but no known documented applications for ISCO using permanganate

In recirculation systems, the presence of MnO_2 particulates must be carefully considered because they will form inevitably during permanganate reduction and may cause problems in fluid-handling piping, well packing, and in the formation itself. Canister cartridge filtration devices with 5 to 1 μm nominal pore sizes can be used for smaller applications (Lowe et al. 2000). For larger applications it may be cost effective to consider other separation methods (e.g., sand filtration, coagulation/flocculation and cross-flow filtration, hydrocyclones).

Process Monitoring and Control

Delivery flow rates and pressures and feed concentrations of oxidant should be monitored at startup and frequently thereafter. Pressure increases during constant flow can indicate particulate formation and deposition within the piping or delivery well system. That may necessitate temporary system shutdown for system maintenance (e.g., piping can be cleaned and/or wells redeveloped). MnO_2 precipitates may be removed with a weakly acidic hydrogen peroxide solution.

The delivery and distribution of permanganate throughout a subsurface region may be evaluated by use of monitoring probes and wells from which samples are collected periodically and analyzed. Very low concentrations of permanganate (< 1 mg/L) can be visually detected in water. As a result, the extent of permanganate migration can be evaluated easily by sampling existing or new monitoring wells. Permanganate groundwater concentrations can be quantified readily by use of a field-portable spectrophotometer following standard methods. Absorbance at 525 nm is measured in an aqueous sample (either groundwater directly or in extraction water from a soil sample), and the permanganate concentration is determined from a 5-point calibration curve made with MnO_4^- standards. Filtration of the aqueous sample (0.22 um) prior to spectrophotometer analysis is important to prevent colloidal MnO_2 from interfering with the MnO_4^- determinations. Remote sensing techniques recently have been explored for delineating permanganate migration. Geophysical methods using DC electrical resistivity have been employed to delineate the migration of a 2 to 4 wt.% solution of permanganate injected into a horizontal well located in a 1.5-m thick aquifer zone overlain by 8 m of silty clay (Nyquist et al. 1998, 1999, West et al. 2000).

To evaluate the extent and rate of treatment, groundwater conditions at the site should be monitored routinely. Periodic sampling and analysis should be accomplished to define the concentrations of COCs within the treatment region and to document treatment effects. At some sites, the appearance of reaction products (e.g., Cl⁻) can be useful as an indicator of contaminant degradation. However, unless there are high concentrations of contaminants and relatively low background concentrations of the product of interest, such monitoring may not prove fruitful. Additional field parameters that should be collected include groundwater head levels and groundwater pH and alkalinity.

Performance Monitoring and Assessment

The major objective of monitoring is to ascertain the performance of the remediation and its compliance with applicable standards. An assessment of the performance requires that different factors be considered in an integrated manner. These include remediation technology implementation, operation and maintenance requirements, risk reduction achieved, and cost.

Overall treatment performance can usually be evaluated by determining the extent of permanganate migration. In most cases, chemical oxidation is so rapid that the presence of permanganate coincides with the complete or near complete destruction of contaminants (West et al. 1998a,b, Siegrist et al. 1999). However, in heterogeneous media the presence of permanganate in the groundwater does not necessarily mean that contaminants in the surrounding soil matrix have been remediated. Thus, soil coring and monitoring well networks, as well as other strategies for performance evaluation, must be considered. Also, beyond monitoring the initial COCs and verifying their disappearance to a target goal, monitoring and assessment should continue beyond the end of the active oxidant delivery phase.

With regard to risk reduction achieved by remediation of DNAPL compounds in the subsurface, it is increasingly recognized that achieving near 100% efficiency with any available, emerging, or conceivable in situ method is virtually impossible (e.g., Freeze and McWhorter 1997). Thus, establishment of cleanup goals for COCs must be based on a risk reduction underpinning. For this purpose, consideration must be given not only to the total mass removed or destroyed, but also to the reduction in mobile mass and the mobility of any residuals that are not removed or treated. Mobile mass in this context is defined as that mass of contaminant that is actively migrating either by advection or diffusion in the liquid or vapor phase. Depending on site conditions, the mobile mass normally can create an unacceptable impact on groundwater by leachate migration vertically downward or to the atmosphere by volatilization and vapor migration upwards.

At a minimum, contaminant reduction, operating conditions, and unit treatment costs must be routinely monitored. Specific contaminants, operating parameters (oxidant dose, flow rates, pressure, etc.) and subsurface properties (soil permeability impacts, soil moisture content, pH, cations, etc.) are dependent on the delivery approach selected. After agreed upon performance goals are met, the monitoring parameters and frequency should be re-evaluated and modified as appropriate.

Maintenance and Closure Criteria

The long-term maintenance procedures and the closure requirements for a site where ISCO is applied must be developed. These are site-specific and will be dependent on the site cleanup goals and the post-ISCO land use.

6.4. LABORATORY TESTING

Subsurface characteristics will vary significantly from site to site. Therefore, it is often essential to conduct laboratory testing on actual soil and groundwater collected from the site to evaluate the application of ISCO with permanganate (Table 6-6). A customized, site-specific approach is required. If the outcome of a particular treatability study is unfavorable, a similar outcome in the field can be expected. Laboratory tests conducted in batch, column, or diffusion cell mode may be performed depending on the site characteristics and the type of information required. Treatability studies generally consist of a series of batch tests conducted to evaluate the oxidant demand associated with soil and groundwater from the site. However, additional tests such as kinetic experiments and one-dimensional column or diffusion cell experiments can also be performed to further characterize matrix-oxidant interactions.

Study Purpose and Data to be Generated

Degradation Efficiencies, Kinetics, and Byproducts. Degradation efficiency is the extent to which a particular target contaminant is destroyed (e.g., 99% destruction) for a given reaction time, the latter usually taken as the time at which changes in COC levels are no longer detectable (i.e., the reaction is complete). The oxidation kinetics describe

| TABLE 6-6. | Summary of laboratory analyses and treatability testing. |

Phase	Parameters	Importance
Preliminary screening	Treatment efficiency	High
	Reaction kinetics	Variable
	Permanganate demand	High
	Reaction products	Variable
Preliminary design for groundwater application	Permanganate demand	High
	pH	Medium
	Alkalinity	Medium
	DO	Low
	COD	High
	TOC	Medium
	Fe^{+2} and Mn^{+2}	High
	Co-contaminant metals	Variable
Preliminary design for soil application	Porosity and bulk density	High
	Grain size distribution	High
	Hydraulic conductivity	High
	Permanganate demand	High
	pH	Medium
	TOC	High
	Co-contaminant metals	Variable

the rate at which it is destroyed (e.g., second-order kinetics for MnO_4^- and TCE with $k_2 = 0.89$ L mol^{-1} s^{-1}). If the target chemical is susceptible to oxidation by permanganate, the reaction generally will be quite rapid (e.g., $t_{1/2} \sim$ minutes). Assuming that a viable reaction mechanism exists, degradation efficiency is dependent on providing an adequate dose of oxidant to react with the target COCs as well as other reductants that may compete for the permanganate. Thus, it is customary to run a series of tests with different initial permanganate concentrations $[MnO_4^-]_o$) and to track the COC disappearance over time. This provides both efficiency data and site specific kinetics data. If only efficiency is of interest, then the time for the reaction to go to completion can be estimated based on an assumed second-order kinetics model and the reaction can be carried out for that period of time plus some excess. To establish a range of oxidant dose concentrations ($[MnO_4^-]_o$) for testing, the NOD of the media as measured by standard methods can provide an estimate of the ultimate demand for oxidant and enable establishment of $[MnO_4^-]_o$ to be used in testing. Alternatively, the concentration of $[MnO_4^-]_o$ can be set based on the target contaminants and the oxidation reaction stoichiometry as well as the estimated NOD (based on a separate test as noted below or an estimate from Table 2-5).

Contaminant byproducts may also be a potential concern, especially when the particular contaminant is complex and the reaction with permanganate has not been well studied. In addition, the production of manganese dioxide, a colloidal byproduct of permanganate reduction, may reduce the permeability of treated soils over time. During the degradation test, samples can be analyzed for production of possible intermediates (e.g., chlorocarbons) as well as filterable solids (for tests in groundwater) as an indicator of MnO_2 production. In systems without high background levels, mineralization of a chlorocarbon can be verified by analyses of chloride production.

Tests of degradation efficiency (and possibly kinetics and byproducts) normally are made in well mixed batch reactors maintained under zero headspace conditions to avoid volatilization losses. If kinetics measurements are not needed, the apparatus is simplified and the number of sample collection time points is reduced. Example reactors are presented in the following section.

Natural Oxidant Demand. The natural oxidant demand exerted by reductants in the soil and groundwater media (excluding that of the target analytes) can account for a significant fraction of the total oxidant demand. As such, the NOD must be determined to enable selection of the oxidant dose concentration and total mass loading rate. It is important to measure the NOD with the range of MnO_4^- dose concentrations that might be used. This is necessary because the apparent NOD is a direct function of MnO_4^- dose concentration (see Chapter 2); that is, under otherwise comparable conditions, the NOD increases as the MnO_4^- dose concentration increases. Examples of oxidant demands were given previously in Table 2-5. In some cases, it is possible to combine the measurement of NOD with the degradation efficiency (and possibly kinetics). In these cases, it is important to monitor the disappearance of the target analytes as well as the reductants in the media (e.g., NOD, TOC) and to determine the MnO_4^- concentrations with time. It is reasonable to assume that the NOD measured in a completely mixed batch test under laboratory conditions may overestimate the demand exerted under field conditions where preferential flow and mass transfer limitations can be significant. However, the exact relationship of NOD measured in a batch test compared to that experienced under field conditions is not yet fully established. The NOD can be measured for the media only (in the absence of target VOCs) by simply purging the groundwater or soil slurry with an inert gas such as nitrogen before the NOD test is run. However, if the target organics are semi-volatile or nonvolatile, then this approach will not work and the measurement of NOD will have to be made along with the target chemical demand.

Oxidant Subsurface Transport. It is important that the rate and extent of transport of MnO_4^- in the subsurface be assessed since these factors are critical to selection and design of the delivery method and the oxidant dose. Models can be used to estimate the transport, but experimental data may often be desired to verify the model results. Laboratory experiments to define transport of permanganate in porous media can be difficult to accomplish in a manner that is representative of field conditions, particularly for heterogeneous and low permeability formations. As a result, pilot testing is likely to provide the best design information at many sites.

Experimental Methods

Batch Reaction Experiments. Laboratory batch experiments can be run relatively quickly and inexpensively. Because they can be conducted with a high degree of control, they are particularly well suited for chemical oxidation degradation efficiency and kinetic studies. Batch experiments are also useful as an initial screening tool to define the NOD of soil and groundwater. Because batch experiments are run typically under homogeneous, completely mixed conditions, the oxidant demand expressed by soil and groundwater is considered the "ultimate" oxidant demand and is not necessarily the demand that would be expressed under field conditions. In naturally structured porous media, only a fraction of the resident pore solution is displaced during an infiltration and flow-through period. This behavior is referred to commonly as preferential flow. Consequently, the oxidant demand expressed under field conditions may be many times smaller than the "ultimate" demand expressed during a batch test.

Generally, a series of batch experiments is performed to determine the oxidant dose concentration (e.g., $mg\text{-}MnO_4^-/L$) and mass loading rate (g-MnO$_4^-$/g media) to achieve a target degradation efficiency and/or satisfy the "ultimate" NOD of the soil and groundwater media. That testing can be accomplished in reactors using groundwater or soil slurries (Figures 6-2 to 6-4). It is emphasized that experiments with VOCs must be performed under zero-headspace conditions to minimize volatilization losses and prevent erroneous interpretations of destruction efficiency. In addition, controls consisting of untreated matrix samples containing the COCs of interest must be included in the experimental design.

One approach for groundwater media involves the use of multiple, glass reaction vials (20 or 40 mL) to enable replicate sampling at one or more time points by sacrificially analyzing vials from the series (Case 1997, Urynowicz 2000) (Figure 6-2). Each of a set of reaction vials is prepared to have the same initial conditions with respect to target contaminant concentration and groundwater matrix. Different oxidant dose

FIGURE 6-2.   **Batch experimental apparatus to measure COC degradation efficiency and NOD using 20- to 40-mL glass reaction vials for groundwater.**

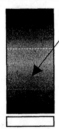

20- to 40-mL clear or amber, glass vial with Teflon sealed solid cap

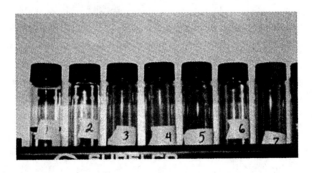

concentrations can be tested. For measurement of NOD without target COCs, groundwater samples can be purged with an inert gas (e.g., N_2) to strip any ambient VOCs or SVOCs from the matrix. The matrix water is added to each vial to fill it almost completely. Then at time zero, a small aliquot of a stock solution that was previously prepared by dissolving permanganate crystals in matrix water (or de-ionized water) is added to each vial to yield the desired initial oxidant concentration $[MnO_4^-]_0$. Reaction vials are tightly capped and then placed on an orbital mixer to ensure complete mixing (e.g., 150 rotations per minute). The reaction vials are sampled in triplicate after the reaction is anticipated to be complete (or for kinetic data, at several time points during the reaction) and then analyzed at a minimum for permanganate concentration by standard methods (e.g., Method 4500-$KMnO_4$ B., APHA 1998). Data on pH, alkalinity, and TOC also can be insightful. It is noted that permanganate analysis with a spectrophotometer requires that aqueous samples are filtered (0.22 μm glass microfibre) prior to analysis to remove any colloidal manganese dioxide solids. It is also important to test the filtering procedure to ensure that the filter material does not also exert a significant oxidant demand. Sample dilution with de-ionized water prior to spectrophotometer analysis may be required also. Controls must be included in the experimental design; they should consist of reaction vials with just MnO_4^- solution and reaction vials with untreated matrix media containing the COCs of interest.

The multiple reaction vial approach can be used for degradation efficiency by using the ambient groundwater with the COCs in it or by spiking the groundwater with target COCs of interest to reach the desired concentrations. This spiking step can be completed by equilibrating the matrix water under zero-headspace conditions overnight. Then during each sampling event, in addition to the water quality parameters noted above, samples can be analyzed for the target organic analytes by use of standard methods for gas chromatography with the appropriate detector. Analyses can be made for chlorides, if appropriate, as well as for organic byproducts (USEPA 1996). Toxicity testing also may be conducted as noted in Chapter 4.

Another approach utilizes a 100-mL reaction apparatus consisting of a glass syringe, mini-inert valve, stainless steel syringe, and micro stir bar (Huang et. al 1999, Urynowicz 2000) (Figure 6-3). The advantage of this approach is that a single bulk solution is prepared and samples can be withdrawn under zero-headspace conditions at different time points during the reaction. In addition to aqueous media only, these reactors can be

FIGURE 6-3.   **Batch experimental apparatus to measure degradation efficiency and NOD using 100-mL syringe reactors for ground water and ground water with 20 wt.% solids added.**

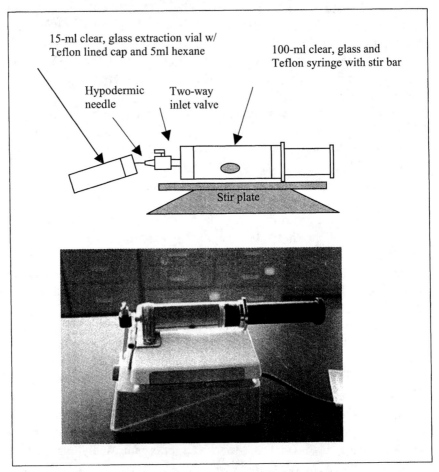

used for groundwater with soil or aquifer solids added at up to 20 wt.% solids. These solids can exert a significant oxidant demand and serve to buffer the aqueous phase pH, thereby better approximating field conditions. The 100-mL reaction syringe apparatus is a zero head-space reactor designed to limit volatilization of the target contaminant(s) during an extended reaction period while enabling periodic sample collection. Either contaminated groundwater from a site or a contaminant stock solution prepared in de-ionized water is used depending on the type of results desired. Syringe vessels are loaded with 90 mL of groundwater or contaminant stock solution (possibly including 5 to 20 g of solids). At time zero, the mini-inert valve on the syringe is opened and 10 mL of concentrated permanganate solution (e.g., ~10 times the target MnO_4^- concentration to allow for dilution) is injected into the reaction syringe. The mini-inert valve then is closed and the reaction syringe apparatus is placed on a stir plate. Samples are collected in extraction vials containing a known volume of solvent (typically hexane or methanol) at various time points depending on the anticipated degradation rate. Immediately upon sample collection, the extraction vials are vigorously shaken to extract residual contaminants from the aqueous phase into the organic solvent phase, thereby causing a cessation of any further contaminant chemical oxidation prior to GC analysis.

The methods used for measurement of NOD without target COCs is analogous to that used with the multiple reaction vials. However, if soil solids are to be used and only NOD is to be determined, the soil can be slurrified with deionized water and then purged with an inert gas to strip any ambient VOCs or SVOCs from the matrix. Air-drying of soil accomplishes the same purpose, but this may be inappropriate if the ambient redox is highly reducing and there are appreciable levels of reduced metals (e.g., Fe^{+2} and Mn^{+2}) in the matrix. The syringe reactor is filled almost completely with matrix water (and possibly a small quantity of solids); a magnetic stir bar is added before the syringe barrel is inserted; the device is now air-tight with zero headspace. Then, an aliquot of stock permanganate solution can be added by connecting the syringe to a vial and simply withdrawing the plunger until the desired volume is added. All volumes can be quantified either by volumetric or gravimetric methods. Then, the syringe is placed on a magnetic stir plate, and the reaction proceeds. At desired time points, a sample of the syringe contents can be removed by opening the syringe needle valve and depressing the plunger. Analyses to be completed are the same as those described above.

For experiments with soil and aquifer solids, 125-mL zero headspace reactors (ZHRs) can be used (Gates and Siegrist 1995, Gates-Anderson et al. 2001) (Figure 6-4). These ZHRs are made of stainless steel or glass and have a plunger apparatus and a sample port with shut-off valve. After the ZHRs are filled and sealed, they can be placed on a mechanical rotator to achieve mixing during a prescribed reaction period. ZHRs can be used to measure degradation efficiency and kinetics and also NOD. The procedures are quite straightforward as described in Gates and Siegrist

FIGURE 6-4. Batch experimental apparatus to measure degradation efficiency and NOD using 125-mL zero headspace reactors for soil slurries.

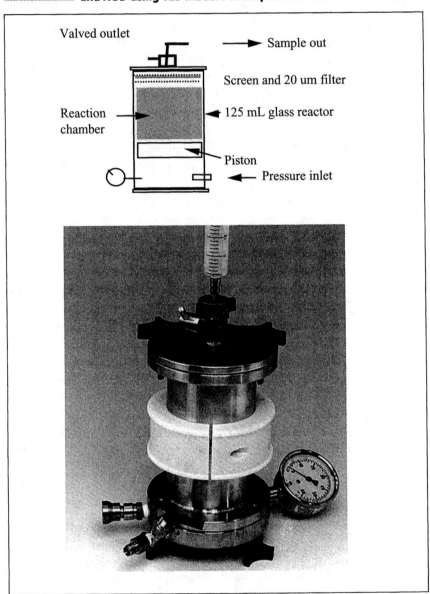

(1995). To measure NOD without target COCs present, 100 g of contaminated air dry soil or aquifer media are quantitatively placed in a precleaned ZHR. Then, 30 mL of deionized water or clean groundwater are added, and the ZHR is sealed and equilibrated for 24 hr during rotation at 30 rpm on a mechanical tumbler. Then, 30 mL of permanganate stock solution are added, to the ZHR via the top sample port (see Figure 6-4) by use of a 50-mL syringe to achieve a desired initial $[MnO_4^-]_o$ concentration in the slurry. At least duplicate ZHRs should be employed for each experimental condition. Following a short period of initial mixing (e.g., 30 min), an initial sample can be taken from each ZHR by slightly pressurizing the chamber below the Teflon piston within the ZHR such that it forces solution out and into a pre-cleaned glass vial. During the reaction period, the ZHRs are mechanically tumbled at 30 rpm at 20C (or at 10C or other temperatures to approximate field conditions). At key time points such as 0.3, 1, 3, 10, and 30 d after oxidant addition, ~5 mL of solution are extracted from the ZHRs for analyses. Solution samples removed from each ZHR at each time point are analyzed for permanganate by standard methods as noted above. Analyses for pH and TOC may also be insightful.

For measurement of COC degradation efficiency with the ZHRs, ambient contaminants can be used, or alternatively background clean (or contaminated but purged) soil or aquifer solids can be spiked with a known concentration of COCs. If ambient contaminants are used, then the methods are quite similar to those described above. However, it is very important to take precautions to avoid loss of VOCs during ZHR preparation and sampling. Also, a set of controls containing the target contaminants and no oxidant must be included in each experiment as a reference. Analyses of the samples withdrawn from the ZHR are made for the analytes noted above plus for the target organics. If spiking of soil slurries is needed (either for control or to test permanganate oxidation at COC levels that exceed ambient levels in samples collected), then the following approach can be taken (Gates and Siegrist 1995, Tyre et al. 1991, Gates-Anderson et al. 2001). For SVOC studies (e.g., PAHs), the spiking solutions are prepared in acetone and added to the soil in the ZHR reaction chamber, covered with a layer of parafilm, and then equilibrated for 24 hr at 20C in a laboratory fume hood. During this time, the acetone will be evaporated while the SVOCs will remain in the soil. Then ~30 mL of groundwater matrix are added and the ZHR is sealed and equilibrated for 24 hr during rotation at 30 rpm on a mechanical tumbler. For VOCs, an aqueous solution is prepared in a gas-tight vessel from which an aliquot can be added to the soil slurry in the ZHR. Then, it is equilibrated for 24 hr. The distribution of organics between the water, soil solids, and NAPL phases within each ZHR can be estimated by using a fugacity-based partitioning model (e.g., Dawson 1997). For SVOCs, the partitioning will be dominated by the NAPL phase (often 90 to 99 wt.%) with lesser mass on the soil solids and very little dissolved in the water. In contrast, the VOCs can be partitioned such that a high fraction (e.g., 75 to 90 wt.%) will be dissolved in the water with the balance sorbed on the soil solids unless sufficient mass is added to create a NAPL phase. Following equilibration, a time zero solution sample should be taken from each ZHR. Permanganate can then be added to achieve a desired oxidant dose to some of the ZHRs, while others must be maintained as controls (i.e., with contaminants but without oxidant). During the reaction period, the ZHRs are mechanically tumbled at 30 rpm at 20C (or at another temperature). At time points such as 0.1, 0.3, 0.5, 1.0, and 2.0 days after oxidant addition, ~5 to 10 mL of solution are extracted from the ZHRs for analyses. The analysis of aqueous subsamples from a ZHR gives an indication of the rate of reaction, but it may not reflect degradation of the total mass of target COCs, since aqueous sample analyses do not include COCs that are sorbed onto porous media in the ZHR. During some runs, repeated oxidant addition can be tested. Solution samples removed from each ZHR at each time point should be analyzed for the target organic chemicals. Analyses should also be completed for permanganate concentrations, and analyses for reaction products (e.g., Cl⁻), pH and TOC can also be useful. To enable a mass balance at the end of the reaction period, an extractant (e.g., hexane) can be added to the ZHR to extract any untreated residual organic contaminants. After that, subsamples of the extractant can be analyzed for the target organics.

Flow-Through Experiments. Although the ability to recreate field conditions in the laboratory is limited, one-dimensional (1-D) column experiments generally provide a better physical simulation of site-specific conditions than well-mixed, batch experiments. One-dimensional column experiments to study single-grain sediments with limited structure (e.g., sands) can be accomplished readily. Field-collected media (clean or contaminated depending on purpose) are hand-packed into mini-columns made from 50 or 100-mL polypropylene syringes or macro-size columns that are 2 to 15 cm in diameter and 30 to 100 cm in length. The columns can be operated in different modes to assess NOD, COC degradation efficiency, and matrix effects (e.g., permeability loss) (see Figures 6-5 and 6-6). In most cases, after preparation, the columns are initially upflow

FIGURE 6-5. Flow-through column apparatus packed with aquifer sediments for laboratory treatability studies of oxidant delivery, NOD, permeability effects and metal mobility studies.

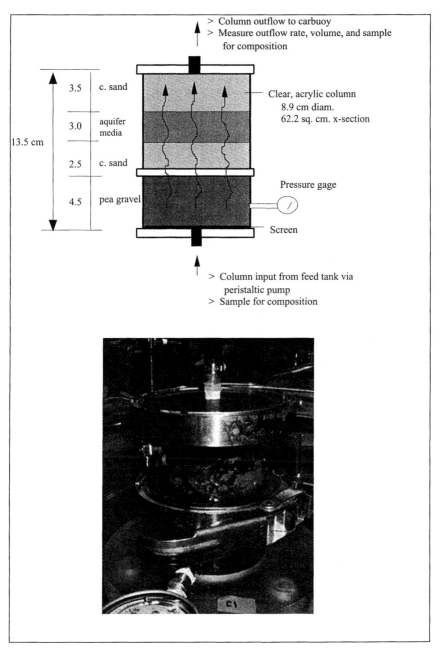

FIGURE 6-6. One-dimensional column apparatus for flow-through experiments to measure COC degradation, NOD, permeability effects, and metal mobility during ISCO.

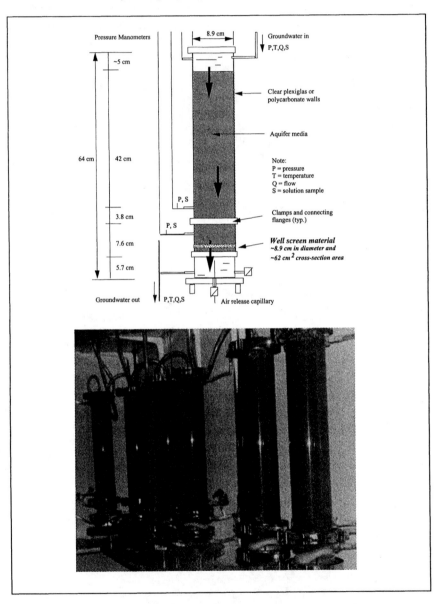

saturated with real site groundwater or a matrix water to simulate the groundwater at the site. NOD in clean media (clean sediment and aquifer media from a background location at the site) can be determined by upflow (saturated) or downflow (saturated or unsaturated) experiments. This is accomplished by monitoring the feed flow rate and composition to the column, and the MnO_4^- concentrations in the pore water with distance from the inlet or just the outflow composition. In a constant flow column operating mode, inlet pressure changes can be observed and used as an indicator of permeability changes possibly due to particle generation and deposition.

In contrast to columns packed with disturbed media collected from a site, column studies performed with relatively intact cores of media are considerably more difficult to accomplish. This is due in part to the greater effort required to preserve the integrity of the subsurface medium's structure during collection, packaging, and transport from the field site to the laboratory. Nevertheless, intact cores of structured soils and sediments (e.g., silts and clays, fractured/porous rock) may be warranted to understand transport and reaction processes. These tests can be completed with advective flow-through columns such as illustrated in Figure 6-5 and 6-6. However, the intact cores are inserted and sealed in the laboratory apparatus, rather than packed into it. For low permeability media where advection does not contribute to oxidant transport, studies can be performed to evaluate diffusive transport processes and matrix interactions. Those studies can be performed by use of small diffusion cells as shown in Figure 6-7 or large monoliths as shown in Figure 6-8. In either case, advective and diffusive transport (with or without chemical reaction) can be observed by tracking the fate of the permanganate in clean media (e.g., from a background location). For degradation studies, as well as NOD measurements in contaminated media (e.g., from the contaminated zone to be treated, or from background media that has been spiked with COCs), sampling and analyses can be completed to define the permanganate fate and that of any target COCs. Matrix interactions also can be studied and changes in media chemistry and permeability can be assessed.

Two-dimensional, intermediate-scale studies have been used to study transport processes and in situ remediation technologies. While generally they are costly and not used to support most site-specific designs, they could be appropriate for some complicated and costly sites as well as for technology development purposes. These tank-scale experiments can

FIGURE 6-7.   **Transport cell apparatus used to experimentally measure the effective diffusion coefficient and tortuosity with intact cores of soil media and to assess COC treatment efficiency (after Struse, 1999).**

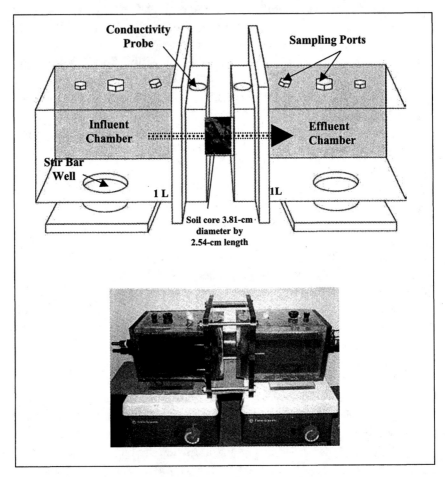

examine the ISCO reaction and transport processes at the pilot-scale, including the effects of ISCO on different DNAPL mass and distribution characteristics (e.g., low and high mass in dispersed ganglia residual vs. isolated entrapment zones) in subsurface conditions of varying hetero-geneity (Figure 6-9). Previous studies have shown that the entrapment morphology (residual, ganglia, or pools) and dimensionality of the flow field can affect the mass transfer coefficients and DNAPL degradation both under natural and treatment-enhanced dissolution (e.g., Saba and

FIGURE 6-8.   Intact cores of soil or aquifer sediments for laboratory treatability studies of ISCO delivery using injection probes.

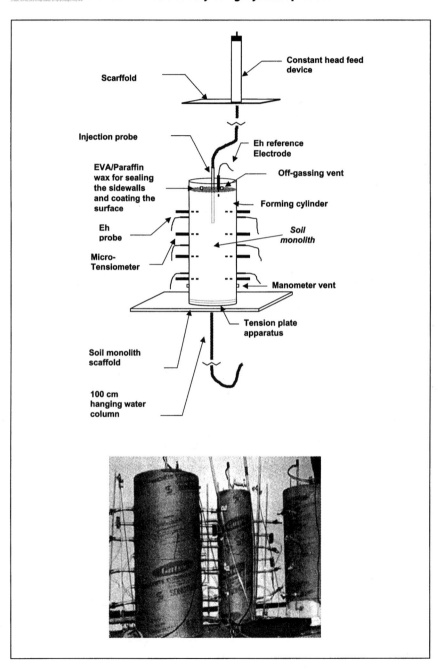

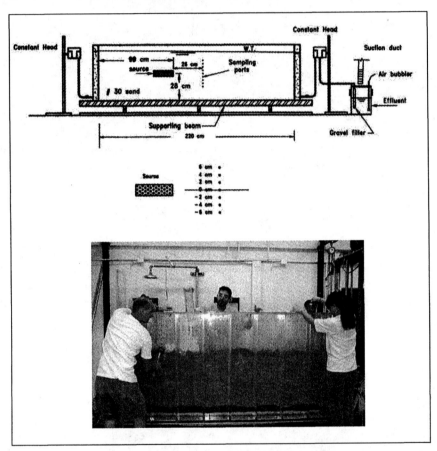

FIGURE 6-9. Two-dimensional tanks at CSM for studying ISCO effects on DNAPLs based on heterogeneity and entrapment morphology (from Saba and Illangasekare 2000).

Illangasekare 2000, Saba et al. 2000). Thus, it can be appropriate to conduct a set of experiments in both 1-D columns and 2-D transport tanks to determine the DNAPL degradation rates as a function of emplacement conditions and ISCO operation (e.g., dose concentration, total loading, delivery density, and flow rates). The intermediate-scale testing methods to be used can follow those developed by Illangasekare et al. (e.g., Held and Illangasekare 1995, Chao et al. 2000, Li and Schwartz 2000, Reitsma and Marshall 2000, Barranco et al. 2001). The columns and tanks can be packed with field materials or well-characterized test sands simulating field heterogeneity (as defined by variances and correlation lengths of a spatially correlated random field) (Chao et al. 2000). The entrapment configurations can be adapted from those used recently by Illangasekare et al. at CSM to examine flow and transport behavior, surfactant flooding, and partitioning tracer utility (e.g., Barranco et al. 2001). The entrapment configurations of interest can include homogeneously distributed DNAPL blobs at residual saturation, and heterogeneously distributed DNAPL pools at relatively high saturation. The two former cases can be created with the use of only one of the sands (thereby creating a uniform intrinsic permeability field), while the latter can be created with the use of two of the sands. Two sands are used in the pooled case to provide accurate control of the spatial limits of the DNAPL within the tank utilizing capillary barrier effects at the coarser/fine sand interfaces.

More manageable and relatively less expensive experiments can be conducted first in 1-D columns packed with media to understand the reaction and transport processes at a smaller scale. Following these experiments, larger scale testing can be performed in 2-D tanks (e.g., as large of 9.8 m long, 1.8 m high, 0.05 m thickness) (Figure 6-9). These types of tests can provide a good understanding of the effects of site and DNAPL conditions on ISCO performance to achieve an overall mass reduction, as well as the effects on mobility of untreated DNAPL compounds.

6.5. FIELD PILOT TESTING

As mentioned in the previous section, laboratory treatability tests provide quantitative information regarding chemical oxidation effectiveness under relatively ideal, controlled conditions. However, a successful implementation of ISCO is highly dependent on the delivery technique used to disperse the permanganate throughout the subsurface. For most applications of ISCO using permanganate, it is highly recommended that field pilot tests using a chosen delivery technique be performed (Figure 6-10). These should be completed after laboratory tests confirm acceptable degradation efficiencies of the target contaminants and before a final decision is made to use the technology for full-scale remediation. Chapter 7 contains five case studies describing field applications of ISCO using permanganate, including information on field pilot testing. Additional examples of field pilot testing can be found in the literature (e.g., Chambers et al. 2000, Huang et al. 2000, IT Corporation and SM Stoller Corporation 2000, Moes et al. 2000, Mott-Smith et al. 2000).

FIGURE 6-10. Photograph of field pilot testing of a 5-spot vertical well recirculation of NaMnO₄ to treat high levels of TCE in groundwater at a site in Ohio.

Note: Center point injection well with extracted groundwater from four perimeter wells transferred in secondary contained lines after amendment with oxidant.

The primary objectives of a field pilot test are to determine the effectiveness of a specific delivery system for dispersing oxidant into the subsurface, to quantify contaminant degradation efficiencies under field conditions, to identify potential problems that may affect full-scale operations, and to determine the effects of the treatment process on areas surrounding the treatment zone. Pilot test design begins with the selection of a representative zone within the target region for remediation. Geochemical characteristics within the pilot test area are then measured to establish pre-test, baseline conditions. For this purpose, geochemical measurements can be made either on groundwater samples collected from monitoring wells, or on discrete soil samples collected from soil borings. It is recommended that soil samples be used for test area characterization rather than groundwater samples alone, particularly where groundwater flow is relatively rapid and may not be in geochemical equilibrium with the surrounding soil. Parameters that should be measured in soil samples include initial contaminant levels, permanganate levels, ph, TOC, and possibly extractable Mn (to evaluate the movement of permanganate through the pilot test zone). Where possible, soil borings can be completed as monitoring wells for minimal additional costs. Hydraulic conductivity should be measured in wells that will be used to deliver the oxidant.

Following pretest characterization, dispersal of the permanganate into the pilot test region is accomplished through the chosen delivery system. Equipment for preparing the oxidant solutions and handling any extracted fluids (i.e., in a recirculation delivery system) planned for the full-scale application should be used during the pilot test so that problems in these systems may be identified and resolved before the full-scale application. Operational parameters are selected such that they are as similar as possible to what is anticipated for the full-scale remediation. For example, oxidant concentrations, total mass loading, and solution volumes (e.g., permanganate mass per unit mass of soil and/or pore volumes of solution) for the pilot test must be comparable to what will be used for a full-scale remediation. Groundwater samples should be collected, if possible, during oxidant delivery to evaluate the migration of the oxidant through the subsurface. At a minimum, permanganate levels should be measured in these groundwater samples. Specific conductivity also can be used as a surrogate measurement for MnO_4^- if levels are sufficiently high (> 0.5%). Other chemical measurements also should be considered such as pH, Cl^-, contaminant levels, total suspended and dissolved solids, TOC, and metals that may be mobilized by the oxidative treatment (e.g., Cr).

Immediately after delivery of the oxidant is ended (i.e., before the influx of upgradient contaminated water into the test area), soil samples should be collected from locations corresponding to the pre-test soil samples. As in the pre-test soil samples, contaminant levels, MnO_4^-, pH, TOC, and extractable Mn may be measured in the post-test soil samples. Detectable levels of MnO_4^- and extractable Mn will indicate whether the oxidant was sufficiently dispersed into the treatment area. Comparisons between pre- and post-treatment contaminant levels will provide an indication of whether sufficient contaminant degradation was achieved with the oxidant concentrations and mass loading. Post-test hydraulic conductivity measurements in the delivery wells will determine whether well plugging will be a problem for the full-scale application.

It is also recommended that monitoring wells down gradient from the pilot test site be installed. Groundwater samples collected from these wells can be used to assess the migration of MnO_4^- away from the treatment region, as well as to define the movement of MnO_4^- and dissolved

Mn after active oxidant delivery is completed. The latter would be of interest if there were any concern regarding the fate of Mn in the treated aquifer.

Tracer tests can be very insightful regarding hydraulics of the region to be treated. Delivery of a slug of tracer (e.g., Br$^-$) at the beginning of the delivery phase before MnO$_4^-$ is delivered can help ascertain the hydraulic control achieved in the system. If combined with the oxidant, conservative tracers can also provide some evidence of the oxidant demand in the region based on the retardation of the permanganate relative to the tracer.

6.6. ISCO SYSTEM MODELING

If well networks will be used for oxidant delivery, the design of such systems must be based on the hydraulic and geologic characteristics of the treatment region that will control how the oxidant is dispersed into the subsurface. At a minimum, modeling of groundwater flow expected during well network operations should be performed. Such modeling would estimate the zone of influence for a given system of wells, and the number of pore volumes that can be flushed through a treatment zone for a given time period. Subsequent to groundwater modeling, one may model MnO$_4^-$ as a conservative tracer and predict its movement through the treatment region assuming only advection and dispersion as the primary transport mechanisms. One can also simulate the movement of a contaminant induced by the well network operations assuming only advection, dispersion, and sorption as the operative transport mechanisms. The preceding modeling approach ignores the reactivity of the MnO$_4^-$ with oxidizable species, including the contaminant. The ultimate goal for modeling MnO$_4^-$ migration would be to incorporate its interaction with the contaminant as well as other oxidizable species. For example, this interaction can be modeled by use of second-order kinetics (see Chapter 2); this approach allows one to simulate the spatial and temporal variations of MnO$_4^-$, COCs, and NOD (Zhang and Schwartz 2000). A more qualitative approach to modeling the movement of MnO$_4^-$ is to assume first-order kinetics of MnO$_4^-$ consumption (see Chapter 5).

Modeling of MnO$_4^-$ transport can also be applied to delivery systems that rely mostly on diffusive transport for oxidant delivery (e.g., permanganate grouts in low permeability media). However, because migration of MnO$_4^-$ in such delivery systems can be significantly slower (i.e., the migration and degradation rate of MnO$_4^-$ are probably of the same order), the reactivity of MnO_4^- with the surrounding soil would be more critical to incorporate into the transport model. A number of existing models can be adapted to model the diffusive transport and fate of MnO_4^- in the subsurface and the degradation of target organics.

ISCO system modeling can also estimate the movement of the contaminants, including those present as a pure phase for sites where DNAPL contamination is suspected. Such modeling would provide some assurance that well pumping does not "push" DNAPL pools outside of the treatment zone. Geochemical equilibrium calculations can also be made to estimate the final Mn speciation after ISCO is completed. Estimates on this can be of interest for sites where soluble Mn levels are of concern.

6.7. DETAILED DESIGN AND IMPLEMENTATION

The detailed design and implementation of remediation technologies employing ISCO with potassium or sodium permanganate solutions or solids are similar in many respects to other in situ technologies. That is, there are pro forma requirements regarding plans and specifications pertaining to the equipment mobilization/demobilization, apparatus and instrumentation installation and startup, process control and performance monitoring, quality assurance, and so forth. There are also regulatory requirements such as an underground injection control permit (or a variance therefrom). For ISCO with permanganate, special attention must be paid to several facets of the design and implementation engineering. These include the aboveground fluid and material handling system, subsurface delivery system (and avoidance of potential problems with plugging of injection wells or trenches), and implementation concerns (e.g., materials of construction, regulations and permitting, health and safety protection, waste management). Materials of construction need to be compatible with permanganate at the concentrations and duration of operation intended. The process control and monitoring needed to generate real- or near-real-time data on process function must be well conceived to enable timely adjustments to locations of oxidant delivery and/or oxidant concentrations and mass loading rates. Performance monitoring that can verify achievement of a goal must be delineated. Finally, waste management must pay attention to the properties of oxidizers and the potential hazards associated with corrosion and combustion. Further information on worker and environmental safety, regulatory considerations and permitting, and waste management are given below.

Worker and Environmental Safety

Potential worker safety risks include those associated with standard construction operations as well as those associated with work at a contaminated site and with potentially hazardous chemicals. All activities must meet guidelines in the Occupational Safety and Health Administration (OSHA) 29 Code of Federal Regulations (CFR) 1910 and 29 CFR 1926. Work at a DOE facility will also require meeting DOE guidelines (U.S. DOE 1984). Generally, a Health and Safety work permit is required with all work conducted under an approved Health and Safety Plan. During preparation of a specific health and safety plan, the topics listed in Table 6-7 should be considered and addressed.

Additional worker risks specific to ISCO are associated with the aboveground handling of the oxidants. The nature and seriousness of the risks will depend on the oxidant form and concentrations used and methods of general handling and subsurface delivery. For example, hydraulic fracturing to deliver reactive solids may involve risk to workers during handling of reactive proppants (e.g., permanganate crystals). Handling of concentrated liquid $NaMnO_4$ inherently requires more care than handling of solid $KMnO_4$ crystals. Once oxidants are delivered to the subsurface, worker risk is negligible because the ground surface is not disturbed by most in situ delivery methods.

Environmental risk is limited because of the nature of materials used. However, delivery of $KMnO_4$ may require additional consideration related to worker and environmental safety due to naturally occurring ^{40}K. During application of ISCO at a DOE facility, concerns were raised related to exposure in excess of the DOE Order 5400.5 derived concentration guidelines (DCG) for ^{40}K (the DCG is 7000 pCi/L which correlates to 280 pCi/L/yr for a standard man dose from ingestion). While minimal leaching of the oxidant into the soil pore water and ultimately into the groundwater could occur, a person would be required to drink several liters of soil pore water from within the treatment area or of the delivery fluid. Adequate engineering controls to prevent this type of exposure are easily implemented.

Regulatory Considerations and Permitting

Local and federal regulations must be considered prior to ISCO implementation. On the federal level, regulations that must be taken into account prior to any remediation scheme may include those listed in

TABLE 6-7.	Topics to be addressed in a health and safety plan for an ISCO application.

Topics to be addressed

✓ Purpose and scope of work

✓ Risk evaluation including a hazard evaluation of the personnel, chemical, biological, physical and potentially radiological hazards

✓ Site-specific concerns

✓ Permits

✓ Generated waste

✓ Confined spaces

✓ Key project personnel and responsibilities

✓ Qualifications and training requirements, particularly the handling of oxidants

✓ HAZCOM program

✓ Control measures and zones

✓ Standard operating procedures

✓ Equipment inspection

✓ Equipment operation potentially including cutting and welding, ground fault protection, hoisting and rigging

✓ Protection during heavy equipment operations

✓ Vehicular safety

✓ Personnel protection equipment (PPE) requirements including air monitoring, PPE levels, limitations of protective clothing, and potential respirator selection, use, and maintenance

✓ Medical surveillance including heat and cold stress monitoring as required

✓ Types and frequency of monitoring including calibration, background readings, and environmental monitoring and recording

✓ Decontamination procedures including personnel and equipment decontamination

✓ Emergency response procedures

Table 6-8. Careful attention should be given to the applicable sections of these regulations. Also, state and local regulations may prove to be more restrictive.

Specific permits depend on site and remediation technology specific application and must be coordinated with the appropriate regulators. A subsurface injection well installation and completion permit may be required by local and state agencies. CERCLA or RCRA permitting may also be required. Finally, a National Environmental Policy Act (NEPA) review may be required. The primary permitting and environmental compliance issues are outlined in Table 6-9.

TABLE 6-8	Local and federal regulations to be considered prior to ISCO implementation.

Regulations and regulated activities to be considered

✓ Safe Drinking Water Act (40 USC 300)

✓ Clean Water Act (33 USC 1251-1376)

✓ U.S. Water Quality Act, 1986

✓ Comprehensive Environmental Response, Compensation and Liability Act (CERCLA)

✓ Resource Conservation and Recovery Act (RCRA)

✓ Land Disposal Restrictions

✓ Corrective Action Management Units

✓ Clean Air Act of 1990 (42 USC 7401-7642)

✓ OSHA (29 USC 651-678 and 29 CFR parts 1910, 1904, and 1926)

✓ Department of Transportation rules for hazardous materials transport (49 USC 1801-1813 and 49 CFR 107, 171-177)

TABLE 6-9	Primary permitting and environmental compliance issues affecting ISCO implementation.

Permitting and environmental compliance issues

✓ An Underground Injection Control (UIC) permit is typically required for injection of any material into the subsurface. Additionally, during site remediation, a UIC may be required and routine reports describing the associated volume of oxidant injected into the subsurface

✓ Air permits will not typically be required for the release of VOCs during in situ chemical oxidation due to the minimal subsurface disruption. However an air permit may be required depending on the selected remediation technology.

✓ A Permit to Install (PTI) and/or a Permit to Operate (PTO) may be required if a surface facility is required. In situ passively operated systems typically would not require a PTI and PTO. However, more aggressively operated systems (e.g., recirculation of fluids) may require a PTI and PTO.

✓ A health and safety work permit is typically required for all field activities and will be issued by the site/facility Health and Safety organization.

✓ Other site/facility specific permits may be required such as a penetration permit prior to any disruption to the ground surface.

Waste Management

Waste management issues are related to type and volume of ISCO generated waste. Disposition of the waste and delivery to the appropriate storage facility is site/facility-specific and must be arranged through appropriate waste management organizations. Possibly the most difficult wastes to manage are the containers, materials, and personal protective equipment that have permanganate oxidant residual in them. Flushing such materials with water can be used to reduce the residual oxidant concentration and mass to acceptable levels. A reducing agent (e.g., sodium thiosulfate) can also be used to neutralize the permanganate, however this will generate MnO_2 solids that need to be dewatered and disposed of appropriately. Addition of reducing agents to residual permanganate solutions should be done carefully and with proper safety protection (see appropriate material safety data sheets). Reducing agents in dispersed crystal or aqueous form should be gradually blended into permanganate solutions that are relatively low in concentrations. Large quantities of solid reducing agents should never be dumped in mass into containers filled with concentrated solutions of MnO_4^- (e.g., 40 wt.% $NaMnO_4$) since a violent and potentially explosive reaction can occur. Care must be taken to avoid disposal of combustible materials that have contacted permanganate in bulk waste containers, as fires can be a risk. Additional generated wastes expected during ISCO are drilling cuttings produced during installation of oxidant delivery access points and other monitoring locations. Driven casing techniques are available to minimize this waste when it is a concern. There can also be groundwater from the site due to dewatering or recirculation, MnO_2 precipitates from extracted and ex situ treated groundwater or from neutralized oxidant solutions, wastewater from equipment decontamination, and miscellaneous sanitary trash. MnO_2 stains that result when permanganate is reduced on a surface (e.g., skin, tools, equipment, pavement) can be effectively removed by using a solution (1:1:1 v/v) of water, vinegar, and 2 wt.% hydrogen peroxide.

6.8   COST ESTIMATION

The cost-effectiveness of ISCO using permanganate is highly dependent on a number of factors related to the contaminant characteristics, site conditions, and performance goals. In general, ISCO applications to sites with dissolved and sorbed organic COCs have been able to achieve high degrees of treatment (e.g., up to 99% mass destruction and/or MCLs) in time periods shorter than those associated with other in situ remediation approaches and at costs that are competitive (e.g., operational periods of weeks to a month or two at costs of $40 per c.y. of subsurface region treated). The costs associated with the oxidant chemicals at these sites are typically only 10 to 20% of the total project costs. Applications to DNAPL source areas at highly contaminated sites can be considerably more expensive, in part due to the duration of treatment and the higher chemical costs required to satisfy the oxidant demand of high COC masses. The time periods and costs required to treat DNAPL source zones to 90% mass destruction can extend to a few months or more and reach costs of $200 per c.y. or more. However, the time and costs of other aggressive DNAPL source zone remediation methods such as surfactant or thermal enhanced recovery have similar or greater timeframes and costs. Information regarding application of ISCO at five sites, including the operation times and costs, are given in Chapter 7.

6.9. SUMMARY

ISCO has been shown to be a promising environmental technology for the remediation of organic chemicals at contaminated sites. However, the feasibility of all environmental technologies is dependent on site-specific conditions, and the performance achieved is based on sound design and implementation for those conditions. In situ remediation using potassium or sodium permanganate should include an initial screening step to determine the validity of the technology as a treatment process. In many cases, much of the site characterization data required for a screening level assessment of the applicability of ISCO oxidation may be available from previous environmental characterization efforts or routine monitoring reports. If after the screening evaluation the site is deemed to be a good candidate for ISCO, existing data again may be used to prepare a preliminary design. However, additional site characterization should be performed if significant data gaps are present or preexisting data are out of date or inadequate. Following the initial screening and preliminary design, a treatability study is often warranted to determine the contaminant degradation efficiency and the natural oxidant demand of soil and groundwater collected from the site. Treatability studies may consist of laboratory screening, bench-scale testing, and/or pilot scale testing.

Implementation activities must couple the technology capabilities with available site-specific information in order to achieve performance goals

(Figure 6-1). Once an oxidant delivery method has been chosen, modeling that includes flow and transport can be used to optimize final design parameters, such as the location of oxidant delivery points, oxidant concentration, and oxidant loading. Implementation requires careful attention to system function over time, process monitoring and control, and health and safety protection. Finally, performance monitoring and post-treatment assessment activities must be carefully developed to enable verification that performance goals have been achieved.

6.10. REFERENCES

APHA (1998). Standard Methods for the Examination of Water and Wastewater, 20th ed. Clesceri, L. S., A. E. Greenberg and R.R. Trussell, eds. APHA-AWWA-WPCF, Washington, DC.

Barranco, F.T., Dai, D., and Illangasekare, T.H. (2001). Partitioning and interfacial tracers for differentiating NAPL entrapment configuration: Column-scale laboratory results. *Environmental Science and Technology* (accepted and in press).

Burlage, R.S., R. Atlas, D. Stahl, G. Geesy, and G. Sayler (1998). Techniques in Microbial Ecology. Oxford University Press, New York.

Carter, M.R. (1993). Soil Sampling and Methods of Analysis. Lewis Publishers, Ann Arbor, MI.

Case, T.L. (1997). Reactive permanganate grouts for horizontal permeable barriers and in situ treatment of groundwater. M.S. Thesis, Colorado School of Mines, Golden, CO.

Chambers, J., A. Leavitt, C. Walti, C.G. Schreier, and J. Melby (2000). Treatability study–Fate of chromium during oxidation of chlorinated solvents. In: Wickramanayake, G.B., A.R. Gavaskar, and A.S.C. Chen (ed.). Chemical Oxidation and Reactive Barriers. Battelle Press, Columbus, OH. pp. 57-66.

Chao, H-C, H. Rajaram and T.H. Illangasekare (2000). Intermediate scale experiments and numerical simulations of transport under radial flow in two-dimensional heterogeneous porous medium, *Water Res. Res.,*36(10):2869-2878.

Dawson, H.E. (1997). Screening level tools for modeling fate and transport of NAPLs and trace organic chemicals in soil and groundwater: SOILMOD, TRANS1D, NAPLMOB. Colorado School of Mines, Office of Special Programs and Continuing Education. Golden, CO.

Environmental Security Technology Certification Program (1999). Technology status review: In situ oxidation. http://www.estcp.gov.

Freeze, R.A. and D.B. McWhorter (1997). A framework for assessing risk reduction due to DNAPL mass removal from low-permeability soils. *Ground Water*, 35(1):111-123.

Gates, D.D., and R..L. Siegrist (1995). In situ chemical oxidation of trichloroethylene using hydrogen peroxide. *J. Environmental Engineering*. Vol. 121, pp. 639-44.

Gates, D.D., R.L. Siegrist and S.R. Cline (1995). Chemical oxidation of contaminants in clay or sandy Soil. Proceedings of ASCE National Conference on Environmental Engineering. Am. Soc. of Civil Eng., Pittsburgh, PA.

Gates-Anderson, D.D., R.L. Siegrist and S.R. Cline (2001). Comparison of potassium permanganate and hydrogen peroxide as chemical oxidants for organically contaminated soils. *J. Environmental Engineering*. 127(4):337-347.

Held, R.J. and T.H. Illangasekare (1995). Fingering of dense nonaqueous phase liquids in porous media 1. Experimental investigation. *Water Resources Res.*, 31(5):1213-1222.

Huang, K., G.E., Hoag, P. Chheda, B.A. Woody, and G.M. Dobbs (1999). Kinetic study of oxidation of trichloroethylene by potassium permanganate. *Environmental Engineering Science*. 16(4): 265-274.

Huang, P. Chheda, G.E. Hoag, B.A. Woody, and G.M. Dobbs (2000). Pilot-scale study of in situ chemical oxidation of trichloroethene with sodium permanganate. In: G.B. Wickramanayake, A.R. Gavaskar, A.S.C. Chen (ed.). Chemical Oxidation and Reactive Barriers: Remediation of Chlorinated and Recalcitrant Compounds. Battelle Press. Columbus, OH. pp. 145-152.

IT Corporation and SM Stollar Corp (2000). Implementation report of remediation technology screening and treatability testing of possible remediation technologies for the Pantex perched aquifer. October, 2000. DOE Pantex Plant, Amarillo, Texas.

Klute, A. et al. (ed.) (1986). Methods of Soil Analysis, Part 1. Physical and Mineralogical Methods. Soil Sci. Soc. Am. Madison, WI.

Li, X.D. and F.W. Schwartz (2000). Efficiency problems related to permanganate oxidation schemes. In: Wickramanayake, G.B., A.R. Gavaskar, and A.S.C. Chen (ed.). Chemical Oxidation and Reactive Barriers. Battelle Press, Columbus, OH. pp. 41-48.

Lowe, K.S., F.G. Gardner, R.L. Siegrist, and T.C. Houk (2000). Field pilot test of in situ chemical oxidation through recirculation using vertical wells at the Portsmouth Gaseous Diffusion Plant. EPA/625/R-99/012. U.S. EPA ORD, Washington, DC. pp. 42-49.

Moes, M. C. Peabody, R. Siegrist, and M. Urynowicz (2000). Permanganate injection for source zone treatment of TCE DNAPL. In: Wickramanayake, G.B., A.R. Gavaskar, and A.S.C. Chen (ed.). Chemical Oxidation and Reactive Barriers. Battelle Press, Columbus, OH. pp. 117-124.

Mott-Smith, E., W.C. Leonard, R. Lewis, W.S. Clayton, J. Ramirez, and R. Brown (2000). In situ oxidation of DNAPL using permanganate: IDC Cape Canaveral demonstration. In: Wickramanayake, G.B., A.R. Gavaskar, and A.S.C. Chen (ed.). Chemical Oxidation and Reactive Barriers. Battelle Press, Columbus, OH. pp.125-134.

NATO (1999). Monitored natural attenuation. NATO/CCMS Pilot Study Special Session, May 1999, Angers, France. EPA/542/R-99/008. www.NATO. int.ccms

Nyquist, J.E., B. Carr, and R.K. Davis (1998). Geophysical monitoring of a demonstration of in situ chemical oxidation. Proc. 10th Annual Symp. of the Application of Geophysics to Environmental and Engineering Problems, Eng. Environ. Geophys. Soc., pp. 583-591.

Nyquist, J.E., B. Carr, and R.K. Davis (1999). DC resistivity monitoring of potassium permanganate injected to oxidize TCE in situ. *J. Environmental and Engineering Geophysics*. September 1999.

Phelps, T.J., D Ringelberg, D. Hedrick, J. Davis, C.B. Fliermans, and D.C. White (1989). Microbial biomass and activities associated with subsurface environments contaminated with chlorinated hydrocarbons. *Geomicrobial Journal.* 6:157-170.

Reitsma, S. and M. Marshall (2000). Experimental study of oxidation of pooled NAPL. . In: Wickramanayake, G.B., A.R. Gavaskar, and A.S.C. Chen (ed.). Chemical Oxidation and Reactive Barriers. Battelle Press, Columbus, OH. pp.25-32.

Saba, T. and T.H. Illangasekare (2000). Effect of groundwater flow dimensionality on mass transfer from entrapped nonaqueous phase liquids, *Water Res. Res.,* 36(4): 971-979.

Saba, T., T.H. Illangasekare and J. Ewing (2000). Surfactant enhanced dissolution of entrapped NAPLs, submitted for publication in J. of Cont. Hydrology (in revision)

Schnarr, M.J. and G.J. Farquhar (1992). An in situ oxidation technique to destroy residual DNAPL from soil. Subsurface Restoration Conference, Third International Conference on Ground Water Quality, Dallas, Texas, Jun 21-24, 1992.

Schnarr, M., C. Truax, G. Farquhar, E. Hood, T. Gonully, and B. Stickney (1998). Laboratory and controlled field experimentation using potassium permanganate to remediate trichloroethylene and perchloroethylene DNAPLs in porous media. *Journal of Contaminant Hydrology*. 29:205-224.

Siegrist, R.L. and J.J. van Ee (1994). Measuring and interpreting VOCs in soils: state of the art and research needs. EPA/540/R-94/506. U.S. EPA ORD, Washington, D.C. 20460.

Siegrist, R.L., K.S. Lowe, D.R. Smuin, O.R. West, J.S. Gunderson, N.E. Korte, D.A. Pickering, and T.C. Houk (1998). Permeation dispersal of reactive fluids for in situ remediation: field studies. Project Report prepared by Oak Ridge National Laboratory for the U.S. DOE Office of Science & Technology. ORNL/TM-13596.

Siegrist, R.L., K.S. Lowe, L.C. Murdoch, T.L. Case, and D.A. Pickering (1999). In situ oxidation by fracture emplaced reactive solids. *J. Environmental Engineering*. Vol.125, No.5, pp.429-440.

Sparks, D.L., A.L. Page, P.A. Helmke, R.H. Loeppert, P.N. Soltanpour, M.A. Tabatabai, C.T. Johnson, and M.E. Sumner (ed.) (1996). Methods of Soil Analysis: Part 3 – Chemical Methods. Soil Sci. Soc. Am. Madison, WI.

Struse, A.M. (1999). Mass transport of potassium permanganate in low permeability media and matrix interactions. M.S. Thesis, Colorado School of Mines, Golden, CO.

Struse, A.M. and R.L. Siegrist (2000). Permanganate transport and matrix interactions in silty clay soils. In: Wickramanayake, G.B., A.R. Gavaskar, and A.S.C. Chen (ed.). Chemical Oxidation and Reactive Barriers. Battelle Press, Columbus, OH. pp. 67-74.

Tan, K.H. (1996). Soil Sampling, Preparation, and Analysis. Marcel Dekker, Inc. New York. 407 pp.

Tratnyek, P.G., T.L. Johnson, S.D. Warner, H.S. Clarke, and J.A. Baker (1999). In situ treatment of organics by sequential reduction and oxidation. Proc. First Intern. Conf. on Remediation of Chlorinated and Recalcitrant Compounds. Monterey, Calif. pp. 371–376.

Tyre, B.W., Watts, R.J., and Miller, G.C. (1991). Treatment of four biorefractory contaminants in soils using catalyzed hydrogen peroxide. *J. Environ. Qual.* 20, pp. 832-38.

Urynowicz, M.A. (2000). Reaction kinetics and mass transfer during in situ oxidation of dissolved and DNAPL trichloroethylene with permanganate. Ph.D. dissertation, Environmental Science & Engineering Division, Colorado School of Mines. May.

Urynowicz, M.A. and R.L. Siegrist (2000). Chemical degradation of TCE DNAPL by permanganate. In: Wickramanayake, G.B., A.R. Gavaskar, and A.S.C. Chen (ed.). Chemical Oxidation and Reactive Barriers. Battelle Press, Columbus, OH. pp. 75-82.

USEPA (1986). Test methods for the evaluation of solid waste, physical/chemical methods. SW-846, 3rd ed. Off. Solid Waste and Emergence Response, Washington, D.C.

USEPA (1990). Second update to SW-846 methods section. Office of Solid Waste. U.S. Environmental Protection Agency, Washington, D.C.

USEPA (1993). Subsurface characterization and monitoring techniques. EPA/625/R-93/003. U.S. Environmental Protection Agency, Office of Research and Development, Washington, D.C. 20460.

USEPA (1996). Method 515.3 Determination of chlorinated acids in drinking water by liquid-liquid extraction, derivatization and gas chromatography with electron capture detector. National Exposure Research Laboratory, Cincinnati, OH. July.

USEPA (1998a). In situ remediation technology: in situ chemical oxidation. EPA 542-R-98-008. Office of Solid Waste and Emergency Response. Washington, D.C.

USEPA (1998b). Technical protocol for evaluating natural attenuation of chlorinated solvents in ground water. EPA 600-R-98-128. http://www.epa.gov/ada/reports.html.

Vella, P.A., G. Deshinsky, J.E. Boll, J. Munder, and W.M. Joyce (1990). Treatment of low level phenols with potassium permanganate. Res. Jour. WPCF. 62 (7): 907-14.

Vella, P.A. and B. Veronda (1994). Oxidation of Trichloroethylene: A comparison of potassium permanganate and Fenton's reagent. 3rd Intern. Symposium on Chemical Oxidation. In: In Situ Chemical Oxidation for the Nineties. Vol. 3. Technomic Publishing Co., Inc. Lancaster, PA. pp. 62-73.

Weaver, R.W., S. Angle, P. Bottomley, D. Bezdicek, S. Smith, A. Tabatabai, and A. Wollum (ed.) (1994). Methods of Soil Analysis: Part 2 – Microbiological and Biochemical Properties. Soil Sci. Soc. Am. Madison, WI.

West, O.R., R.L. Siegrist, T.J. Mitchell, and R.A. Jenkins (1995). Measurement error and spatial variability effects on characterization of volatile organics in the subsurface. Environ. Sci. & Technol. 29(3): 647-656.

West, O.R., S.R. Cline, W.L. Holden, F.G. Gardner, B.M. Schlosser, J.E. Thate, D.A. Pickering, and T.C. Houk (1998a). A full-scale field demonstration of in situ chemical oxidation through recirculation at the X-701B site. Oak Ridge National Laboratory Report, ORNL/TM-13556.

West, O.R., S.R. Cline, R.L. Siegrist, T.C. Houk, W.L. Holden, F.G. Gardner, and R.M. Schlosser (1998b). A field-scale test of in situ chemical oxidation through recirculation. Proc. Spectrum '98 International Conference on Nuclear and Hazardous Waste Management. Denver, Colorado, Sept. 13-18, pp. 1051-1057.

West, O.R., R.L. Siegrist, S.R. Cline, and F.G. Gardner (2000). The effects of in situ chemical oxidation through recirculation (ISCOR) on aquifer contamination, hydrogeology and geochemistry. Oak Ridge National Laboratory, Internal report submitted to the Department of Energy, Office of Environmental Management, Subsurface Contaminants Focus Area.

Yan, Y.E. and F.W. Schwartz (1999). Oxidative degradation and kinetics of chlorinated ethylenes by potassium permanganate. *Journal of Contaminant Hydrology.* 37, pp. 343-365.

Zhang, H. and F. Schwartz (2000). Simulating the in situ oxidative treatment of chlorinated ethylenes by potassium permanganate. *Water Resources Research*, (36):3031-3042.

CHAPTER **7**

Field Applications and Experiences

7.1. APPLICATIONS AND PERFORMANCE

Field applications of in situ chemical oxidation (ISCO) using permanganate are increasing in the U.S. and abroad. As of this writing there are more than 100 field applications that have been completed, are in progress, or are in design. The most common applications to date appear to have been for treatment of chlorinated solvents in soil and groundwater to reduce the mass of contaminant in a source area and thereby aid the remediation of a site by longer-term natural attenuation. Other applications have attempted to reduce the mass of solvents in an industrial area to facilitate regulatory approvals and property transfers.

The choice of in situ chemical oxidation over other source treatment methods (e.g., surfactant flushing, thermally enhanced recovery) has been motivated by the ability of chemical oxidation to (1) be engineered to accommodate site specific conditions, (2) be implemented quickly with commercially available equipment and materials, and (3) yield measurable results in weeks to a few months. The choice of permanganate over other oxidants, such as hydrogen peroxide or ozone, often has been due to contaminant and site conditions such as the fact that permanganate (1) can be effective over a pH range that is wider than that in which other oxidants are effective, (2) effects oxidation by electron transfer rather than free radical processes and thus has a slower rate of reaction in the subsurface which enables enhanced delivery and transport, and (3) has lower gas and heat evolution and a lower risk of fugitive emissions.

247

Based on experiences to date at contaminated sites, MnO_4^- has been injected to yield groundwater concentrations that typically range from 200 to 20,000 mg/L (~1 to 150 mM MnO_4^-) depending on site and COC conditions. Delivery and dispersal into the subsurface has been achieved by injection probes, flushing by vertical and horizontal groundwater wells, and reactive zone emplacement by hydraulic fracturing (Schnarr et al. 1998, West et al. 1998a,b, USEPA 1998, Siegrist et al. 1999, Lowe et al. 2000, Siegrist et al. 2000a,b). Five applications of ISCO using permanganate are highlighted in Table 7-1 and described in the balance of this chapter. Those example applications were selected for inclusion because the authors have been actively engaged in them, and they illustrate a variety of delivery methods, oxidant loadings, treatment results, and lessons learned. It is recognized that there are a large number of other applications that have been carried out previously or are currently ongoing as described elsewhere (e.g., USEPA 1998, ESTCP 1999). The following five field applications are described in this chapter:

Site 1–Horizontal Well Recirculation with Potassium Permanganate

Site 2–Vertical Well Recirculation with Sodium Permanganate

Site 3–Deep Soil Mixing With Potassium Permanganate

Site 4–Multipoint Injection and Permeation of Potassium Permanganate

Site 5–Hydraulic Fracturing with Potassium Permanganate Oxidative Particle Mixture

These and other field applications have yielded encouraging results with respect to implementing permanganate ISCO at a contaminant site and the ability to reduce the mass of organics in soil and groundwater.

7.2. HORIZONTAL WELL RECIRCULATION WITH POTASSIUM PERMANGANATE (SITE 1)

Introduction

An oxidant delivery technique has been developed wherein the treatment solution is made by adding an oxidant to extracted groundwater. The oxidant-laden groundwater is then injected and recirculated into a contaminated aquifer through multiple horizontal and/or vertical wells. This technique, referred to as "in situ chemical oxidation through recirculation" (ISCOR), can be applied to saturated and hydraulically conductive

TABLE 7-1.	Examples of field applications of permanganate for soil and groundwater remediation.			
Application	Oxidant loading–delivery method	Treatment media	Efficiency	Lessons learned
1. DOE Portsmouth Gaseous Diffusion Plant, Piketon OH	Recirculation via horizontal wells Groundwater was taken from a portion of the site and mixed with $KMnO_4$, then injected without pressurization into a well so flow between horizontal wells was opposite direction of ground water flow	Permeable aquifers; hydraulic conductivity $>10^{-4}$ cm/s	TCE concentrations dropped to non-detectable or low ppb levels	In situ chemical oxidation through recirculation is a viable approach to chemical oxidation at remote sites where transport of pre-mixed oxidant solutions may be logistically difficult. Amorphous manganese oxides formed during oxidation may plug flow. The flow system must be designed to include a means for removing manganese oxide particles from extracted ground water. Non-uniform distribution of oxidant may result from soil heterogeneity Diffusion of oxidant into less conductive layers is possible Oxidant injection through a vertical well was more efficient than through a horizontal well. It may also be more cost-effective.

continued

TABLE 7-1. (continued)

Application	Oxidant loading–delivery method	Treatment media	Efficiency	Lessons learned
2. DOE Portsmouth Gaseous Diffusion Plant, Piketon OH	Recirculation via vertical wells An injection well and four extraction wells were installed to accommodate recirculation flow rates of 2-20 gpm supplemented with $NaMnO_4$ to maintain delivery of 250 mg $NaMnO_4$ per L of re-injected ground water	Permeable saturated subsurface media with an underlying aquifer.	~92% reduction in TCE within 3 days, and ~97% reduction 10 days.	Tracer studies should be performed in advance to determine preferential flow paths. Oxidant remaining in pore spaces after shutdown can continue to oxidize remaining contaminant. Oxidant injection using vertical wells in a 5-spot pattern is a delivery technique applicable to relatively permeable, saturated subsurface media contaminated with dissolved and sorbed phase contaminants and potentially ganglia of non-aqueous phase liquids (NAPLs). An underlying aquitard is required to prevent spread of contamination during injection. Application to situations with large masses of NAPLs (e.g., pools) may require substantially higher oxidant loadings and potentially modified hydraulic control approaches.

continued

TABLE 7-1. (continued)

Application	Oxidant loading–delivery method	Treatment media	Efficiency	Lessons learned
3. DOE Kansas City Plant, Kansas City, MO	Deep Soil Mixing The contaminated area was mixed by a crane-mounted vertical rotating 8 ft diameter mixing blade. The oxidant was injected through the hollow shaft and into the subsurface through orifices located on the mixing blades. Air was supplied to the mixed region during the initial pass of the apparatus through the soil column.	Low-permeability soils	An average TCE removal of 67% was computed. Total VOC removal may have been higher since 1,2-DCE was present and has been shown to be more easily oxidized than TCE.	The selection of a reagent delivery system to be used is closely tied to site conditions/parameters beyond initial contaminant concentration. Site characterization is an important step in the implementation of a chemical treatment process. Air mixing created some problems during operations, particularly while mixing the shallow cells until the air flow rate was reduced. Up to 30% of $KMnO_4$ added ponded on the surface due to difficulties delivering such an oxidant volume into the tight clay soil. Assessment of COC destruction from large test regions is often difficult since data interpretation is based on a very limited number of pre- and post-treatment soil samples. 60% reduction in the oxidant loading used in the field compared to laboratory testing still resulted in acceptable TCE reductions. Residual $KMnO_4$ in the vicinity of treatment cells did not pose an environmental concern due to the relatively high soil organic matter content of the soils. Site soils must be adequately characterized for other parameters in order to select the most promising oxidant delivery system.

TABLE 7-1. (continued)				
Application	Oxidant loading–delivery method	Treatment media	Efficiency	Lessons learned
4. DOE Portsmouth Gaseous Diffusion Plant, Piketon OH	Injection probes for permeation Several different treatment agents were injected at a volumetric loading rate of 5% v/v and pressures of 2-5 atm. using a four-lance vertical injector system.	Silty clay vadose zone –low permeability	Not applicable since this was a clean test site.	Volumes of treatment agents can be introduced into a relatively low permeability deposit. The nature and extent of the biogeochemical changes observed suggested that permeation dispersal could facilitate the transformation and degradation/ immobilization of many contaminants despite their presence in low permeability media. Spatial variability on the effects of biogeochemical properties was observed and believed to be due to the following of preferential pathways in the subsurface. $KMnO_4$ reacted more mildly, permeated further, and persisted longer then hydrogen peroxide. Soil total organic carbon concentrations decreased close to the ground surface, but little change was observed at greater depths. Manganese oxide solids deposition was observed on soil surfaces adjacent to major preferential pathways.

continued

TABLE 7-1.	(continued)			

Application	Oxidant loading– delivery method	Treatment media	Efficiency	Lessons learned
5. Portsmouth Gaseous Diffusion Plant, Piketon OH	Hydraulic Fracturing Test cells were comprised of 5 horizontal fractures stacked one over the other within a subsurface region. A new KMnO₄ mixture (OPM) was used to create permanganate-filled fractures. The cells were left in passive mode of operation for 15 months.	Low permeability silty clay vadose zone	Degradation efficiency > 99% after 2 hours of contact for 0.5 and 1.2 mg dissolved TCE per g of media.	Permanganate-filled fractures yielded a broad zone of reactivity within the low permeability media. Degradation potential for high levels of TCE was sustained even after 10 months of fracture emplacement. The use of fractures may curtail TCE release to the atmosphere or an underlying aquifer. Diffusive transport is slow, and the rate and extent are highly dependent on the physical and chemical properties of the formation. OPM had a TCE degradation rate that was equal to or greater than that of permanganate alone. Oxidant demand of natural organic matter or the advective loss of oxidant out of the treatment region could diminish permanganate in fractures. Further work is necessary and appropriate to provide needed design, implementation, and performance data for a range of site and contamination conditions.

formations and used with relatively stable oxidants such as $KMnO_4$. A field-scale test of ISCOR using $KMnO_4$ was completed at the DOE Portsmouth Gaseous Diffusion Plan in Piketon, OH, as part of the environmental restoration program at that site. Oak Ridge National Laboratory (ORNL) coordinated the demonstration in collaboration with the site operating contractor, Lockheed Martin Energy Systems (LMES). Assistance with facets of the demonstration was provided by Carus Chemical and Schumacher Filters America.

The field-scale ISCOR test was implemented at the X-701B Plume by use of a pair of previously installed parallel horizontal wells (Figure 7-1) (West et al. 1998a,b). The wells were 90 ft apart and had 200-ft screened sections consisting of 5-in. diameter, high-density polyethylene porous filters (500 mm pore size). The horizontal wells were drilled such that the screened sections were on top of the Sunbury shale and within the Gallia water-bearing unit (Figure 7-2).

The goal of the field test was to determine the efficacy of introducing $KMnO_4$ into the Gallia using an ISCOR approach. The field test duration was preset at 4 weeks. As such, clean up of the entire zone between the horizontal wells was not a test objective. ISCOR performance was

FIGURE 7-1. Conceptual schematic of ISCOR flow system at the X-701B site at the Portsmouth Gaseous Diffusion Plant (after West et al. 1998a,b).

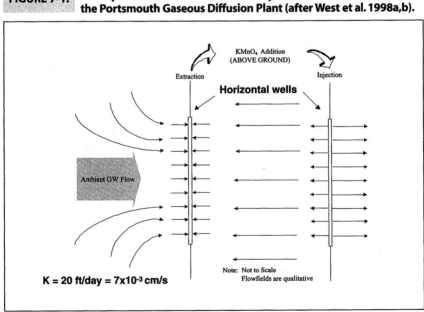

FIGURE 7-2.	Subsurface conditions and contamination levels at the X-701B site where ISCOR was applied (after West et al. 1998a,b).

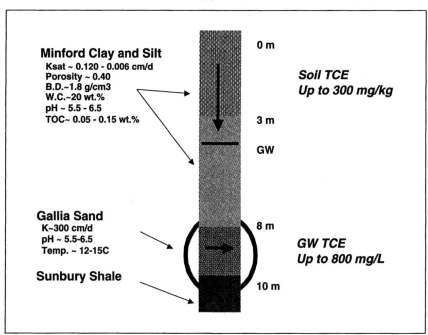

evaluated by comparing pre- and post-treatment soil and groundwater contamination in areas that were permeated by the oxidant. The advantages of ISCOR when compared to well injections alone include better control of oxidant and contaminant migration within the treatment zone, and the introduction of potentially higher volumes of oxidant solution because existing soil pore water is extracted prior to oxidant injection. ISCOR could be a viable approach to chemical oxidation at remote sites where transport of pre-mixed oxidant solutions may be logistically difficult.

Site Description and Treatability Testing (Site 1)

The stratigraphy underlying the X-701B site consists of the following layers proceeding downward from the surface (Figure 7-2): (1) Minford silt and clay with a thickness of 25 to 30 ft, (2) Gallia silty gravel with a thickness varying from 2 to 10 ft, (3) Sunbury shale, which is the first bedrock layer and consists of a 10 to 15-ft thick, moderately hard shale that often exhibits an upper weathered zone of gray, highly plastic clay, and (4) Berea sandstone, which is present at an approximate depth of 47 ft in this area. The Minford silt and Sunbury shale layers have very low conductivities, while the Gallia and Berea layers are the main water-bearing units. Past characterization has shown TCE contamination to be mainly in the Minford, Gallia, and upper weathered region of the Sunbury shale. The highest risk for offsite contaminant migration is associated with the relatively permeable Gallia layer. Hence, the ISCOR field test was targeted at treating contamination in the Gallia. Soil samples from the Minford, Gallia and Sunbury shale layers exhibited a wide range of TCE concentrations below 20-ft (Table 7-2).

Prior to design and implementation of the field application, a column-scale treatability test was completed at ORNL to test the recirculation and flushing of $KMnO_4$ solution through a sandy media contaminated with DNAPL TCE (Figure 7-3). During the test, a total of 2.0 L of solution with an initial concentration of ~4 wt.% $KMnO_4$ was recirculated through the column at a rate of 1.1 pore volumes/min for a total of 550 pore volumes (PV) over 8.3 hr (Figure 7-4). The final $KMnO_4$ concentration in the flushing solution was 1.8 wt.%. The effluent TCE concentration averaged 56 mg/L during the initial 30 PVs followed by a decrease to 0.13 mg/L for the balance of the flushing. The post-treatment TCE within the column apparatus was ~0.3 wt.% of the initial TCE mass added indicating a treatment efficiency of 99%. The $KMnO_4$ consumption was 3.6 g-$KMnO_4$/g-TCE.

TABLE 7-2.	TCE concentrations in soil samples from ISCOR pretreatment characterization at Site 1 (West et al. 1998a,b).

Subsurface Layer	No. of Samples	Trichloroethylene concentration (μg/kg)**				
		Average	Std. Dev.	Median	Min.	Max.
Minford *	90	19,493	21,770	10,002	nondetect	80,471
Gallia	163	53,596	52,713	43,320	nondetect	302,237
Sunbury*	13	132,405	269,791	46,932	32	1,048,174

*Minford: based on samples collected at depths > 20 ft. Sunbury: based on samples from the top weathered region.
**Based on wet soil weight, nd = not detected at an approximate detection limit of 5 μg/kg. Avg. moisture content = 16%.

FIGURE 7-3. Column-scale treatability test completed at ORNL prior to design and implementation of the full-scale ISCOR application.

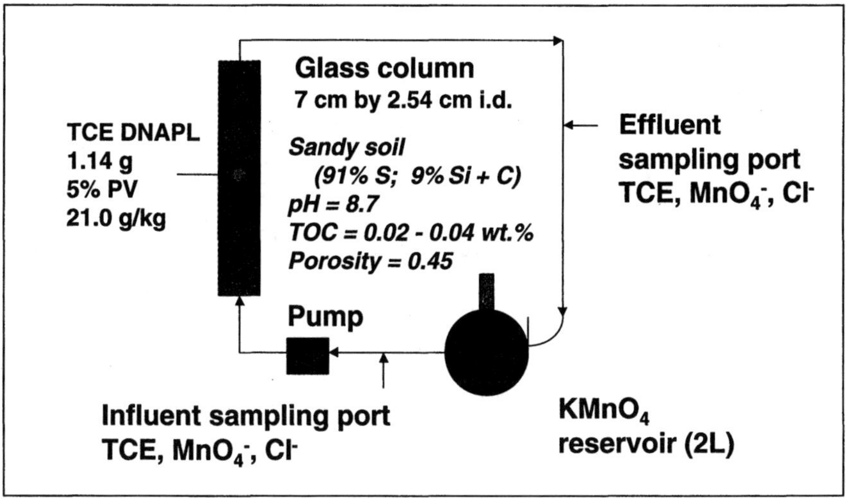

System Implementation (Site 1)

A conceptual schematic for the flow system used during the ISCOR field test is shown in Figure 7-1; photographs of the field site during operation are shown in Figure 7-5. During the implementation of ISCOR at the X-701B site, water extracted from the west (upgradient) horizontal well was sent through a nearby groundwater treatment facility before the addition of KMnO₄. This modification to the original ISCOR concept was to ensure that the process met the regulatory requirement that TCE in the re-injected groundwater be <5 ppb.

Groundwater extracted from the west horizontal well was delivered to the nearby groundwater treatment facility, where the water was treated by use of air strippers and activated carbon. Water for the oxidant injection solution was taken from a portion of the groundwater treatment facility effluent, and mixed with crystalline KMnO₄ by use of a solids feeder. The oxidant-laden water then was injected without pressurization into the east horizontal well. As such, flow between the horizontal wells was opposite the direction of ambient groundwater flow. Extraction from the west horizontal well was set to ~10 gpm by flow regulators. The target injection flow rate at the east horizontal well was 10 gpm. However, this well could

FIGURE 7-4. Results of column-scale treatability test in support of ISCOR using KMnO₄ at the DOE Portsmouth site.

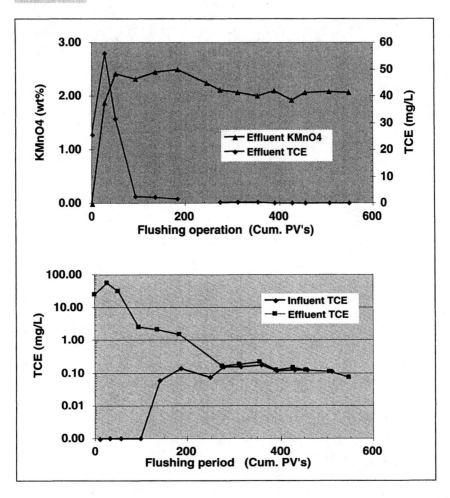

only take a maximum of 6 gpm as water backed up to the ground surface when higher injection flow rates were attempted.

ISCOR between the horizontal wells began operation on July 26, 1997, and continued through August 21, 1997. Simultaneous injections in the east horizontal well and a nearby vertical well (74G) were initiated on August 20, 1997, to attempt greater coverage of permanganate injection (Figure 7-6). Well 74G was selected because it is centrally located within

FIGURE 7-5. Photographs of the X-701B site during the ISCOR application at the DOE Portsmouth site near Piketon, Ohio.

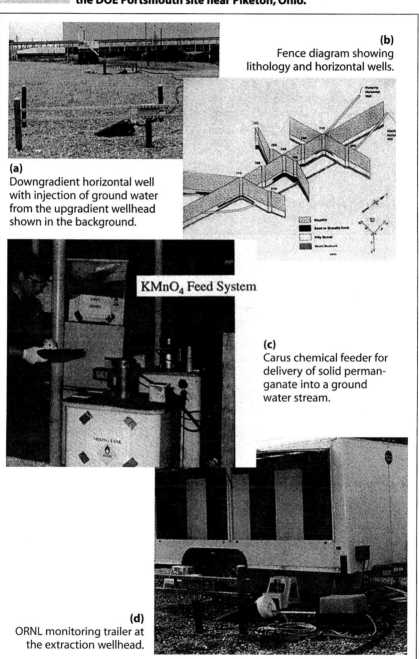

(b)
Fence diagram showing lithology and horizontal wells.

(a)
Downgradient horizontal well with injection of ground water from the upgradient wellhead shown in the background.

KMnO₄ Feed System

(c)
Carus chemical feeder for delivery of solid permanganate into a ground water stream.

(d)
ORNL monitoring trailer at the extraction wellhead.

FIGURE 7-6. **Vertical well locations in the vicinity of the ISCOR field test site.**

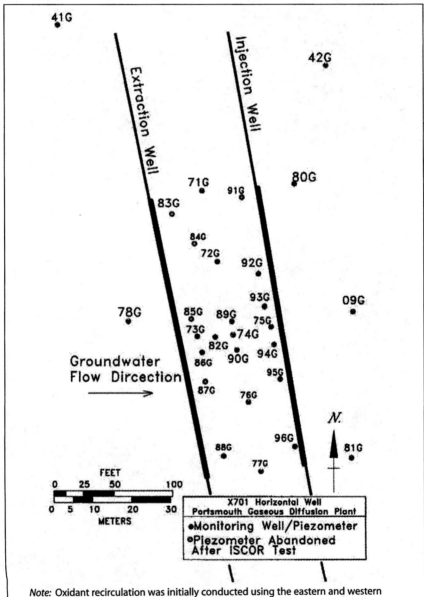

Note: Oxidant recirculation was initially conducted using the eastern and western horizontal wells for injection and extraction, respectively. Oxidant was injected into Well 74G towards the latter part of the field test. Well 21G, which is located ~300 ft downstream of the horizontal wells and not shown in the figure above, was sampled after the ISCOR field test.

the treatment region and had high levels of TCE (734 ppm in ground-water) prior to the beginning of the ISCOR field test.

Injection and extraction from the horizontal wells was halted on August 21, 1997; oxidant injection into well 74G was continued through August 28, 1997, approximately 4 weeks after the ISCOR field test was initiated. As mentioned previously, the field test duration was preset at 4 weeks. The oxidant flow rate into 74G was 2 gpm. Thus, injection into well 74G, which had a screened section of 5 ft, was significantly more efficient than injection into the 200-ft horizontal well that could only be sustained at a maximum of 6 gpm.

Recirculation between the horizontal wells was halted 3 weeks into the field test because of increasing amounts of colloidal particles coming from the extraction well, which the groundwater treatment facility was not prepared to handle. These particles, on the order of 1 μm in size, were identified by use of a scanning electron microscope and energy dispersive x-ray as amorphous manganese oxides. In future applications of this technology, the flow system must be modified to include a means for removing manganese oxide particles from extracted groundwater. Such techniques have been developed for wastewater treatment where manganese oxides are a by-product of dissolved iron removal using $KMnO_4$.

A total of ~12,700 kg of $KMnO_4$ was delivered to the treatment region, 1960 kg of which was introduced through vertical well 74G. Of the 206,000 gallons of oxidant solution, 14,000 gallons were delivered through well 74G.

The performance of the ISCOR system was monitored through collection of water samples from the influent and effluent streams (daily), and from monitoring wells (daily to every three days) in the vicinity of the treatment region. The MnO_4^- concentration in these water samples was quantified in the field by measuring absorbance of filtered samples with a Hach DR2000 spectrophotometer at 525 nm. TCE concentrations were quantified by hexane extraction followed by GC/ECD analysis of the extracts. TCE analyses were done within 7 days of groundwater sample collection.

Operation and Performance Results (Site 1)

The delivery of oxidant solution through the east horizontal well was not uniformly rapid through the entire treatment region. However, after 32 days, the oxidant had been detected throughout most of the region between the two horizontal wells. The oxidant was detected in the central area of the treatment zone after oxidant injection in vertical well 74G was initiated. The non-uniform distribution of MnO_4^- probably was due to spatially variable hydraulic conductivities that ranged from 24 to 411 ft/day as measured in monitoring wells within the treatment zone. The highest conductivity of 411 ft/day corresponds with the rapid detection of $KMnO_4$ in the wells in the southern portion of the treatment zone (Figure 7-7), although pressure drop along the 200-ft well screen also could have caused more oxidant solution to flow out near the wellhead. Finally, the injection well screen may have had a variable efficiency along its length.

Overall, whenever permanganate was detected in groundwater samples from the monitoring wells, TCE concentrations dropped to non-detectable or low ppb levels. That reduction in groundwater TCE concentration may be due to the degradation or removal of TCE from associated sediments or to displacement of contaminated groundwater from the pore space and non-equilibrium between the pore water and residual TCE in the sediments. Approximately two weeks after the ISCOR field test ended, boreholes were drilled in locations where MnO_4^- was detected during treatment operations. All post-treatment boreholes revealed that TCE levels in the low permeability Minford silt and Sunbury shale did not change significantly with treatment. However, this post-treatment sampling showed that TCE soil concentrations were reduced significantly in the Gallia layer wherever MnO_4^- was able to permeate. That is illustrated in Figures 7-8 and 7-9 where pre- and post-treatment TCE concentrations in soil are compared for boreholes drilled near two monitoring wells located approximately 40 ft. and 10 ft., respectively, from the injection well. Soil TCE levels remained elevated in the upper layer of the Gallia at some locations probably due to the presence of the silt layer that may have interfered with oxidant dispersion. Because MnO_4^- is still detected at up to 93 mg/L in some of the monitoring wells more than 3 years after the ISCOR field test (AIMTech 2001), diffusion of the oxidant into the Minford and Sunbury layers is possible.

Concentrations of TCE and MnO_4^- (as $KMnO_4$) versus time in groundwater are shown in Figures 7-10 and 7-11. MnO_4^- was detected in both of these wells 4 months after the field test, and TCE was below the 5-ppb detection limit. Five months after the field test (January 1998), MnO_4^- was no longer present at one location and corresponding TCE levels increased (Figure 7-11). Because TCE was not detected in post-treatment soil samples (Figure 7-9), the increase in TCE concentration indicates that recontamination from groundwater upgradient and/or from the adjacent Minford silt and Sunbury shale layers may be occurring.

FIGURE 7-7. Distribution of KMnO₄ in groundwater at the ISCOR field test site 32 days after initiation of oxidant recirculation (West et al. 1998a,b).

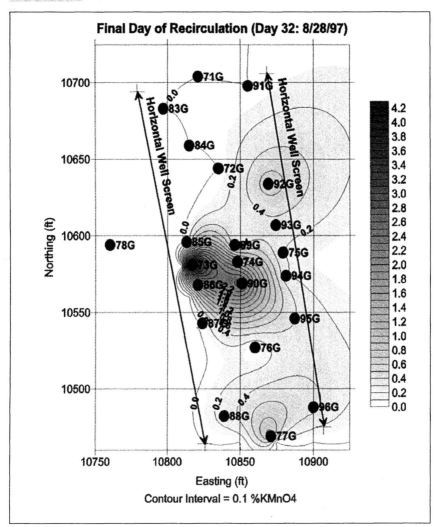

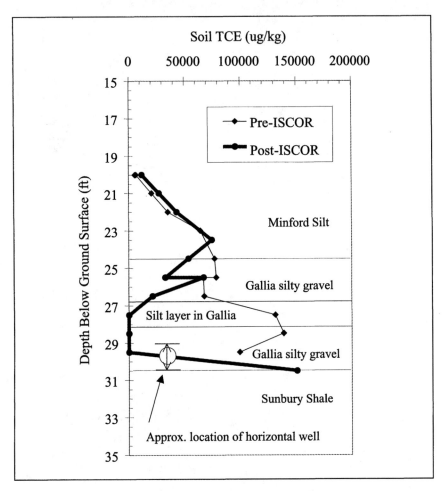

FIGURE 7-8.   **Pre-and post-treatment TCE concentrations in soil samples collected from boreholes near well 90G approximately 40 feet from the KMnO$_4$ injection well (West et al., 1998a,b).**

The total horizontal well recirculation demonstration cost was approximately $562K with the majority of the costs being related to the operation of the recirculation system and pre-demonstration characterization. Of the total demonstration costs: generator and oxidant feed system rental accounted for approximately 2%, materials (well materials, sampling supplies, etc.) accounted for approximately 6%, KMnO$_4$ accounted for approximately 7%, and recirculation system operation accounted for approximately 85%. The unit cost for the demonstration

FIGURE 7-9. Pre- and post-treatment TCE concentrations in soil samples collected from boreholes near well 95G adjacent to the oxidant injection well (West et al. 1998a,b).

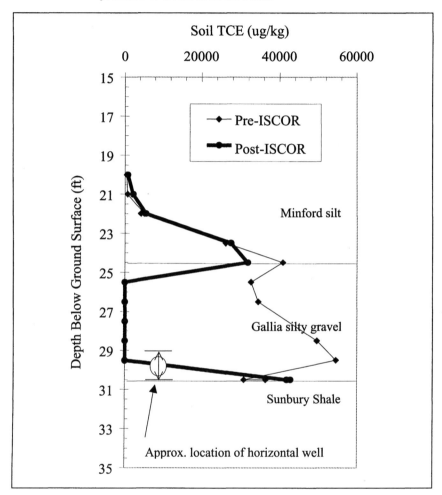

was $101 per c.y. of soil, including oxidant costs of $21/c.y. and O&M costs of $80/c.y. The estimated unit cost per pound of TCE treated ranged from $363 to $778 based on three different mass scenarios (8,000 lbs TCE, 16,000 lbs TCE, and 25,000 lbs TCE). The lower unit cost was associated with the highest total mass of TCE scenario due to leveraged well installation and pre-characterization costs. Based on this demonstration, a treatment rate of ~70 m³/d (24kL/d) is expected.

FIGURE 7-10.   TCE and KMnO$_4$ concentrations in groundwater samples collected from well 75G adjacent to the KMnO$_4$ injection (east) horizontal well (West et al. 1998a,b).

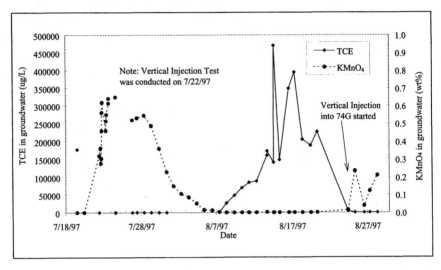

FIGURE 7-11.   TCE and KMnO$_4$ concentrations in groundwater samples collected from well 95G adjacent to the KMnO$_4$ injection (east) horizontal well (West et al. 1998a,b).

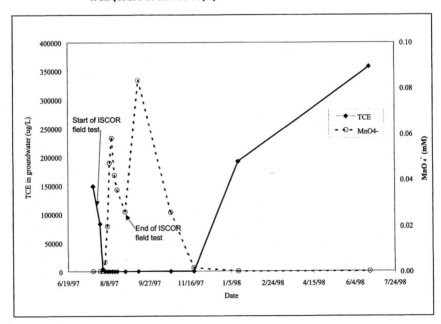

Conclusions (Site 1)

The field test described above showed that ISCOR is a viable remediation technique for saturated subsurface media. Lateral and vertical heterogeneity impacted the uniform delivery of oxidant through the horizontal wells. Where the oxidant was able to permeate the Gallia, significant reductions in TCE were measured in both groundwater and soil samples. Oxidant injection through the vertical well was significantly more efficient than injection through the horizontal well.

While TCE concentrations in the less permeable Minford and Sunbury layers adjacent to the Gallia water-bearing unit did not decrease within 2 weeks following ISCOR, diffusion of the oxidant into the less conductive layers is possible, given that MnO_4^- was still present in some monitoring wells 2 years after the ISCOR field test. Although oxidant delivery/recirculation was impacted by heterogeneity within the demonstration area, where the oxidant was able to permeate the aquifer, significant and sustained reductions in TCE were measured in both soil and groundwater.

7.3. VERTICAL WELL RECIRCULATION WITH SODIUM PERMANGANATE (SITE 2)

Introduction

Based on results from the horizontal well ISCOR treatability test described in Section 7.2, a second ISCOR treatability test with vertical wells in a 5-spot pattern was conducted at the DOE Portsmouth Gaseous Diffusion Plant. Oak Ridge National Laboratory coordinated the demonstration with the operating contractor for the site, Bechtel Jacobs Company, LLC. The demonstration was funded jointly by the DOE Subsurface Contaminants Focus Area and the DOE Portsmouth Site Office.

The test was conducted in an industrial area with buildings in close proximity and with overhead and underground utilities present (Figure 7-12). The test design was comprised of a 5-spot vertical well pattern with the central well serving as the injection point (Figure 7-13). The four corner extraction wells were placed in a square grid surrounding the injection point on a fixed radius of 45 ft. The system was designed to accommodate recirculation flow rates of 2 to 20 gpm supplemented with

FIGURE 7-12. Photograph of the field site during 5-spot vertical well recirculation of NaMnO₄.

FIGURE 7-13. Conceptual process diagram of 5-spot vertical well injection and recirculation of permanganate.

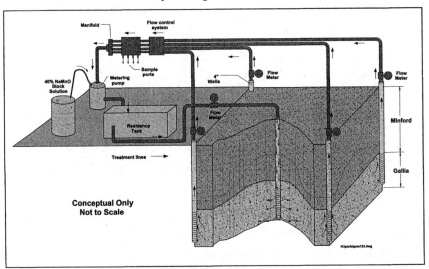

$NaMnO_4$ to maintain delivery of the oxidant concentration at ~250 mg $NaMnO_4$ per L of re-injected groundwater.

The goal of this project was to test an alternative oxidant delivery approach using recirculation via vertical wells in a 5-spot pattern (Lowe et al. 1998, Lowe et al. 2000). Several monitoring wells also were installed throughout the region to monitor changes in contaminant concentrations, oxidant concentrations, water-quality parameters, and water levels during operation. The potential advantages of the use of liquid $NaMnO_4$ and a vertical well recirculation approach include: (1) better control of oxidant and contaminant migration within the treatment zone as compared to single well injections, (2) the introduction of higher volumes of oxidant solutions in a given time because existing soil pore water is extracted concurrent with oxidant injection, (3) easier oxidant handling, and (4) potentially lower overall cost for treating larger volumes.

Site Description and Treatability Testing (Site 2)

The field test was conducted within the source area of the 5-Unit Investigative Area, within the central portion of the Portsmouth Gaseous Diffusion Plant. The geology under the pilot test area is comprised of 23–26 ft of Minford sediments underlain by 3 to 8 ft of Gallia (see Figure 7-2). Both units are unconsolidated, Quaternary age members of the Teays formation. The Gallia lithology typically is comprised of a saturated silty-sandy gravel unit of higher permeability overlying a drier lithified unit of coarse angular gravel in a silty clay matrix. The upper more permeable member of the Gallia varies in thickness from 0.5 to 2.5 ft. The Gallia is thicker and more permeable across the southern two thirds of the test cell area. In general, groundwater in the 5-Unit Area flows horizontally from north to south, and the hydraulic gradient is very low because of the flat valley floor, the presence of thicker and more permeable Gallia deposits, and the proximity of the east-west-trending groundwater divide along the central part of the facility. Groundwater samples collected prior to system operation indicated TCE concentrations ranged from 133 ug/L to 2148 ug/L with higher concentrations detected in the northern and western monitoring wells. Soil samples ranged from 3 to 4527 ug/kg, typically with concentrations increasing with depth.

Prior to implementing the field system, laboratory treatability tests were done to verify that the oxidation kinetics of the $NaMnO_4$ were essentially the same as that of the $KMnO_4$. These tests were conducted by use of 100-mL syringe reactors loaded with simulated groundwater spiked with TCE and dosed with oxidant. The initial concentration of TCE was 4.0 mg/L and the permanganate concentration was either 250 mg/L of $NaMnO_4$ or $KMnO_4$. During reaction at 20C, the efficiency and kinetics of degradation of TCE were essentially the same for both oxidants. The reaction was pseudo-first order with a $t_{1/2} < 5$ min and complete destruction was achieved before 50 min.

System Implementation (Site 2)

The field test required the installation of wells, development and hydraulic testing of the wells, baseline soil and groundwater characterization, and assembly of the oxidant delivery system. One 6-in-diameter injection well, four 4-in-diameter extraction wells, and nine monitoring wells to monitor the pilot test operations were installed (Figures 7-13 and 7-14). Soil and groundwater characteristics—physical and chemical—were obtained throughout the area prior to recirculation. Hydraulic tests (pressure injection test and single well tests) were performed to select the recirculation pumping and injection rate and to determine the distribution of permeability within the test area.

During injection and recirculation, oxidant solution was injected into the Gallia through the central well and extracted at equal rates (each of four at 25% of the injection rate) from the 4 perimeter wells located at 45 ft from the central injection well. Groundwater was extracted from the four perimeter wells at a maximum combined rate of 18 gpm (4.5 gpm from each extraction well). Groundwater from the four extraction wells was pumped through a manifold system into a single line where the $NaMnO_4$ solution was added with an oxidant-resistant chemical metering pump. The metering pump was adjusted to feed the concentrated stock solution at a rate sufficient to maintain 250 mg/L of $NaMnO_4$ in the injection water. The oxidant-laden water was then pumped into a 120 gal holding tank providing approximately 7 minutes of residence time for the oxidant with the contaminated groundwater. This ensured that only treated water was re-injected into the aquifer. Injection of the extracted water was permissible as long as the extracted water had been "treated".

A 250 mg/L $NaMnO_4$ solution was recirculated at 18 gpm for the 1st pore volume (approximately 82,000 gals or 3 days). Then the recirculated water with residual $NaMnO_4$ was supplemented with approximately 100 mg/L of $NaMnO_4$ solution to maintain a delivery of ~200 to 250 mg/L of oxidant into the region at 18 gpm for the 2nd and 3rd pore volumes. Approximately 78 gals of 40 wt% $NaMnO_4$ (162.4 kg $NaMnO_4$) were

FIGURE 7-14. Site plan for the field site where vertical well recirculation of NaMnO$_4$ was deployed using a 5-spot network (center injection and perimeter extraction wells).

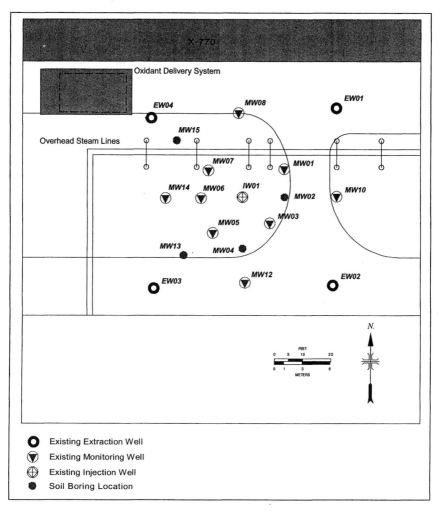

delivered throughout the test region within 10 days. Approximately 240,660 gals of treated groundwater were recirculated. The recirculation flow rates remained steady throughout the test duration with ~1 gpm reduction in total recirculation flow rate over the course of the 10-day test. This gradual but minor decline in the total recirculated flow rate is attributed to injection-or-extraction well fouling, matrix plugging, pump performance, or clogging/plugging within the system lines (at in-line valves, gauges, etc). Pressures within the system and at the injection well gradually increased over time. The injection pressure at the well head nearly doubled, from 10 psi to 18.5 psi, but remained within safe operating ranges based on optimum flow out of the selected well screen, potential fracturing of the formation, and potential damage to the well construction. It is likely that the gradual increase in system pressures and decline in extraction-well flow rates is, in part, a result of injection-well performance or potential redistribution of fine particles in the formation near the well.

Performance assessment was based on TCE reduction as determined by characterizing contaminants in the aqueous phase before and after treatment. Groundwater was recirculated through the pilot test area until sufficient oxidant had been delivered or breakthrough of the residual oxidant at the injected concentration (indicating that the $NaMnO_4$ was not being consumed in the aquifer) was observed. The TCE concentrations in the test area were monitored during recirculation and for 24 hrs after recirculation was terminated. Finally, post-treatment groundwater concentrations were monitored weekly for approximately 1 month and once at 2 months after recirculation to determine the potential aquifer rebound effects.

Operation and Performance Results (Site 2)

A pressure injection test conducted at the injection well indicated that the injection well could sustain recirculation rates up to 26 gpm without exceeding optimum pressures at the well head. Observation of water level fluctuations during the pressure injection test also indicated significant variability in hydraulic conductivity across the 64-ft by 64-ft test area. Single well tests confirmed the variability and provided a relative measure of permeability with values ranging from approximately 26 to >300 ft/d (Table 7-3). Bromide tracer test results confirmed the hydraulic tests and indicated that the Gallia is heterogeneous throughout the test region. The bromide curves from the northern wells tend to flatten out indicating a more heterogeneous flow system and greater dispersion while the curves to the west and south indicate a more pronounced (i.e., sharp) bromide front, suggesting a more homogeneous flow system with less dispersion. Water levels, measured continuously by a data acquisition system, indicate that the system reached hydraulic equilibrium within a few hours after recirculation was initiated and remained constant through-

TABLE 7-3.	Hydraulic conductivities measured during single well aquifer tests for Site 2.		
Well Number	Average Pre-injection ft/day	Average Post-injection ft/day	Ratio of Post-injection to Pre-injection
X770-EW02	26.5	26.7	1.00
X770-EW03	86.5	87.3	1.00
X770-EW04	44	54.3	1.23
X770-MW01	148	94	0.64
X770-MW03	297.5	271.7	0.91
X770-MW05	691.5*	532.3	0.77
X770-MW07	202.5	187.7	0.93
X770-MW08	121	143	1.18
X770-MW10	145	113.7	0.78
X770-MW12		366.3	–
X770-MW14	601.5*	337	0.56

*Limited data points used in analysis.

out the test. The bromide tracer test also provided insight into the preferential flow paths within the region.

Oxidant injection by use of vertical wells in a 5-spot pattern was capable of providing sufficient hydraulic control to deliver oxidant throughout the permeable zones of the Gallia in the test area within 3 days or less. The pre-test TCE concentrations throughout the region were reduced to <10 ug/L throughout all but the lower permeable eastern edge of the test region within 3 days, indicating an apparent reduction in contaminant levels of ~92% (Figure 7-15). Continued recirculation for 10 days, provided uniform oxidant delivery throughout the entire area excluding the northeast extraction well (Figure 7-16). As of two hours after recirculation was terminated, TCE concentrations were reduced to below detection limits (5 ug/L) at all but one location, (from ~1500 ug/L to 650 ug/L at

274   In Situ Chemical Oxidation

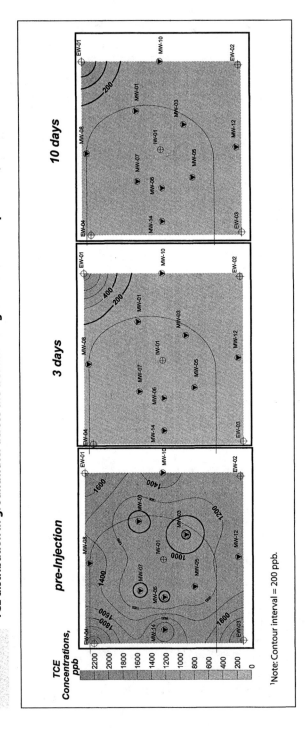

FIGURE 7-15. TCE distribution in groundwater across the treatment region with time of operation (Lowe et al. 1998, 2000)[1].

¹Note: Contour interval = 200 ppb.

FIGURE 7-16. NaMnO$_4$ distribution in groundwater across the treatment region with time of operation (Lowe et al. 1998, 2000)[1].

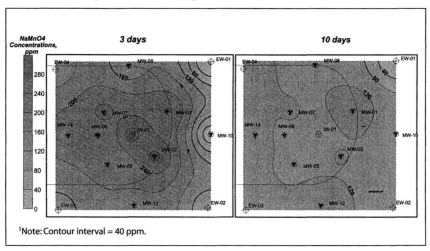

[1]Note: Contour interval = 40 ppm.

the northeast extraction well), indicating an apparent ~97% reduction in TCE. The slower arrival of oxidant to the northeast portion of the region also was noticeable and attributed to the lower permeability of this region. It is important to note that, while the permeability of the Gallia varied by over an order of magnitude in the region (Table 7-3), oxidant was delivered effectively throughout the area in a relatively short time frame (i.e., 3 days).

To evaluate the short-term effects of the residual oxidant and TCE within the region, groundwater samples were collected at 1, 2, 4, 6, 9, 21, and 27 hrs after system shut down. Additionally, groundwater samples were collected weekly for one month after recirculation. Although low levels of TCE were present along the upgradient edge of the test area (14 ug/L at X770-MW08, 39 ug/L at X770-EW04 and up to 738 ug/L at X770-EW01) when the system was shut down, the TCE concentrations continued to decline after recirculation was terminated. This decline in concentrations may be due to the oxidation of TCE by the oxidant remaining in the pore spaces.

TCE remained near or below detection limits (<20 ug/L) for approximately two weeks after recirculation at which point the TCE concentrations along the upgradient northern edge (X770-EW01, X770-MW08, and X770-EW04) and lower permeable eastern edge (X770-MW10) of the test area began to increase. The higher concentrations detected at these locations can be attributed to the migration of untreated upgradient plume groundwater into the test region (northern edge), as well as TCE diffusion from the less permeable zones of the test area (eastern edge). It is important to note that near the injection well (15-ft-radius monitoring wells) where the delivered oxidant mass was the highest, TCE concentrations remained below detection limits one month after recirculation except at the most upgradient location (35 ug/L at X770-MW01).

One month after system shut down, the oxidant concentrations within the region had declined to less than 1 mg/L, while a gradual increase in TCE concentrations was observed. However, the apparent TCE mass reduction within the mobile groundwater fraction of the test cell remained at ~83%. The oxidant consumption over time is as expected and presumed to be due to a combination of factors, including oxidation of the natural organic material present and any TCE that was diffusing from the finer grained less permeable zones in the Gallia or the overlying Minford and advecting from the upgradient contamination into the test area. Soil and groundwater samples collected 2 years after injection indicate persistence of $NaMnO_4$ localized to the area immediately adjacent (within 2 feet) to the injection point in the soil and groundwater at this site (AIMTech 2001).

Samples of groundwater were also analyzed by use of biotoxicity test procedures. A Microtox Model 500 analyzer (AZUR Environmental, Carlsbad, CA) was used to analyze groundwater samples for acute toxicity to a luminescent bacteria *(Vibrio fischeri)*. Toxicity of samples collected after the active period of oxidant flushing was compared to that of background samples collected at the same time but from outside the treated region. When exposed to the highest concentration of the groundwater within the total sample of the test media (91%), no toxicity was measured for post-oxidant flushing samples or for background samples.

The costs of the 5-spot ISCOR vary depending on the scale of the application and the performance goals required. Avoidance of costs associated with treatment and discharge of extracted groundwater is a major benefit of the recirculation delivery method; during this field test, ~240,000 gal. of groundwater were extracted, amended with oxidant, and then re-injected (rather than being treated and disposed of ex situ). The estimated cost per gallon of treated groundwater during the pilot test at a DOE facility was ~$1.70/gal, including well installation costs, but excluding cost of site support (e.g., health and safety, health physics, construction engineering, waste management, and project management oversight). While it is important to recognize that specific site conditions will greatly influence cost, the most significant costs associated with the 5-spot ISCOR approach are well installation and the oxidant. The cost of the injection wells and extraction wells will be dependent on the size, depth, location, and materials, as well as the drilling subcontractor. The cost of the oxidant will depend on the concentration delivered to the sub-surface over some period of time based on performance goals and initial contaminant concentrations.

Conclusions (Site 2)

Oxidant injection by use of vertical wells in a 5-spot pattern was capable of providing sufficient hydraulic control to deliver oxidant throughout the permeable zones of the Gallia within the pilot test area within 3 days. Pre-test baseline TCE concentrations were reduced to <20 ug/L throughout all but the lower permeable eastern edge of the test region within 3 days, indicating an apparent reduction in contaminant levels of ~92% within 3 days and ~97% reduction at 10 days (2 hours after the end of the test). This oxidant delivery technique is applicable to relatively permeable, saturated subsurface media contaminated with dissolved and sorbed phase contaminants and potentially to ganglia of non-aqueous phase liquids. An underlying aquitard is required in order to prevent spread of contamination during the injection phase. It is noted that application to situations in which large masses of DNAPLs (e.g., pools) occur may require substantially higher oxidant loadings and potentially modified hydraulic control approaches.

7.4. DEEP SOIL MIXING WITH POTASSIUM PERMANGANATE (SITE 3)

Introduction

A project to evaluate ISCO of chlorinated solvents using potassium permanganate delivered via deep soil mixing (DSM) was completed at the DOE Kansas City Plant (KCP), in Kansas City, MO, as part of the DOE environmental restoration program at that site (Cline et al. 1997, Gardner et al. 1998). In addition to chemical oxidation, the demonstration included calcium oxide enhanced mixed region vapor stripping and bioaugmentation, both of which used the same DSM delivery method. Oak Ridge National Laboratory coordinated the demonstration with the operating contractor for the site, Allied Signal. GeoCon, Inc., provided the DSM equipment, and Carus Chemical provided the permanganate. The demonstration was funded jointly by the DOE Subsurface Contaminants Focus Area and the DOE KCP environmental program.

The ISCO system involved a crane-mounted vertical rotating blade system designed to mix the subsurface by use of 8 to 10-ft diameter blades (Figure 7-17). During the in situ mixing process, treatment agents were injected through a vertical, hollow shaft and into the soil through orifices at the rear of the horizontal soil-mixing blades.

The selection of DSM technology for the DOE KCP site was based on the low permeability, wet conditions within the subsurface, and the need to reduce TCE and other chlorinated solvents in a source area for a groundwater plume (DOE 1996a,b). DSM was chosen for effective introduction and mixing of the oxidant within the low permeability contaminated media. The goal of the demonstration was to achieve >70% treatment of TCE. Performance was evaluated by pre- and post-treatment soil samples, water samples, and off-gas monitoring.

Site Description and Treatability Testing (Site 3)

The field site for application of $KMnO_4$ with DSM delivery was located at an old land treatment unit where the sediments consisted of low permeability silty clays and clayey silts with TCE concentrations as great as 3,800 mg/kg. Contamination existed down to 45-ft below ground surface (bgs), which included ~35 ft. in the saturated zone. The subsurface soil media were characterized as having a dry density = 1.3 to 1.4 g/cm^3; water content = 19 to 33 wt.%; plastic limit = 18 to 20%; liquid limit = 33-50%; and TOC = 0.4 to 0.7 wt.%.

A substantial laboratory treatability study was conducted to support the design and implementation of the field application (Gardner et al. 1998). The purpose of the lab tests was to determine if oxidation treatment could be used to degrade TCE and DCE in contaminated soil from the DOE KCP site. The study included three series of tests conducted at ORNL to determine (1) the effect of oxidant type and concentration, (2) the effect of oxidant volume, and (3) oxidant persistence and effect on soil composition. The first two test series were conducted by use of zero headspace reactors (ZHRs) following the methods of Gates and Siegrist (1995). The ZHRs were loaded with either field or artificially contaminated soil from the KCP. Oxidant solution(s) were added to contaminated soil under gastight conditions and pre- and post-treatment contaminant concentrations were measured to determine treatment efficiency. A final series of tests

FIGURE 7-17. Photographs of the permanganate DSM project at the DOE Kansas City Plant (after Cline et al. 1997, Gardner et al. 1998).

(a) Deep soil mixing equipment.

(b) Auguring blade with delivery outlets opposite of the cutting teeth.

(c) Surface spoil and permanganate bleed off during deep soil mixing to 47 ft. in fine-grained soils.

were conducted in a laboratory pilot-scale soil mixing apparatus designed to replicate reagent injection and deep soil mixing (Gates and Siegrist 1995).

Three oxidants ($KMnO_4$, H_2O_2, or H_2O_2 supplemented with $FeSO_4$) were tested and compared initially. The $KMnO_4$ concentration was set at 4 wt.% while the H_2O_2 was set at 8.5 wt.%. During the first series of tests, each oxidant solution was added to field moist soil, and the decrease in TCE was observed. The greatest decrease in TCE was achieved with $KMnO_4$ (96%) compared to 40% and 72% for H_2O_2 and H_2O_2 plus $FeSO_4$, respectively. Based on the relatively better results with $KMnO_4$, all subsequent tests were completed with $KMnO_4$ only. Contaminated soil slurries (1:1 soil:water) from the field site were treated with different concentrations of $KMnO_4$. The slurry tests indicated that the TCE treatment efficiency increased as the oxidant concentration increased (35% TCE reduction with 0.5 wt.% $KMnO_4$ to 96% TCE reduction with 4 wt.% $KMnO_4$). DCE treatment of ~100% from an initial 12 mg/kg level was observed at all $KMnO_4$ concentrations.

A second series of tests was conducted to determine if the volume of oxidant solution added to contaminated soil could be minimized or eliminated by treating the soil with > 4 wt.% $KMnO_4$ solutions or with solid $KMnO_4$ crystals. Field moist soil (not slurries) were treated in ZHRs with 5, 8, and 12 wt.% $KMnO_4$ such that the oxidant solution volume and solids mass added provided an equivalent dose of 20 g $KMnO_4$/kg soil. In each test, >85% treatment of TCE was achieved from an initial level of 14 mg/kg and 536 mg/kg. For DCE, 100% treatment was achieved from an initial level of 9.7 mg/kg and 124 mg/kg. Those results suggested that, at an equivalent mass dose, lower volumes of oxidant solution could achieve adequate treatment results and reduce the volume of fluid added.

A final series of tests was conducted to determine the persistence of $KMnO_4$ in field moist soil. In these tests, uncontaminated background soil from the site was mixed with three different volumes of 4 wt.% $KMnO_4$ by use of a pilot-scale mixing apparatus. The volumes of oxidant solution added were equivalent to 0.05, 0.1, and 0.15 mL/g soil (or 0.6, 1.2, and 1.8 gal/ft^3 or 5, 10, and 15 wt.% oxidant/ soil). Soil samples were collected periodically after oxidant addition for analyses of moisture content, pH, and oxidant concentration. The pH increased slightly. In soils treated with 0.05 and 0.1 mL/g, the $KMnO_4$ concentration decreased by over 98% within 20 min of oxidant addition and 100% after 24 h, indicating rapid consumption by the soil (without contaminants present).

Based on the treatability studies, it was concluded that >90% treatment of TCE and near 100% treatment of DCE could be achieved with an oxidant concentration of 4 wt.% and an oxidant mass loading rate equal to ≥ 16 g $KMnO_4$/kg soil (Gardner et al. 1998).

System Implementation (Site 3)

During field application of the DSM process in spring 1995, various treatment agents were used during mixing of six test cells (Figures 7-17 and 7-18). Of these six test cells, one test cell was mixed with $KMnO_4$ added to a depth of 25-ft bgs in the unsaturated zone, and a second test cell was mixed with $KMnO_4$ added to final depth of 47-ft bgs that was within the regional groundwater table. The contaminated area was mixed by a crane-mounted vertical rotating 8-ft-diameter mixing blade. The oxidant was injected through the hollow shaft and into the subsurface through orifices located on the mixing blades. Air was supplied to the

---

FIGURE 7-18. Plan view of test cell layout and sampling locations (after Cline et al. 1997).

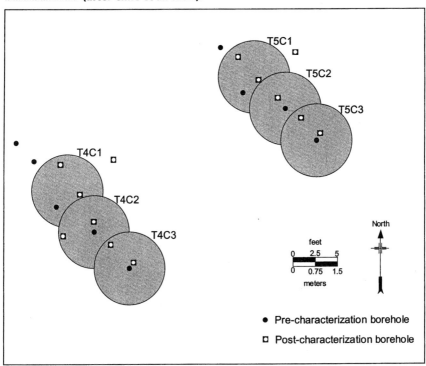

mixed region during the initial pass of the apparatus through the soil column. This pre-mixing step helped break up the tight clay soil to enhance the distribution of the oxidant in the subsurface. The area immediately above the mixed region was covered by a shroud and maintained under negative pressure. Some removal via volatilization was expected; hence, the resulting off-gas was passed through activated carbon canisters prior to atmospheric release. Once the target mixing depth was reached, $KMnO_4$ solution (5 wt.% maximum) was pumped through the Kelly bar and injected into the subsurface from the mixing blades. Several up and down passes were made before moving to the next mixing location. Crystalline, technical grade $KMnO_4$ was used to prepare the oxidant solution on-site.

The reagent volumes and masses injected into each mixed soil column are presented in Table 7-4. During the field test, 6,800 gal of $KMnO_4$ were applied to the three soil columns in the shallow test cell to yield a total $KMnO_4$ mass of 2,450 lb. However, up to 30% of the $KMnO_4$ added to the shallow test cell ponded on the surface due to difficulties delivering such an oxidant volume into the tight clay soil. Hence, the average effective loading rate for the shallow test cell may be as low as 4.9 g $KMnO_4$/kg soil. To avoid excessive reagent ponding, the oxidant loadings were reduced for the columns mixed to the 47-ft bgs.

TABLE 7-4.	DSM $KMnO_4$ chemical oxidation operational data for Site 3 (after Cline et al. 1997).				
Treatment column	Mix date	$KMnO_4$ added (gallons)	# of Passes	Conc. (wt %)[1]	Effective loading (g $KMnO_4$/ kg soil)[2]
T5C1	7/13/96	1950	3	3.7	4.13
T5C2	7/13/96	1125	4	4.2 (Assumed)	6.95
T5C3	7/12/96	3700	5	4.7	9.96
T4C1	7/15/96	3850	3	3.1	3.64
T4C2	7/16/96	3088	4	3.4	6.08
T4C3	7/15/96	4000	4	4.9	5.97

[1]Measured using spectrophotometry, unless otherwise noted.
[2]Effective loading calculations for the C2 overlap columns takes into account contributions from the C1 and C3 columns which overlapped the middle column by 30% each.

An oxidant volume of 10,950 gal of $KMnO_4$ (3,510 lb) was delivered to the three columns in the deep test cell. This reagent volume and mass resulted in an effective in situ oxidant loading rate of 5.3 g $KMnO_4$/kg soil. Significant volumes of oxidant were not observed on the ground surface following the treatment of the deep test cells.

Since sampling and monitoring of the subsurface is difficult during mixing operations, evaluation of the effectiveness of the process was based primarily on the collection and analysis of both pre- and post-treatment soil borings. The collected soil borings were 1.5 inches in diameter and were collected along the entire depth of the treatment region; the borings were subsampled at discrete depth intervals. Sampling locations are presented in Figure 7-18. For baseline characterization, soil samples also were collected from a background soil boring located approximately 150 ft west of test cells.

Soil analyses were performed for TCE and $KMnO_4$ concentrations, soil moisture, microbial populations, total and inorganic carbon determinations, and manganese oxide content. Samples for TCE analysis were extracted with hexane immediately upon collection and analyzed on-site.

Operation and Performance Results (Site 3)

Initially, a large amount of TCE appeared to be located in the 15- to 20-ft and 3- to 35-ft bgs intervals. A typical plot of pre-and post-treatment TCE concentrations is presented in Figure 7-19. Comparison of all pre- and post-treatment borings indicate a total TCE mass treatment efficiency of 70%. However, it is recognized that quantitative assessment of VOC destruction from large test regions involving DSM is often difficult because data interpretation is based on a very limited number of pre- and post-treatment soil samples. For example, in regions with low initial contaminant concentrations, soil analyses indicate that TCE increases as a result of interactions/mixing with the adjacent, overlapping columns having greater initial TCE concentrations. The range of pre- and post-treatment TCE concentrations in soil core samples obtained for the 25-ft deep test cell are presented in Figure 7-20 (note log scale). In addition to the magnitude of TCE removed, the effect of mixing/homogenization also is illustrated.

The concentration of $KMnO_4$ was determined immediately after sample collection via color spectrophotometry. Residual $KMnO_4$ was not detected in the post-treatment soils samples 3-5 days following oxidant

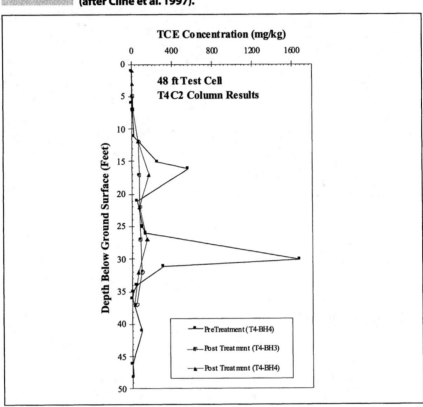

FIGURE 7-19. Typical pre- and post-treatment soil TCE distribution with depth (after Cline et al. 1997).

injection. That was expected based on treatability study data indicating nearly complete oxidant degradation after 24 h.

Analyses for manganese oxides (ammonium oxalate extraction) in the background and post-treatment samples were also performed. Those soil borings indicated average Mn soil concentration within the treatment zone approximately 3 to 4 times higher than background concentrations.

Other soil analyses indicated that (1) soil pH did not decline after mixing with $KMnO_4$ but rather increased, (2) the soil TOC within the treated region remained in the same range as that of background soil (3 to 9 g-TOC/kg soil), and (3) soil microbial analyses indicated that aerobic and anaerobic organisms still persisted in the soil following $KMnO_4$ addition.

For the demonstration, the soil mixing unit cost ranged from \$130 to \$200/c.y., including equipment and labor costs of \$43/c.y. and material

FIGURE 7-20. Typical plot of soil TCE distribution (a) before and (b) after soil mixing/oxidation (after Cline et al. 1997).

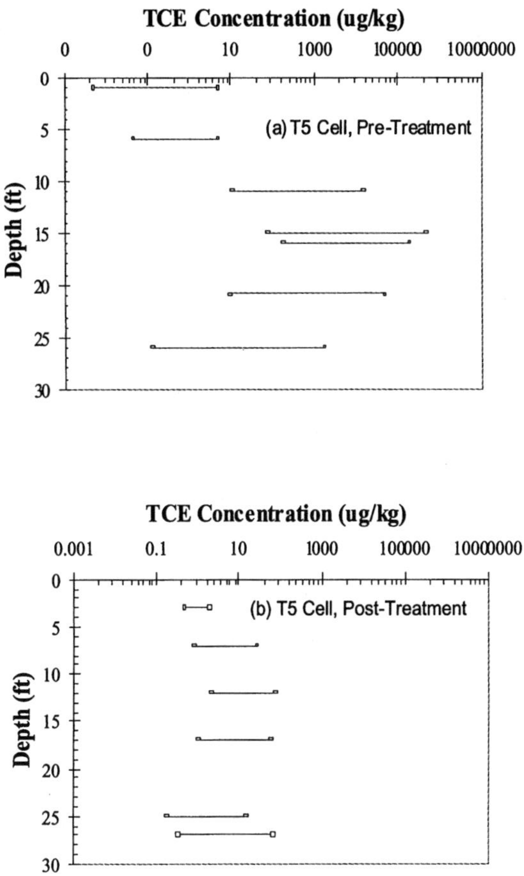

TCE Concentration (ug/kg)

(a) T5 Cell, Pre-Treatment

(a) Pre-treatment TCE levels.

TCE Concentration (ug/kg)

(b) T5 Cell, Post-Treatment

(b) Post-treatment TCE levels.

costs of $46/c.y. (for a 5% slurry of $KMnO_4$). The majority of the costs are for equipment operation, which can be as high as ~$20K/day. Based on this demonstration, a treatment rate of ~50 to 100 m^3/d is expected.

Conclusions (Site 3)

During the field application of ISCO using $KMnO_4$ and DSM technology, 500 tons of soil were treated with nearly 18,000 gal of $KMnO_4$

(6,000 lb). While performance assessment of in-situ treatment applications are difficult, comparisons of pre-and post-treatment soil properties with depth suggest that the DSM process effectively mixed and homogenized the treatment region and achieved the goal to remove at least 70% of the initial TCE from the treatment zone. The selection of a reagent delivery system to be used is closely tied to site conditions/ parameters beyond initial contaminant concentrations. Hence, site characterization is an important step in the implementation of a chemical treatment process.

7.5. MULTIPOINT INJECTION / PERMEATION OF POTASSIUM PERMANGANATE (SITE 4)

Introduction

Vertical injection probes can be used to inject treatment agents such as chemical oxidants, reductants, or bionutrients to accomplish in situ treatment of the vadose and saturated zones (Siegrist et al. 1996, 2000, Urynowicz and Siegrist 2000). That approach is conceivable to depths of 50 m or more, particularly if small hot spots are to be treated. However, practically, it is constrained to shallower depths if the area to be treated is larger (e.g., hectares). A field trial was conducted at the DOE Portsmouth Gaseous Diffusion Plant to demonstrate permeation dispersal of reactive fluids in a shallow, silty clay vadose zone. The test was conducted by ORNL in collaboration with LMES and the environmental restoration program at the DOE Portsmouth site. Hayward Baker Environmental, Inc. (HBE) provided the multipoint injection system used during the demonstration. Supporting experimentation and modeling were conducted at ORNL and CSM.

Site Description and Treatability Testing (Site 4)

At the DOE Portsmouth Plant, in situ remediation technologies were being evaluated to determine their viability for full-scale application at the Plant (Siegrist et al. 1995, Siegrist et al. 1999). As part of this program of evaluation, a test area was established in an uncontaminated but representative location near several contaminated land treatment units (Figures 7-21, 7-22). The use of a clean test site was deemed appropriate to permit controlled hydrodynamic and biogeochemical studies of subsurface manipulation methods without the health and safety, waste management, and environmental concerns associated with a contaminated site.

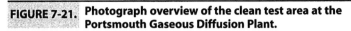

FIGURE 7-21. Photograph overview of the clean test area at the Portsmouth Gaseous Diffusion Plant.

FIGURE 7-22. Site plan of the field test area used for the MPIS trials at the DOE Portsmouth site.

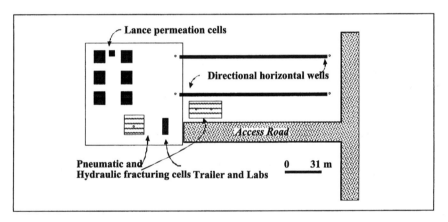

Subsurface features of the test area were representative of several contaminated land treatment sites at the Portsmouth Plant.

The test area was 5 hectares in size and underlain by a silty clay fluvio-
lacustrine deposit from ground surface to 8-m depth (Minford member).
The bulk permeability of this silty clay deposit was ~10^{-7} to 10^{-5}
cm s^{-1} although a fine platy structure with cross-cutting fractures provided
preferential transport pathways that permitted vertical migration of
contaminants released during land treatment of wastes. The air-filled
porosity was estimated at 5% v/v with a zone of groundwater saturation
at 3- to 4-m depth. Underlying the Minford is a moderately permeable

sand known as the Gallia member that consists of pebbles and gravel in a fine-grained silty-sand matrix that is 1 to 2.5-m thick (see also Figure 7-2). The water table is located within the Minford member at about 4-m bgs. Groundwater flows vertically through the Minford to the Gallia member, where flow is predominantly horizontal.

The goal of field testing activities was to explore the potential for rapid but relatively non-disruptive injection and dispersal of treatment fluids into and throughout the shallow subsurface to achieve in situ treatment of organic and metal contaminants by physiochemical or biological processes. If successful, this permeation technology could enable rapid and extensive treatment in source areas and plumes associated with land treatment sites, impoundments, and tank and transfer line areas. A multipoint injection system (MPIS) developed by HBE was chosen as the best available technology based on its relative simplicity and low cost, as well as its potential for effective performance (Figure 7-23). As described below, treatability studies were completed in parallel with the field application of multipoint injection and permeation dispersal of treatment agents.

FIGURE 7-23. **Photograph of the multipoint injection system in operation during field trials at the DOE Portsmouth Plant.**

System Implementation (Site 4)

The field trial included injections of different agents into seven un-confined test cells (six cells were 60 m^2 and 4-m deep; one was 3 m^2 and 4-m deep) (Figure 7-22, Table 7-5) (Siegrist et al. 1999). The following agents were selected to achieve the stated purposes: (1) tracers to evaluate uniformity of delivery, (2) an alkaline slurry to increase soil pH and immobilize metals, (3) $KMnO_4$ and H_2O_2 to oxidize organics, (4) zero valent iron metal to reductively degrade chlorinated solvents, (5) bionutrients to stimulate indigenous microbes and enable bioremediation of organics, and (6) compressed air to increase pneumatic permeability and facilitate soil vapor extraction. The agents were injected at a volu-metric loading rate of in the range of 5% by volume and at pressures of 30–75 psig by use of a four-lance, vertical injector system (Figure 7-23). Injections were made on a 60-cm horizontal spacing to 3.2-m depth. Each test cell was injected with either (1) a mixture of solute and colloidal trac-ers (i.e., bromide and ice nucleating bacteria), (2) H_2O_2 (~10% by wt.), (3) $KMnO_4$ (~4% by wt.), (4) calcium hydroxide (~12% by wt.), (5) bionutrients, (6) colloidal iron (~10% by wt.), or (7) compressed air (~17 m^3 min^{-1}). The operation of the injection system was monitored, and the effects of the injected reagents on subsurface biogeochemical charac-teristics were monitored through analyses of pre- and post-treatment soil cores and solution samples. The injections were completed during fall 1994, and monitoring of test cell characteristics was completed for several years thereafter.

Monitoring devices included vadose zone sensors for temperature and water content, suction lysimeters for soil pore water composition, and observation wells for monitoring water quality in the underlying saturated zone beneath each cell (Figure 7-24). Soil core samples were collected before and after and analyzed for pH, Eh, temperature, water content, organic carbon content, and elemental composition. (Figure 7-25). Micro-bial analyses also were conducted in the bionutrient cell and tracer cell.

To evaluate the permeation dispersal of $KMnO_4$ and H_2O_2 to destroy TCE in the silty clay soil, intact soil cores recovered from the test area were used for laboratory experiments conducted at CSM (Urynowicz and Siegrist 2000). The soil cores were collected by use of intact core pedestal techniques to preserve preferential pathways vital to transport processes. They were transported to the pilot laboratory at CSM where each was setup with a tension plate and porous polyethylene permeation tube, instrumented with soil moisture tensiometers, Eh electrodes, and an Eh

TABLE 7-5. Summary of injection system operations for each test cell at Site 4.

Characteristic	Tracer (T1)	Lime (T4)	Peroxide (T2)	KMnO$_4$ (T5)	Biotreat (T3)	Air (T6)	Iron (T7)
Cell area (ft^2)	24 x 24	24 x 24	24 x 24	24 x 24	24 x 24	24 x 24	8 x 8
Cell depth (ft)	10.4	10.4	10.4	10.4	10.4	10.4	10.4
Injection date	Nov. 12	Nov. 13	Nov. 15	Nov. 16	Nov. 18	Nov. 19	Nov. 19
Injection start time	13:54	11:21	09:50	13:50	13:35		
Injection end time	17:46	16:04	13:09	17:35	17:39		
Injection time (min)	232	283	199	225	245		
Injection vol. (gal)	2870	2330	2420	2260	2220	~5520 cf	310
Loading rate (v/v)	0.064	0.052	0.054	0.050	0.050	0.90	0.47
Injection passes	1	1	1	1	1	1	1
Injector spacing (ft.)	2	2	2	2	2	12	2
Injector setups	36	36	36	36	36	4	4
Time/setup (min)	4-5	4-5	4-5	4-5	4-5	1	4-5
Interval volume (gal)	2.5	2.0	2.0	2.0	2.0	N/A	
Setup volume (gal)	79	64	68	62	64	1380 cf	77
Injector Q (gpm)	4-5	4	4	4	8	~10 cfs	
Injector P (psig)	40-100	40-100	40-100	40-100	40-100	100-120	40-100
Batch size (gal)	6 @ 500	5 @ 500	Bulk tank	5 @ 500	2 @ 1000 1 @ 500	Air comp.	1 @ 500
Batch makeup:							
Tapwater (gal)	500	500	—	500	—	—	500
KBr (kg)	0.72	—	—	—	0.42	—	—
Snomax (kg)	0.27	—	—	—	0.22	—	—
Lime (lb)	—	500	—	—	—	—	—
H$_2$O$_2$ (~10 wt.%; gal)	—	—	Bulk	—	—	—	—
KMnO$_4$ (~4 wt.% ; kg)	—	—	—	~100	—	—	—
Biotreat (~5 wt.%; gal)	—	—	—	—	500	—	—
Air (600 cfm)	—	—	—	—	—	Air comp.	—
Iron (5 um; lb)	—	—	—	—	—	—	400
Guar gum (lb)	—	—	—	—	—	—	25

FIGURE 7-24. **Monitoring vadose zone sensors and lysimeters in the zero valent iron cell.**

FIGURE 7-25. **Intact soil core collected from the permanganate cell illustrating presence of oxidant (Eh along the entire core length was elevated above 800 mv).**

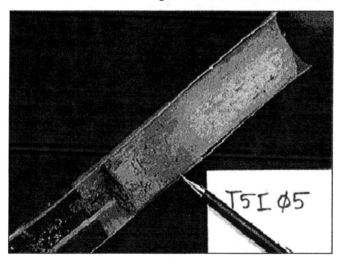

reference electrode, and encased with an ethyl vinyl acetate/paraffin wax thermoplastic (see Figures 7-26 and 7-27).

The laboratory study involving intact cores had four distinct phases: (1) bromide tracer study to determine the hydrologic properties of the

FIGURE 7-26. Intact core monolith prepared for collection from the shallow vadose zone by perimeter carving and wax encapsulation.

FIGURE 7-27. Intact soil monoliths established in the laboratory at CSM for in situ chemical oxidation studies using $KMnO_4$ or H_2O_2 delivered by injection probes.

media; (2) contamination of each core with TCE dyed red with 1 ppm Sudan IV (a nonvolatile dye that does not visibly partition from TCE to the water phase); (3) lance injection of cores M1 and M2 with approximately one air-filled pore volume (~1000 mL) of 1.0 wt.% $KMnO_4$ and 3.0 wt.% H_2O_2 solutions, respectively (note that M3 remained untreated for the purpose of experimental control); and (4) core dismantling and sampling to compare treatment efficiencies and evaluate the media specific geochemical effects of each reagent.

Operation and Performance Results (Site 4)

During the field trial, all reagents were successfully delivered with commercially available equipment to a total depth of approximately 4 m bgs. The effects within the vadose zone varied with the different agents, because of differences in their reactivity as well as their amenability to permeation away from the point of injection from the lance. An illustration of this is shown in Figure 7-28, which depicts the redox potential

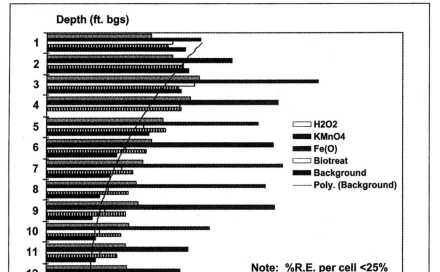

FIGURE 7-28. Redox potential changes within two months of injection probe delivery of different treatment agents in a silty clay vadose zone.

with depth in each cell. The effects of the oxidants are clear, with permanganate having a greater effect than the peroxide. The H_2O_2 reacted rapidly and degraded within a few days to 2 weeks. The $KMnO_4$ appeared to react more mildly, permeate further, and persist for several months.

During the field-scale tests, soil sample TOC concentrations showed a significant decrease within the first 2 ft of ground surface as a result of hydrogen peroxide injection (~2.5 g H_2O_2/kg soil); however, little or no change was observed at greater depths. Peroxide injection also increased dissolved oxygen (DO) concentrations in lysimeter water samples. Post-injection field test soil sample results for $KMnO_4$ indicated a dramatic increase in Eh throughout the entire soil profile (~ 1.3g $KMnO_4$/kg soil). Lysimeter water sample results showed a slight pH decrease and a significant increase in manganese concentrations. Manganese concentrations remained elevated for approximately five weeks. $Ca(OH)_2$ injection effectively elevated the soil and groundwater pH to over 10, and this was sustained for > 12 months.

During the post-experimental sampling and characterization phase of the laboratory study, manganese dioxide solids (MnO_2) deposition was observed on soil surfaces adjacent to major preferential pathways such as the top of the monolith, the injector borehole, and large fractures. Permeation dispersal of H_2O_2 within M2 appeared to be more limited. Visual evidence of soil pore alteration was only observed on soil surfaces adjacent to the top of the core and injector borehole. Migration of H_2O_2 along large fractures does not appear to have occurred. Extreme TCE and Cl⁻ residual variability was observed in each core. Based on the sample volume (~1 mL) and frequency (168 discrete samples and 21 composite samples), it appears that core heterogeneity was on the order of millimeters or less. Although approximately 2,628 mg of TCE was injected into the intact cores M1 and M2 during the contamination phase, the post treatment mass balances accounted for less than 20 percent of the total TCE. It appears that core heterogeneity led to a biased low measure of TCE and Cl⁻ residual. Treatment efficiencies were determined based on monolith effluent and residual TCE and Cl⁻ concentrations, assuming 70% of the TCE degraded by $KMnO_4$ and H_2O_2 was converted to chloride ion. The treatment efficiencies were between 68 to 84% and 61 to 76% for M1 ($KMnO_4$) and M2 (H_2O_2), respectively. The treatment efficiency range for each monolith was the result of discrete and composite sample data discrepancies.

Conclusions (Site 4)

The results of the field trial and laboratory core studies indicated that volumes of different treatment agents could be introduced into a relatively low permeability deposit and could dramatically impact subsurface properties (e.g., raising Eh to >800 mV or elevating pH to >10). While pronounced effects were observed in the test cells, there was substantial short-range spatial variability in the effects on biogeochemical properties. That variability is believed to be due to the injected agents following existing preferential pathways in the subsurface and then, depending on fluid reactivity, their permeation to varying degrees into the finer pore and matrix structure. The nature and extent of the biogeochemical changes observed suggested that permeation dispersal could facilitate the transformation and degradation/immobilization of many contaminants despite their presence in low permeability media. However, questions remain regarding the uniformity of distribution that can be achieved as a function of subsurface characteristics, fluid properties, and dispersal system operation; the mobility and risk of any untreated contaminants; and the risk-reduction requirements for uniformity of dispersal and effect in order to ensure adequate in situ treatment effectiveness.

7.6. HYDRAULIC FRACTURING WITH POTASSIUM PERMANGANATE OPM (SITE 5)

Introduction

A field trial was completed to evaluate in situ remediation by use of hydraulic fracturing to emplace zero valent iron metal and potassium permanganate solids (i.e., OPM) in a silty clay vadose zone to chemically treat TCE (Siegrist et al. 1999) (Figure 7-29). The work was performed by ORNL and FRx, Inc., in collaboration with LMES (Murdoch et al. 1997a,b, Siegrist et al. 1999). Supporting experimentation and modeling were conducted at ORNL and CSM (Case 1997, Struse 1999).

Site Description and Treatability Testing (Site 5)

The field trial was conducted at the X-231A land treatment unit located at the DOE Portsmouth Gaseous Diffusion Plant (Table 7-6, Figure 7-30). The 2.2 hectare site was used for disposal of waste oils and solvents during the 1970s. Its use was terminated around 1980, when it was capped with a temporary geomembrane. The X-231A site was located adjacent to

FIGURE 7-29. Chemical treatment zones installed using hydraulic fracturing to deliver KMnO₄ OPM solids into silty clay soils.

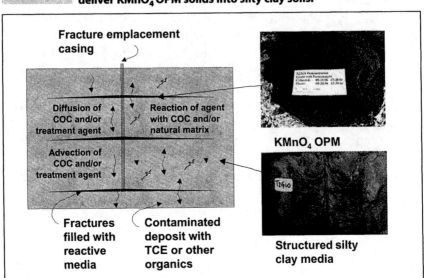

another similar site (X-231B) where in situ soil mixing and mixed region vapor stripping had been demonstrated earlier (Siegrist et al. 1995). Comprised of unconsolidated Quaternary-age deposits, the X-231A site is underlain by 6 to 8 m of low-permeability clays and silts ($K_{sat} < 10^{-6}$ cm/s) known as the Minford member (see also Figure 7-2).

The Minford has a naturally occurring platy structure and vertically dipping fractures that provide pathways for vertical migration of contaminants released during land treatment of wastes. Underlying the Minford is a moderately permeable sand known as the Gallia member, which consists of pebbles and gravel in a fine-grained silty-sand matrix that is 1 to 2.5-m thick. Previous characterization activities at the site revealed total VOCs in soil (dominantly TCE) ranging to as much as 300 mg/kg, whereas groundwater at the site contained variable TCE concentrations including DNAPL globules.

Treatability studies were conducted in the laboratory during development of the permanganate OPM to determine its properties and oxidative degradation potential (Case 1997). That work showed that the OPM had an equal or higher degradation efficiency when delivered at a mass loading rate equivalent to that of a KMnO₄ solution. Before the soil fracturing

TABLE 7-6.	Summary of the test site characteristics at Site 5.	

Characteristic	Units	Conditions
Soil type and genesis:		Minford silty clay deposit of fluvio- lacustrine origin. Typically with a 4.6 m thick upper clay unit (CH) transitioning to a lower 3.0 m thick silt unit (CL).
Soil particle size distribution: - Sand (0.050 - 2.000 mm) - Silt (0.002 - 0.050 mm) - Clay (<0.002 mm)	 dry wt.% dry wt.% dry wt.%	 ~05 ~60 to 85 ~10 to 35
Soil mineralogy:		In the Minford, the sand fraction consists of mainly quartz with minor geothite. The silt fraction consists of quartz and minor feldspars but no geothite. The clay fraction is a mixture of illite (~33%), quartz (~29%), kaolinite (~26%), and smectite (~12%).
Soil properties:		
- Bulk density	g/cm^3	1.8
- Water content	wet wt.%	20%
- Liquid limit, plastic index	%	~60 and 35
- Total fractional porosity	v/v	0.40
- Water-filled porosity	% pores	90
- pH (in water)	-	6.0
- Organic carbon	mg/kg	500 to 1500
- Iron oxides - free	mg/kg	23000
- Iron oxides - amorphous	mg/kg	1350
- Cation exchange capacity	meq/100g	17.5
- Total bacteria	org./g	100 to 10000

was implemented at the X-231A site, a test cell comprised of three stacked fractures was installed at the PORTS clean test site. The test fractures were installed to verify that the subsurface formation at PORTS was conducive to hydraulic fracturing. The test fractures were installed in a fashion similar to that to be implemented at the X-231A site, but sand was used as the proppant rather than the OPM. Laboratory analyses and treatability tests also were conducted in parallel with monitoring of the X-231A field site over more than 2 years following the initial soil fracturing and delivery of oxidative particles into the subsurface.

FIGURE 7-30. **Photographs of the X-231A site at the DOE Portsmouth Gaseous Diffusion Plant during in situ treatment of TCE using hydraulic fracturing with permanganate OPM or iron metal.**

(a) Overview of the X-231A land treatment unit.

(b) FRx, Inc. soil fracturing equipment.

(c) Test cell installation with access casings for fracture emplacement.

System Implementation (Site 5)

Hydraulic fracturing was used to emplace zero valent iron metal in one cell and potassium permanganate OPM in a second cell. Each of the test cells was comprised of five horizontal fractures stacked one over the other within a subsurface region roughly 6 m in diameter and 5 m deep (Table 7-7). Sand-filled fractures were propagated at nominal depths of 1.2 and 4.9 m with fractures containing reactive media propagated at depths of 1.8, 2.4, and 3.6 m. The saturated zone was encountered at 3.6-m depth bgs. In one test cell, iron metal in the form of 0.2-mm diameter Fe^0 particles (Master Builder) was suspended in guar gum gel, and this slurry was used to create three iron-filled fractures. In another test cell, an OPM consisting of 0.1 to 0.3 mm diameter $KMnO_4$ particles (Carus Chemical) suspended in a mineral-based gel was used to create three permanganate-filled fractures. In each test cell, the shallowest fractures were created first, followed by successively deeper ones. For each fracture, steel casing was driven to depth and used to inject the reactive slurry; the casing was left in place for future access following procedures similar to those described in U.S. EPA (1993) and Murdoch et al. (1995, 1997a,b) (Figure 7-29). Forms of the fractures were estimated by measuring ground surface displacement during fracturing as well as via direct observation of intact cores collected from 8 boreholes made after fracture emplacement.

TABLE 7-7.	Test cell installation features at Site 5.	

Test cell characteristic	Iron-filled fractures for dechlorination	Permanganate-filled fractures for oxidation
Method and time of installation	Iron metal and guar gel; 2 to 3 hr	$KMnO_4$ OPM; 2 to 3 hr
Fracture depth -proppant - amount	1.2 m - Sand - 0.14 m³ 1.8 m - Fe^0 - 1000 kg 2.4 m - Fe^0 - 1300 kg 3.6 m - Fe^0 - 2600 kg 5.0 m - Sand - 0.57 m³	1.2 m - Sand - 0.14 m³ 1.8 m - $KMnO_4$ - 400 kg 2.4 m - $KMnO_4$ - 600 kg 3.6 m - $KMnO_4$ - 600 kg 5.0 m - Sand - 0.57 m³
Test cell size	6 m diam. by 5 m deep	6 m diam. by 5 m deep

The test cells were established during September 1996 and left in a passive mode of operation for the next 15 months. Since there was a geomembrane over the site, there was limited infiltration, although minor amounts of moisture did enter through perforations made in the cover for instrumentation and soil coring. Fracture emplacement was monitored and soil and groundwater conditions were characterized. After 3, 10, and 15 months of emplacement, continuous cores were collected, and morphologic and geochemical data were taken across the fracture zones. Controlled degradation tests were completed using site groundwater with TCE concentrations near 53, 144, and 480 mg/L, equivalent to 0.5, 1.2, and 4.1 g TCE per kg media, respectively. Laboratory and modeling studies were focused on permanganate oxidation chemistry and delivery/transport processes since a body of research was already available and/or in progress regarding Fe^0 metal reduction. Laboratory studies included (1) batch tests to define permanganate oxidation kinetics, (2) development and testing of a permanganate OPM, and (3) intact core studies to quantify permanganate diffusive transport and matrix interactions.

Operation and Performance Results (Site 5)

In general, the fractures were flat-lying around the point of initiation but gradually climb upwards to form a gently bowl-shaped form, in some cases interacting with overlying fractures. The iron-filled fractures formed a discrete reactive seam less than 1 cm thick, wherein the Eh decreased and reductive dechlorination could occur, but effects in the adjacent silty clay soils were negligible (Figures 7-31a and 7-32). While the emplaced iron exhibited some surface corrosion after extended emplacement in the subsurface, the reactivity was unaffected. Iron from the fractures degraded TCE at efficiencies of as much as 36% after 24 to 48 hr of contact, which is consistent with Fe^0 packed bed degradation half-lives of 1 to 2 hr.

The permanganate-filled fractures yielded a diffuse reactive zone that expanded over time, reaching 40 cm in thickness after 10 months (Figures 7-31b and 7-33). Throughout this oxidizing zone, the degradation efficiency was >99% after 2 hr of contact for dissolved TCE at 0.5 and 1.2 mg TCE per g of media. When exposed to higher TCE loadings (i.e., 4.1 mg per g), degradation efficiencies after 10 months dropped to 70% as the TCE load exceeded the oxidant capacity remaining. These efficiencies and rates are consistent with oxidation stoichiometry and previously determined half-lives of <2 min. for permanganate oxidation of TCE. In both

FIGURE 7-31. Photographs of intact cores from the locations of hydraulic fractures filled with permanganate OPM or Fe⁰ metal collected 10 months following initial emplacement.

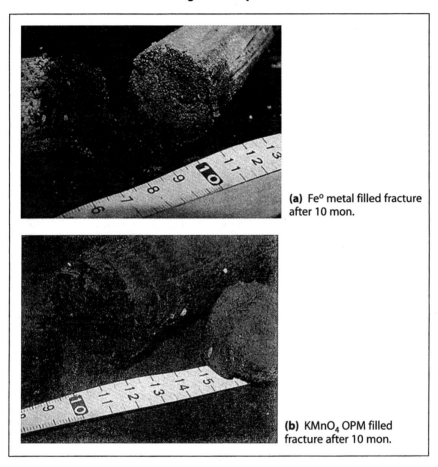

(a) Fe⁰ metal filled fracture after 10 mon.

(b) KMnO₄ OPM filled fracture after 10 mon.

test cells, there were no marked effects on the chemistry or contamination levels in the groundwater beneath the cells (Siegrist et al. 1999).

Results of laboratory work revealed that the OPM had a TCE degradation rate that was equal to or greater than that of permanganate alone. The rate of oxidative destruction of TCE was second-order with respect to TCE and MnO_4^- yielding a reaction rate constant in the range of 0.6 to 0.9 L mol⁻¹s⁻¹ (at 20C without appreciable NOM). Lower temperatures reduced the rate of organic chemical destruction as will the presence of other oxidant-demanding substances in the system (e.g., NOM). Diffusive

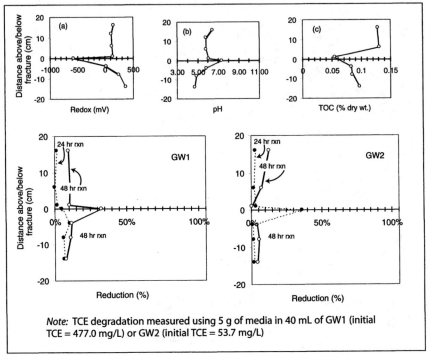

FIGURE 7-32. Geochemical properties and TCE degradation potential of Fe⁰ metal zones 10 mon. after emplacement (after Siegrist et al. 1999).

Note: TCE degradation measured using 5 g of media in 40 mL of GW1 (initial TCE = 477.0 mg/L) or GW2 (initial TCE = 53.7 mg/L)

transport of potassium permanganate from a 5 g-$KMnO_4$/L source zone through the silty clay soil was studied in the laboratory. In uncontaminated cores, diffusing permanganate did oxidize some, but not all of the NOM in the soil (~10 to 30% destruction), and did yield MnO_2 deposits but they did not alter system tortuosity. In contaminated cores (~100 to 1000 mg/kg TCE), permanganate transport was retarded due to reaction with the TCE. Oxidation of residual TCE in the silty clay soil was almost complete, but tortuosity appeared to be increased, possibly due to more focused and intense MnO_4^- oxidation of residual TCE. With a MnO_4^- concentration of 40 g/L in a fracture upon emplacement, the enveloping reactive zone was modeled and predicted to have extended to 60 cm in total width after 6 months of emplacement (Struse 1999). These experimental and modeling results compared well with the field observations made during the field trial: at 24 months after emplacement, the oxidizing

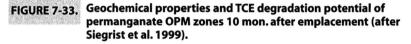

FIGURE 7-33. Geochemical properties and TCE degradation potential of permanganate OPM zones 10 mon. after emplacement (after Siegrist et al. 1999).

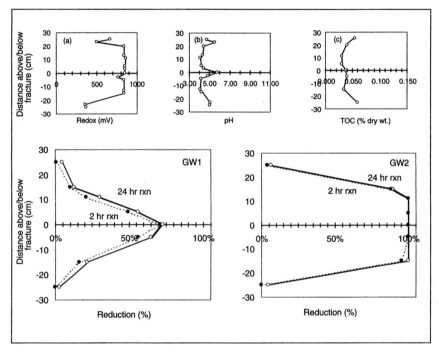

zone was continuous between fractures, but the degradation efficiency continued to decline.

The total hydraulic fracturing demonstration cost was approximately $1,259K, which included all test cells (steam injection, hot air injection, zero valent iron, and OPM). Of the total demonstration costs: equipment shakedown at the clean test site accounted for approximately 4%, materials (iron, permanganate, well materials, sampling supplies, etc.) accounted for approximately 9%, comparative demonstration operations and monitoring (excluding the vendor subcontract) accounted for approximately 23%, and the vendor subcontract accounted for approximately 27%. The typical costs of creating hydraulic fractures involve labor, materials, equipment usage, and mobilization expenses. The cost per fracture will decrease as the size of the job increases, largely because mobilization costs are distributed over several fractures and monitoring efforts can be shared. The cost generally ranges from $850 to $1500 per fracture using current methods for sand-propped fractures. Estimated unit costs for

OPM reactive barriers is ~$28/c.y. of soil for installation of OPM propped fractures (cost of permanganate, ~$1.60/lb., compared to sand, ($.10/lb.) and approximately $275 per day operational costs for resource consumption (e.g., power), sampling and analysis (e.g., no off-gas), and manpower requirements. Based on this demonstration, a treatment rate of ~300 m^3/d is expected.

Conclusions (Site 5)

Hydraulic fracturing equipment and methods were used to create reactive zones of Fe^0 metal or $KMnO_4$ OPM in horizontally oriented layers within a silty clay vadose zone at depths up to 5 m. The Fe^0-filled fractures produced a reactive seam with limited effect on the surrounding LPM, while the $KMnO_4$-filled fractures yielded a broad zone of reactivity within the LPM. With both types of fracture zones, degradation potential for high levels of TCE was sustained even after 10 months of emplacement in the subsurface. Both types of horizontal treatment zones may reduce risks associated with exposure to TCE from a contaminated site. Although the system that uses iron-filled fractures may leave immobile contaminants in the ground untreated, data from this study suggest that it is capable of degrading mobile TCE and thus may reduce risk by effectively eliminating TCE release from a low permeability unit to the atmosphere or an underlying aquifer. The system that uses permanganate-filled fractures, where MnO_4^- ions are diffusively distributed through a broad region, offers the possibility to both curtail TCE release to the atmosphere or an underlying aquifer and destroy TCE throughout a low permeability formation. However, diffusive transport is slow, and the rate and extent are highly dependent on the physical and chemical properties of the formation.

The feasibility of in situ remediation of TCE and other organic compounds in low permeability vadose zones was demonstrated at the X-231A site. However, general application of this technology requires consideration of the horizontal continuity, degradation capacity, and longevity of the treatment agents (Siegrist et al., 1999).

7.7. SUMMARY

ISCO is rapidly emerging as a viable remediation technology for mass reduction in source areas as well as for plume treatment. The oxidants most commonly employed to date include peroxide, permanganate, and ozone systems, with subsurface delivery to groundwater by vertical wells, horizontal wells, or sparge points and to soil by lance injectors or hydraulic fracturing. The potential benefits of in situ oxidation include the rapid and extensive reactions with various COCs, applicability to many biorecaciltrant organics and subsurface environments, ability to tailor treatment to a site from locally available components and resources, and facilitation of property transfers and Brownfields development projects. There are some potential limitations, including: potential need for large quantities of reactive chemicals to be introduced due to the oxidant demand of the target organics and the unproductive oxidant consumption of the formation; resistance of some COCs to chemical oxidation; and potential for process-induced detrimental effects, including gas evolution, permeability loss, and mobilization of redox sensitive and exchangeable sorbed metals. Full-scale deployment is accelerating, but care must be taken to avoid poor performance and unforeseen adverse effects. Matching the *oxidant and delivery system* to the *COCs and site conditions* is the key to achieving performance goals. More development work is proceeding in many areas.

7.8. REFERENCES

Advanced Infrastructure Management Technologies (AIMTech) (2001). Manganese in the Environment Evaluation. Department of Energy Report, DOE/OR/11-3068&D1, Grand Junction, CO.

Case, T.L. (1997). Reactive Permanganate Grouts for Horizontal Permeable Barriers and In situ Treatment of Groundwater. M.S. Thesis, Colorado School of Mines, Golden, CO.

Cline, S.R., O.R. West, N.E. Korte, F.G. Gardner, R.L. Siegrist, and J.L. Baker (1997). $KMnO_4$ Chemical Oxidation and Deep Soil Mixing for Soil Treatment. *Geotechnical News*. December. pp. 25-28.

DOE (1996a). In Situ Enhanced Soil Mixing. Innovative Technology Status Report. DOE Office of Science & Technology. Washington, D.C. 25 pp.

DOE (1996b). In Situ Remediation of DNAPL Compounds in Low Permeability Media: Transport/Fate, In Situ Control Technologies, and Risk Reduction. Siegrist, R.L. and K.S. Lowe (ed.). Oak Ridge National Laboratory Report, ORNL/TM-13305. August, 1996.

Environmental Security Technology Certification Program (1999). Technology Status Review: In situ Oxidation. http://www.estcp.gov.

Gardner, F.G., N.E. Korte, J. Strong-Gunderson, R.L. Siegrist, O.R. West, S.R. Cline, and J. Baker (1998). Implementation of Deep Soil Mixing at the Kansas City Plant. Final project report by Oak Ridge National Laboratory for the Environmental Restoration Program at the DOE Kansas City Plant. ORNL/TM-13532.

Gates, D.D., R.L. Siegrist, and S.R. Cline (1995). Chemical Oxidation of Contaminants in Clay or Sandy Soil. Proceedings of ASCE National Conference on Environmental Engineering. Am. Soc. of Civil Eng., Pittsburgh, PA.

Lowe, K.S., F.G. Gardner, R.L. Siegrist, and T.C Houk (1998). In situ Chemical Oxidation Recirculation Pilot Test at the 5-Unit Investigative Area Using Vertical Wells. DOE/OR/11-3010&D1. Prepared by Oak Ridge National Laboratory for Bechtel Jacobs Company, LLC, DOE Portsmouth Gaseous Diffusion Plant, Piketon, OH. October 1998.

Lowe, K.S., F.G. Gardner, R.L. Siegrist, and T.C. Houk (2000). Field pilot test of in situ chemical oxidation through recirculation using vertical wells at the Portsmouth Gaseous Diffusion Plant. EPA/625/R-99/012. U.S. EPA Office of Research and Development, Washington, D.C. 20460. pp. 42-49.

Murdoch, L.C. (1995). Forms of Hydraulic Fractures Created During a Field Test in Fine-Grained Glacial Drift. Quarterly Journal of Engineering Geology. 28:23-35.

Murdoch, L., W. Slack, R. Siegrist, S. Vesper, and T. Meiggs (1997a). Advanced Hydraulic Fracturing Methods to Create In situ Reactive Barriers. Proc. International Containment Technology Conference and Exhibition. February 9-12, 1997, St. Petersburg, FL.

Murdoch, L., B. Slack, B. Siegrist, S. Vesper, and T. Meiggs (1997b). Hydraulic Fracturing Advances. Civil Engineering. May 1997. pp. 10A-12A.

Schnarr, M., C. Truax, G. Farquhar, E. Hood, T. Gonully, and B. Stickney (1998). Laboratory and controlled field experimentation using potassium permanganate to remediate trichloroethylene and perchloroethylene DNAPLs in porous media. *Journal of Contaminant Hydrology*, 29:205-224.

Siegrist, R.L., O.R. West, M.I. Morris, D.A. Pickering, D.W. Greene, C.A. Muhr, D.D. Davenport, and J.S. Gierke (1995). In Situ Mixed Region Vapor Stripping of Low Permeability Media. 2. Full Scale Field Experiments. *Environmental Science & Technology*. 29(9): 2198-2207.

Siegrist, R.L., N.E. Korte, D. Smuin, O.R. West, D.D. Gates, and J.S. Gunderson (1996). In situ Treatment of Contaminants in Low Permeability Soils: Biogeochemical Enhancement by Subsurface Manipulation. Proc. First Int. Conf. on Contaminants in the Soil Environment in the Australasia-Pacific Region. February 1996, Adelaide.

Siegrist, R.L. (1998). "In Situ Chemical Oxidation: Technology Features and Applications." Invited presentation at the Conference on Advances in Innovative Groundwater Remediation Technologies. 15 December 1998. U.S. EPA Technology Innovation Office, Washington, D.C.

Siegrist, R.L., K.S. Lowe, L.C. Murdoch, T.L. Case, and D.A. Pickering (1999). In Situ Oxidation by Fracture Emplaced Reactive Solids. *J. Environmental Engineering*. Vol.125, No.5, pp.429-440.

Siegrist, R.L., M.A. Urynowicz, and O.R. West (2000a). An Overview of In Situ Chemical Oxidation Technology Features and Applications. EPA/625/R-99/012. U.S. EPA Office of Research and Development, Washington, D.C. 20460. pp. 61-69.

Siegrist, R.L., D.R. Smuin, N.E. Korte, D.W. Greene, J. Strong-Gunderson, D.A. Pickering, and K.S. Lowe (2000b). Permeation Dispersal of Treatment Agents for In situ Remediation: 1. Field Studies in Unconfined Test Cells. Final project report for U.S. Department of Energy by Oak Ridge National Laboratory, Oak Ridge, TN. ORNL/TM-13596.

Struse, A.M. (1999). Mass Transport of Potassium Permanganate in Low Permeability Media and Matrix Interactions. M.S. Thesis, Colorado School of Mines, Golden, CO.

Urynowicz, M. and R.L. Siegrist (2000). Permeation Dispersal of Treatment Agents for In situ Remediation: 2. Intact Core Experiments with Chemical Oxidants. Final project report for Oak Ridge National Laboratory by Colorado School of Mines, Golden, CO. Oak Ridge, TN. ORNL/TM-13597.

USEPA (1985). Methods for Measuring the Acute Toxicity of Effluents to Freshwater and Marine Organisms. EPA 600/4-85-013. Environmental Monitoring and Support Laboratory, Cincinnati, OH.

USEPA (1993). Hydraulic Fracturing Technology - Technology Evaluation Report. EPA/540/R-93/505. Office of Research and Development, Cincinnati, OH.

USEPA (1998). In situ remediation technology: In situ chemical oxidation. EPA 542-R-98-008. Office of Solid Waste and Emergency Response. Washington, D.C.

West, O.R., S.R. Cline, W.L. Holden, F.G. Gardner, B.M. Schlosser, J.E. Thate, D.A. Pickering, and T.C. Houk (1998a). A Full-scale field demonstration of in situ chemical oxidation through recirculation at the X-701B site. Oak Ridge National Laboratory Report, ORNL/TM-13556.

West, O.R., Cline, S.R., Siegrist, R.L., Houk, T.C., Holden, W.L., Gardner, F.G. and Schlosser, R.M. (1998b). A Field-scale test of in situ chemical oxidation through recirculation. Proc. Spectrum '98 International Conference on Nuclear and Hazardous Waste Management. Denver, Colorado, Sept. 13-18, pp. 1051-1057.

CHAPTER 8

Status and Future Directions

In situ chemical oxidation (ISCO) is a viable remediation technology for contaminant mass reduction in source areas, as well as for plume control and treatment. The oxidants most commonly employed to date include hydrogen peroxide, ozone, and permanganate. Subsurface delivery to groundwater is implemented by vertical or horizontal wells and sparge points and to soil by injection probes, hydraulic fracturing, and soil mixing. As a site-remediation technology, the goal of ISCO has been to destroy target organic chemicals present in soil and groundwater systems and thereby reduce the mass, mobility, and/or toxicity of contamination. The choice of ISCO over other in situ treatment methods (e.g., bioremediation, surfactant flushing, thermally enhanced recovery) has been motivated by the ability of chemical oxidation (1) to be engineered to accommodate site-specific conditions, (2) to be implemented quickly with commercially available equipment and materials, and (3) to yield measurable results in weeks to a few months. The choice of permanganate over other oxidants such as hydrogen peroxide or ozone often has been due to contaminant and site conditions. For example, compared to other oxidants, permanganate (1) can be effective over a wider pH range, (2) effects oxidation by electron transfer rather than free radical processes and thus has a slower rate of reaction in the subsurface which enables delivery and transport, and (3) has lower gas and heat evolution and a lower risk of fugitive emissions.

As of this writing there are scores of field applications that have been completed, are in progress, or are in design. The number of full-scale applications is growing at an increasing pace. The most common applications to date appear to have been for treatment of chlorinated solvents in soil and groundwater. In these applications, the goals have been to reduce the mass of contaminant in a source area and quickly achieve in situ remediation and site closure, or to enable eventual remediation of a site by longer-term natural attenuation. Applications have also been focused on cleanup of a groundwater plume near a source or down-gradient from a source. The cost of in situ chemical oxidation with permanganate is dependent on site-specific conditions and the remediation performance goals. Remediation costs of ISCO using permanganate normally are competitive with other in situ remediation technologies, and costs are generally not an over-riding factor in technology selection and implementation.

ISCO system selection, design, and implementation using permanganate should rely on an integrated effort based on typical site-characterization data, ISCO-specific laboratory tests, laboratory and field pilot tests, and reaction and transport modeling. As illustrated in Figure 1-8, effective engineering of ISCO technologies must be done carefully with due attention being paid to both reaction chemistry and oxidant delivery. The aim is to achieve a desired performance goal for a target COC without adverse oxidation-induced secondary effects. Matching the permanganate oxidant and delivery system to the COCs and site conditions is critical to achieving performance goals.

During the evaluation, design, and implementation of ISCO, a number of key considerations must be carefully addressed to help ensure an effective application that meets a performance goal without causing adverse secondary effects. Several considerations are common to all of the common chemical oxidants during in situ application, but they have different levels of importance depending on the oxidant and site conditions. Considerations can include: (1) amenability of the target organic COCs to oxidative degradation, (2) determination of the optimal oxidant loading (dose concentration and delivery rate) for a given organic in a given subsurface setting, (3) minimizing unproductive oxidant consumption, (4) managing potential adverse effects (e.g., mobilizing metals, forming toxic byproducts, reducing formation permeability, generating off-gases and heat), (5) regulatory and permitting requirements, and (6) health and safety precautions during chemical handling. In addition, some applications can require explicit consideration of the potential for ISCO to be coupled with other remediation technologies (e.g., bioremediation). For sites with common organics of concern that are present in the dissolved and sorbed phases in relatively permeable groundwater formations with limited heterogeneity, current understanding of ISCO is adequate to support rational design and effective application with limited treatability studies. Understanding is not complete for other more complex sites, such as those with heterogeneous conditions (e.g., low permeability silt and clay zones), fractured bedrock, or in source areas with DNAPLs. For those types of sites, well-conceived and often extensive treatability studies at the laboratory-scale and field-scale are still warranted.

Research continues in academia, national laboratories, and private industry to confirm and extend the current understanding of permanganate oxidation reactions, transport processes, secondary effects, and delivery technologies. The research includes the development of robust remediation system models, the development of advanced methods of implementation including enhancing the oxidant stability during subsurface delivery (e.g., encapsulation), and making the technology of delivery more effective and predictable. In addition, work continues on increasing the understanding of how ISCO using permanganate can be cost-effectively coupled with other treatment processes, including physicochemical methods (e.g., thermally enhanced recovery), biological methods (e.g., in situ bioremediation), and containment (e.g., permeable reactive barriers). The results of this research will enhance the potential of ISCO using permanganate and the ability to achieve predictable, cost-effective applications.

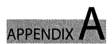

APPENDIX A

Manufacturer's Information on Potassium and Sodium Permanganate

This appendix contains information concerning the general properties and use of permanganate chemicals. The information presented was obtained from Carus Chemical Company, the world's largest manufacturer of potassium permanganate and the materials are reproduced herein with their permission. While the information presented in this appendix was current as of June 2001, the reader is cautioned to contact Carus Chemical Company or other sources as appropriate for current and complete information such as presented herein.

CONTENTS

1. CAIROX technical fact sheet for technical grade potassium permanganate (2 pages)

2. CAIROX technical fact sheet for free-flowing grade potassium permanganate (2 pages)

3. LIQUOX technical fact sheet on liquid sodium permanganate (2 pages)

4. Materials compatibility chart for potassium permanganate (2 pages)

5. Protective equipment, drum disposal and first aid recommendations for CAIROX (2 pages)

6. Material safety data sheet for CAIROX potassium permanganate (6 pages)

7. Material safety data sheet for LIQUOX sodium permanganate (6 pages)

313

CAIROX®
Potassium Permanganate
CAS* Registry No. 7722-64-7

DATA SHEET

Technical Grade is recommended where potassium permanganate is fed as a solution and where particle size is not critical.

TECHNICAL GRADE

Assay
Guaranteed 98% KMnO₄, Typical 98.5%

Particle Size
On request

Standards and Specifications
CAIROX potassium permanganate is certified by the National Sanitation Foundation (NSF) to meet Standard 60: Drinking Water Treatment Chemicals—Health Effects.

Technical Grade meets:
AWWA Standard B603-88
Military Specifications MIL-P-11970-C dated
14 October 1983
Water Chemical Codex RMIC values

CHEMICAL AND PHYSICAL DATA

Formula . . . . KMnO₄

Formula Weight 158.0

Form . . . . Granular Crystalline

Specific Gravity

Solid . . . . 2.703 g/cm³

6% Solution . . . . 1.039 by weight, 20°C/4°C

Bulk Density . . . . Approximately 100 lb/ft³

Decomposition may start at 150°C/302°F

Solubility in Distilled Water

Temperature		Solubility	
°C	°F	g/L	wt oz/gal
0	32	27.8	3.7
20	68	65.0	8.6
40	104	125.2	16.7
60	140	230.0	30.7
70	158	286.4	38.3
75	167	323.5	43.2

Refer to the Solubility Fact Sheet, Form #130, for more information.

*Chemical Abstract Service

Lithographed Full Open Top Metal, with Lever Lock Closure, Single-Trip Drums meet DOT-37A or 37B

25kg pail (55.125 lb) net, with handle, made of 24-gauge steel, weighs 3.8 lb. It is tapered to allow nested storage of empty drums, stands approximately 13½ inches high and has a maximum diameter of 12 inches.

50kg drum (110.250 lb) net, made of 24-gauge steel, weighs 8.25 lb. It is tapered to allow nested storage of empty drums, stands approximately 17¾ inches high and has a maximum diameter of 14¾ inches.

150kg drum (330.750 lb) net, made of 22-gauge steel, weighs 22.4 lb. It stands approximately 29½ inches high and is approximately 19¾ inches in diameter.

Special Packages will be considered on request.

DESCRIPTION

Crystals or granules are dark purple with a metallic sheen, sometimes with a dark bronze-like appearance. Free-Flowing Grade is gray due to an additive. Potassium permanganate has a sweetish, astringent taste and is odorless.

HANDLING

Protect containers against physical damage. When handling potassium permanganate, respirators should be worn to avoid irritation of or damage to mucous membranes. Eye protection should also be worn when handling potassium permanganate as a solid or in solution. Refer to the Protective Equipment and First Aid Recommendations Fact Sheet, Form #171.

STORAGE

Potassium permanganate is stable and will keep indefinitely if stored in a cool, dry area in closed containers. Concrete floors are preferred to wooden decks.
To clean up spills and leaks follow the steps recommended in our MSDS Form #170. Be sure to use goggles, rubber gloves, and respirator when cleaning up a spill or leak.

INCOMPATIBILITY

Avoid contact with acids, peroxides, and all combustible organic or readily oxidizable materials including inorganic oxidizable materials and metal powders. With hydrochloric acid, chlorine gas is liberated.
Potassium permanganate is not combustible but will support combustion. It may decompose if exposed to intense heat. Fires may be controlled and extinguished by using large quantities of water. Refer to the MSDS, Form #170, for more information.

CORROSIVE PROPERTIES

Potassium permanganate is compatible with many metals and synthetic materials. Natural rubbers and fibers are often incompatible. Solution pH and temperature are also important factors. The material must be compatible with either the acid or alkali also being used.

In neutral and alkaline solutions, potassium permanganate is not corrosive to iron, mild steel or stainless steel. However, chloride corrosion of metals may be accelerated when an oxidant such as permanganate is present in solution. Plastics such as polypropylene, polyvinyl chloride Type I (PVC I), epoxy resins, fiberglass reinforced plastic (FRP), Penton, Lucite, Viton A, and Hypalon are suitable. Teflon FEP and TFE, and Tefzel ETFE are best. Refer to Material Compatibility Chart, Form #180.

Aluminum, zinc, copper, lead, and alloys containing these metals may be (slightly) affected by potassium permanganate solutions. Actual studies should be made under the conditions in which permanganate will be used.

SHIPPING

Potassium permanganate is classified by the Hazardous Materials Transportation Board (HMTB) as an oxidizer. It is shipped under Interstate Commerce Commission's (ICC) Tariff 19.

Proper Shipping Name: Potassium Permanganate
(RQ-100/45.4)

Hazard Class: Oxidizer

Identification Number: UN 1490

Label Requirements: Oxidizer and Corrosive

Packaging Requirements: 49 CFR Parts 100 to 199
Sections: 173.152, 173.153, 173.154, 173.194

Shipping Limitations:
Minimum quantities:
Rail car: See Tariff for destination
Truck: No minimum
Postal regulations:
Information applicable to packaging of oxidizers for shipment by the U.S. Postal Service to domestic and foreign destinations is readily available from the local postmaster.
United Parcel Service accepts 25 lb as largest unit quantity properly packaged; consult United Parcel Service.

Regulations concerning shipping and packing should be consulted regularly due to frequent changes.

REPACKING

When potassium permanganate is repacked, the packing, markings, labels, and shipping conditions must meet applicable Federal regulations. See Code of Federal Regulations – 49, Transportation, parts 100-199, and the Federal Hazardous Substances Act, 15 U.S.C. 1261.

APPLICATIONS

These are some of the many applications of potassium permanganate. Permanganate is a powerful oxidizing agent and the condition under which it is to be used should be completely investigated to minimize risks of violent reaction, fire, and explosion.

Oxidation and Synthesis
Organic chemicals and intermediates manufacture. Oxidizes impurities in organic and inorganic chemicals.

Water Treatment
Oxidizes iron, manganese, and hydrogen sulfide; controls taste and odor; and is an alternate pre-oxidant for THM control.

Municipal Wastewater Treatment
Destroys hydrogen sulfide in wastewater and sludge. Improves wastewater treatment.

Industrial Wastewater Treatment
Oxidizes hydrogen sulfide, phenols, iron, manganese, and many other organic and inorganic contaminants; resultant manganese dioxide aids in removing heavy metals.

Metal Surface Treatment
Conditions mill scale and smut to facilitate subsequent removal by acid pickling in wire manufacture and jet engine overhaul.

Equipment Cleaning
Assists in cleaning organic and inorganic residues from refining and cooling towers and other processing equipment.

Purification of Gases
Removes trace impurities of sulfur, arsine, phosphine, silane, borane, and sulfides from carbon dioxide and other industrial gases.

Mining and Metallurgical
Aids in separation of molybdenum from copper; removes impurities from zinc and cadmium; oxidizes flotation compounds.

Radioactive Decontamination
Used with other ingredients to decontaminate nuclear power plants, submarines, and reactors; also to purify heavy water.

Slag Quenching
Controls hydrogen sulfide and acetylene emissions during quenching of hot slag.

Other
Bleaches beeswax and some plasticizers. Increases swelling of sodium bentonite clays. Used in manufacture of printed circuit boards.

 CHEMICAL COMPANY

Division of Carus Corporation
P.O. Box 1500 • Ottawa, Illinois 61350
Telephone: (815) 433-9070 Cable: Carchemco
Telex: 797551

Copyright © 1986, Revised 1989, Carus Chemical Company

Form #100T

CAIROX®
Potassium Permanganate
CAS® Registry No. 7722-64-7

DATA SHEET

Free-Flowing Grade

Free-Flowing Grade is recommended where the potassium permanganate is subjected to high humidity and where the material is to be dry fed, through a chemical feeder or stored in a bin or hopper. This is recommended for use in water or wastewater treatment plants. This is the only grade available in bulk.

FREE-FLOWING GRADE

Assay
Guaranteed 97% KMnO₄, Typical 98%

Particle Size
20% maximum retained on #425 U.S. Standard Sieve
(formerly #40)
7% maximum through #75 U.S. Standard Sieve
(formerly #200)

Standards and Specifications
CAIROX® Potassium Permanganate is certified by the National Sanitation Foundation (NSF) to ANSI/NSF Standard 60: Drinking Water Treatment Chemicals–Health Effects.

Free-Flowing Grade meets:
AWWA Standard B603
West German DIN-19619
Water Chemical Codex RMIC values

CHEMICAL AND PHYSICAL DATA

Formula . . . . KMnO₄

Formula Weight 158.0

Form . . . . Granular Crystalline

Specific Gravity
Solid . . . . 2.703 g/cm³
6% Solution . . . . 1.039 by weight, 20°C/4°C

Bulk Density . . . . Approximately 100 lb/ft³

Solubility in Distilled Water

Temperature		Solubility	
°C	°F	g/L	wt oz/gal
0	32	27.8	3.7
20	68	65.0	8.6
40	104	125.2	16.7
60	140	230.0	30.7
70	158	286.4	38.3
75	167	323.5	43.2

Refer to the Solubility Fact Sheet, Form #130, for more information.

Chemical Abstract Service

SHIPPING CONTAINERS

25kg pail⁽¹⁾ (55.125 lb) net, with handle, made of 24-gauge steel, weighs 3.8 lb. It is tapered to allow nested storage of empty drums, stands approximately 13½ inches high and has a maximum diameter of 12 inches.

50kg keg⁽¹⁾ (110.250 lb) net, made of 24-gauge steel, weighs 8.25 lb. It is tapered to allow nested storage of empty drums, stands approximately 17¾ inches high and has a maximum diameter of 14¾ inches.

150kg drum⁽¹⁾ (330.7500 lb) net, made of 22-gauge steel, weighs 22.4 lb. It stands approximately 29½ inches high and is a approximately 19¾ inches in diameter.

1500kg Cycle-Bin⁽²⁾ (3307 lb) net.

Bulk, up to 48,000#

Special Packages will be considered on request.

(1) All containers meet DOT-37A or 37B Specifications and are Full Open Top Metal containers with Lever Lock Closures.

(2) The Cycle-Bin meets DOT 56 Specifications.

DESCRIPTION
Crystals or granules are dark purple with a metallic sheen, sometimes with a dark bronze-like appearance. Free-Flowing Grade is gray due to an additive. Potassium permanganate has a sweetish, astringent taste and is odorless.

HANDLING
Protect containers against physical damage. When handling potassium permanganate, respirators should be worn to avoid irritation of or damage to mucous membranes. Eye protection should also be worn when handling potassium permanganate as a solid or in solution.

STORAGE
Potassium permanganate is stable and will keep indefinitely if stored in a cool, dry area in closed containers. Concrete floors are preferred to wooden decks.
To clean up spills and leaks follow the steps recommended in our MSDS Form #170. Be sure to use goggles, rubber gloves, and respirator when cleaning up a spill or leak.

INCOMPATIBILITY
Avoid contact with acids, peroxides, and all combustible organic or readily oxidizable materials including inorganic oxidizable materials and metal powders. With hydrochloric acid, chlorine gas is liberated.
Potassium permanganate is not combustible but will support combustion. It may decompose if exposed to intense heat. Fires may be controlled and extinguished by using large quantities of water. Refer to the MSDS, Form #170, for more information.

CORROSIVE PROPERTIES

Potassium permanganate is compatible with many metals and synthetic materials. Natural rubbers and fibers are often incompatible. Solution pH and temperature are also important factors. The material must be compatible with either the acid or alkali also being used.

In neutral and alkaline solutions, potassium permanganate is not corrosive to iron, mild steel or stainless steel. However, chloride corrosion of metals may be accelerated when an oxidant such as permanganate is present in such solution. Plastics such as polypropylene, polyvinyl chloride Type I (PVC I), epoxy resins, fiberglass reinforced plastic (FRP), Penton, Lucite, Viton A, and Hypalon are suitable. Teflon FEP and TFE, and Tefzel ETFE are best. Refer to Material Compatibility Chart, Form #180.

Aluminum, zinc, copper, lead, and alloys containing these metals may be (slightly) affected by potassium permanganate solutions. Actual studies should be made under the conditions in which permanganate will be used.

SHIPPING

Potassium permanganate is classified by the Hazardous Materials Transportation Board (HMTB) as an oxidizer. It is shipped under Interstate Commerce Commission's (ICC) Tariff 19.

Proper Shipping Name: Potassium Permanganate
(RQ-100/45.4)

Hazard Class: Oxidizer

Identification Number: UN 1490

Label Requirements: Oxidizer and Corrosive

Packaging Requirements: 49 CFR Parts 100 to 199

Shipping Limitations:
Minimum quantities:
Rail car: See Tariff for destination
Truck: No minimum
Postal regulations:
Information applicable to packaging of oxidizers for shipment by the U.S. Postal Service to domestic and foreign destinations is readily available from the local postmaster.
United Parcel Service accepts 25 lb as largest unit quantity properly packaged; consult United Parcel Service.

Regulations concerning shipping and packing should be consulted regularly due to frequent changes.

REPACKING

When potassium permanganate is repacked, the packing, markings, labels, and shipping conditions must meet applicable Federal regulations. See Code of Federal Regulations – 49, Transportation, parts 100-199, and the Federal Hazardous Substances Act, 15 U.S.C. 1261.

APPLICATIONS

These are some of the many applications of potassium permanganate. Permanganate is a powerful oxidizing agent and the condition under which it is to be used should be completely investigated to minimize risks of violent reaction, fire, and explosion.

Oxidation and Synthesis
Organic chemicals and intermediates manufacture. Oxidizes impurities in organic and inorganic chemicals.

Water Treatment
Oxidizes iron, manganese, and hydrogen sulfide; controls taste and odor; and is an alternate pre-oxidant for THM control.

Municipal Wastewater Treatment
Destroys hydrogen sulfide in wastewater and sludge. Improves wastewater treatment.

Industrial Wastewater Treatment
Oxidizes hydrogen sulfide, phenols, iron, manganese, and many other organic and inorganic contaminants; resultant manganese dioxide aids in removing heavy metals.

Metal Surface Treatment
Conditions mill scale and smut to facilitate subsequent removal by acid pickling in wire manufacture and jet engine overhaul.

Equipment Cleaning
Assists in cleaning organic and inorganic residues from refining and cooling towers and other processing equipment.

Purification of Gases
Removes trace impurities of sulfur, arsine, phosphine, silane, borane, and sulfides from carbon dioxide and other industrial gases.

Mining and Metallurgical
Aids in separation of molybdenum from copper; removes impurities from zinc and cadmium; oxidizes flotation compounds.

Radioactive Decontamination
Used with other ingredients to decontaminate nuclear power plants, submarines, and reactors; also to purify heavy water.

Slag Quenching
Controls hydrogen sulfide and acetylene emissions during quenching of hot slag.

Other
Bleaches beeswax and some plasticizers. Increases swelling of sodium bentonite clays. Used in manufacture of printed circuit boards.

carus CHEMICAL COMPANY

Division of Carus Corporation
315 5th Street
Peru, IL 61354

Telephone: 800 435-6856
815 223-1500
Fax: 815 223-4105

Copyright © 1986 Rev. 7/93

Form #100F

LIQUOX™

Sodium Permanganate

CAS No. 10101-50-5

Fact Sheet

LIQUOX™ sodium permanganate is a liquid oxidant recommended for applications that require a concentrated permanganate solution.

Product Specifications

Assay	40% minimum as $NaMnO_4$
pH	6.0 - 7.0
Specific Gravity	1.36 - 1.39
Solubility in Water	Miscible with water in all proportions.

Chemical/Physical Data

Formula	$NaMnO_4$
Appearance	Dark Purple Solution
Insolubles	100 - 1900 ppm
Potassium	1000 - 2200 ppm
Stability	> 18 Months

Applications

• Printed Circuit Board Desmearing
• Pharmaceutical Synthesis Reactions
• Metal Cleaning Formulations
• Acid Mine Drainage
• Hydrogen Sulfide Odor Control
   - Remote Locations
   - Unheated Locations

Benefits

• Concentrated liquid oxidant is easily stored and handled. Feed equipment is simplified (no need to transfer and dissolve crystalline product).

• Dust problems associated with handling dry oxidants are eliminated.

• High solubility at room temperature. Reactions requiring a concentrated permanganate solution can be conducted without having to raise the temperature.

• Can be used instead of potassium permanganate whenever the potassium ion cannot be tolerated, or if dusting is a critical issue.

Shipping Containers

5 gallon (18.9L) Tight Head HDPE Jerrican
(UN Specification: 3H1) made of High Density Polyethylene (HDPE), weighs 3.5 lb (1.6 kg). Net weight is 57 lb (25.7 kg). Dimensions, 15.33 in. tall, 10.2 in. wide and 11.4 in. long (38.94 cm tall, 25.91 cm by 28.96 cm).

5 gallon (18.9L) Tight Head Steel Drum
(UN Specification: 1A1) made of 12 gauge, mild steel, weighs 5 lb (2.3 kg). Net weight is 57 lb (25.7 kg). The drum is 13.75 in. tall and 11.5 in. in diameter. (34.93 cm tall, 29.21 cm diameter)

55 gallon (208.2L) Closed Head Steel Drum
(UN Specification: 1A1) made of 16 gauge, mild steel, weighs 53.7 lb (24.4 kg). Net weight is 550 lb (249.5 kg). The drum is 34.6 in. tall, outside diameter 23.5 in., inside diameter 22.5 in. (87.9 cm tall, OD 59.7 cm, ID 57.2 cm).

Handling and Storage

Like any potent oxidant, LIQUOX™ sodium permanganate should be handled with care. Protective equipment during handling should include face shields and/or goggles, rubber or plastic gloves, rubber or plastic apron. If clothing becomes spotted, wash off immediately; spontaneous ignition can occur with cloth or paper. In cases where significant exposure exists, use of the appropriate NIOSH-MSHA dust or mist respirator or an air supplied respirator is advised.

The product should be stored in a cool, dry area in closed containers. Concrete floors are preferred. Avoid wooden decks. Spillage should be collected and disposed of properly. Contain and dilute spillage to approximately 6% with water and reduce with sodium thiosulfate, a bisulfite, or ferrous salt. The bisulfite or ferrous salt may require dilute sulfuric acid to promote reduction. Neutralize any acid used with sodium bicarbonate. Deposit sludge in an approved landfill or, where permitted, drain into sewer with large quantities of water.

As an oxidant, the product itself is non-combustible, but will accelerate the burning of combustible materials. Therefore, contact with all combustible materials and/or chemicals must be avoided. These include, but are not limited to: wood, cloth, organic chemicals, and charcoal. Avoid contact with acids, peroxides, sulfites, oxalates, and all other oxidizable inorganic chemicals. With hydrochloric acid, chlorine is liberated.

Shipping

LIQUOX™ sodium permanganate is classified as an oxidizer. Sodium permanganate is shipped domestically as Class 70 and has a Harmonized Code for export of 2841.69.0000.

Proper Shipping Name: Permanganates, Inorganic, Aqueous solution, n.o.s. (Contains Sodium Permanganate)

Hazard Class: 5.1

Identification Number: UN 3214

Packaging Group: II

Label Requirements: Oxidizer, 5.1

Special Provisions: T8-Intermodal transportation in IM 101 portable tanks

Package Requirement: 49 CFR Parts 171 to 180 Sections: 173.152, 173.202, 173.242

Quantity Limitations: 1 liter net for passenger aircraft or railcar. 5 liters net for cargo aircraft.

Vessel Stowage: D-material must be stowed "ondeck" on a cargo vessel, but is prohibited on a passenger vessel. Other provisions, stow "separated from" ammonium compounds, hydrogen peroxide, peroxides and superperoxides, cyanide compounds, and powdered metal.

Repackaging

When LIQUOX™ sodium permanganate is repackaged, the packaging, markings, labels, and shipping conditions must meet applicable federal regulations. See Code of Federal Regulations-49, Transportation, parts 171-180, and the Federal Hazardous Materials Transportation Act (HMTA).

Corrosive Properties

LIQUOX™ sodium permanganate is compatible with many metals and synthetic materials. Natural rubbers and fibers are often incompatible. Solution pH and temperature are also important factors. The material selected for use with sodium permanganate must also be compatible with any acid or alkali being used.

In neutral and alkaline solutions, sodium permanganate is not corrosive to carbon steel and 316 stainless steel. However, chloride corrosion of metals may be accelerated when an oxidant such as sodium permanganate is present in solution. Plastics such as teflon, polypropylene, HDPE and EDPM are also compatible with sodium permanganate.

Aluminum, zinc, copper, lead, and alloys containing these metals may be slightly affected by sodium permanganate solutions. Actual corrosion or compatibility studies should be made under the conditions in which the permanganate will be used prior to use.

Carus Value Added

LABORATORY SUPPORT

Carus Chemical Company has technical assistance available to its potential and current customers to answer questions or perform laboratory and field testing including:

*Feasibility Studies   * Toxicity Evaluations   *Treatability Studies   *Analytical Services   *Field Trials

CARUS CHEMICAL COMPANY

During its more than 80-year history, Carus' ongoing reliance on research and development, as well as its emphasis on technical support and customer service, have enabled the company to become the world leader in permanganate, manganese, oxidation, and catalyst technologies.

Carus Chemical Company
315 Fifth Street
P.O. Box 599
Peru, IL 61354
Tel. (815) 223-1500
Fax (815) 224-6697
Web: www.caruschem.com
E-Mail: salesmkt@caruschem.com

CARUS

ResponsibleCare®
A Public Commitment

Form #LX1501   Copyright® 1998

CAIROX®

Potassium Permanganate

CAS Registry No. 7722-64-7*

FACT SHEET

The following two tables, *"Metals Compatibility"* and *"Plastics and Other Non-Metals Compatibility"* show materials of construction and whether or not they may be used with solutions of potassium permanganate. In using the information provided in this chart, it should be understood that some of the data was gathered from in-plant and field experiences of engineers and plant operators using permanganate solutions. Over half of the data is from laboratory experiments only. The manufacturers literature was also consulted in the preparation of the charts. In each case, the results are specific to the conditions under which the permanganate was being applied. Use these tables as a guide, but not as a guarantee.

METALS

Potassium permanganate compatibility with metal products will depend upon the solution pH, and for some metals, on the solution temperature.

When adjusting the solution pH, always be certain that the metal is also compatible with the acid or alkali being used.

Ferrous Metals	SOLUTION pH			Metal Alloys	SOLUTION pH		
	Acidic	Neutral	Basic		Acidic	Neutral	Basic
Carbon Steel	NO	YES	YES	Brass	NO	YES	YES
Black Iron	NO	YES	YES	Bronze	NO	YES	YES
Galvanized Steel	NO	NO	NO	Hastalloy			
Stainless Steel[1]				B & D	NO	NO	NO
304	YES	YES	YES	C	*	YES	YES
316	YES[2]	YES	YES	Incoloy			
420	YES[2]	YES	YES	800	*	YES	YES
12% Cr	*	YES[3]	YES[3]	825	*	YES	YES
17% Cr	*	YES[3]	YES[3]	840	*	YES	YES[8]
Carpenter 20	YES	YES	YES	Monel[7]			
Non-Ferrous Metals				400	*	YES	YES
Aluminum	NO	YES[4]	NO[5]				
Copper	NO	YES[4]	NO[5]				
Lead	NO	YES[4]	NO[5]				
Nickel	YES	YES	YES				
Tantalum	*	*	*				
Tin[6]	YES	YES	YES				
Titanium	*	YES	YES				
Zinc	NO	NO	NO	* Information not available			
Zirconium[9]	YES	YES	YES				

SPECIAL NOTES FOR METALS COMPATIBILITY

1. Stainless Steels have a high corrosion rate when chlorides are present in the permanganate solution. They are not compatible with hydrochloric acid (HCl).
2. An accelerated corrosion rate was found when nitric acid is used to acidify permanganate solutions.
3. Compatible at room temperature only.
4. Only "FAIR" or "MODERATE" life when used with permanganate solutions. Short-term use would be acceptable.

5. Unsuitable with alkali, such as sodium hydroxide or potassium hydroxide. Should not be used with alkaline permanganate solutions.
6. "FAIR" with permanganate solutions.
7. "MODERATE" life below 100°F.
8. Incoloy 840 failed when used as the "sheath" material for an immersion heater in a 2% to 4% permanganate solution.
9. Tested at pH 3, 7, and 9.

PLASTICS AND OTHER NON-METALS

Potassium permanganate solutions can affect the strength, flexibility, surface appearance, or color of plastics. The chemical attack that could cause these changes might include: (1) oxidation of the polymer chain, (2) oxidation of the functional groups in or on the chain, or (3) depolymerization.

Fibers

Acetates	YES
Acrylics	YES
Cotton	NO
Nylon (polyamides)	NO
Orlon	NO
Paper	NO
Polyesters	YES[1]
Silk	NO
Wool	NO

Tank, Tank Linings, Pump, and other Equipment Construction Materials

ABS Plastic	YES
Asphaltic Resin	NO
Ceramic	YES
Epoxy Resin	YES[1]
Furan Resin	YES[1]
Glass	YES
Lucite (acrylic resin)	YES
Phenol-Formaldehyde Resin	NO
Phenolic Resin	YES[1]
Styrene Copolymers	YES[1]
Polyallomer	YES
Polybutylene	YES
Polycarbonate	NO
Polyethylene	YES[2]
Polypropylene	YES[3]
Polystyrene	NO
Polysulfone	YES
Polyurethane	YES
Polyvinyl Chloride I	YES[4]
Polyvinyl Chloride II	NO

Hose, Tubing, Pipe, and Gasket Materials

Asbestos	NO
Chlorinated Polyvinyl Chloride (CPVC)	YES[3]
Ethylene Propylene Rubber (EPR)	NO
Ethylene Propylene Terpolymers (EPT)	YES[6]
Hycar	NO
Hypalon	YES
Natural Rubber	NO
Nitrile Butadiene Rubber (NBR)	NO[6]
Neoprene	NO
Norprene	YES
Penton	YES
Polyphenylene Oxides (PPO)	YES
Polyvinylidene Chloride (Tygon)	YES
Polyvinylidene Fluoride (PVDF)	YES[7]
Styrene Butadiene Rubber (SBR)	
Buna N	NO
Buna S	NO
Teflon FEP	YES[7]
Teflon TFE	YES
Viton	YES[5]

Oils, Greases, and Lubricants

All oils, greases, and lubricants must be tested for compatibility with potassium permanganate.

When unknown, assume that potassium permanganate will react with these compounds resulting in fire and/or explosion.

SPECIAL NOTES FOR PLASTICS AND OTHER NON-METALS

1. Temperatures up to 200° F
2. Discolored at 140° F
3. Temperatures of 68° – 176° F
4. Temperatures up to 140° F
5. Temperatures of 68° – 140° F
6. Temperatures to 68° F
7. Temperatures of 68° – 248° F

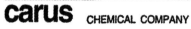

carus CHEMICAL COMPANY

Division of Carus Corporation
1001 Boyce Memorial Drive
Ottawa, IL 61350

Telephone: 800 435-6856
815 433-9070
Fax: 815 433-9075

Copyright © 1984 Rev. 7/93

Form #180

CAIROX®
Potassium Permanganate

FACT SHEET

Protective Equipment
Drum Disposal & First Aid
Recommendations

When handling CAIROX® potassium permanganate, either in form of crystals or in form of solution, avoid contact with eyes, skin, and clothing. Avoid inhalation of dust or mist.

PROTECTIVE EQUIPMENT

(a) Face shield or safety goggles should be worn.

(b) Rubber or plastic gloves should be worn.

(c) Normal work clothing covering arms and legs, and rubber apron should be worn.

(d) Respirators meeting general industry standards or NIOSH-MSHA approved dust and mist respirators (such as TC-21C-287) advised.

(e) Wash hands with soap and water after handling and before eating or smoking.

STAIN REMOVAL

Brown stains often form on skin exposed to potassium permanganate. These stains, manganese dioxide, usually wear off with time. A simple mixture of 30 parts of 3% USP hydrogen peroxide and 40 parts 5% food-grade white vinegar with 30 parts of water can be applied to speed up the process. Before applying this solution, clean the skin with soap and water. Next, apply the solution to the stained area, repeating the application until the stain is gone. Then, wash again with soap and water.

If the skin becomes red or irritated from using this solution, wash the affected area thoroughly with water, but DO NOT repeat the treatment.

NEVER use this solution on sensitive tissue such as eyes, mucous membranes, open wounds, and burns.

FIRST AID

EYES:

Immediately flush eyes with large amounts of water for at least 15 minutes, holding lids apart to ensure flushing of the entire surface. Do not attempt to neutralize chemically. Seek medical attention immediately.

SKIN:

Immediately wash contaminated areas with plenty of water. Remove contaminated clothing and footwear. Wash clothing and decontaminate footwear before use. Seek medical attention immediately if irritation is severe or persistent.

INHALATION:

Get person out of contaminated area to fresh air. If breathing has stopped, resuscitate and administer oxygen, if readily available. Seek medical attention immediately.

INGESTION:

Never give anything by mouth to an unconscious person. If conscious, give large quantities of water. Seek medical attention immediately.

 carus CHEMICAL COMPANY

Division of Carus Corporation
1001 Boyce Memorial Drive
Ottawa, IL 61350

Telephone: 800 435-6856
815 433-9070
Fax: 815 433-9075

Copyright © 1994   Rev. 7/93

CAIROX®
Potassium Permanganate

FACT SHEET

Empty Drum Disposal

REGULATIONS

D.O.T. INFORMATION

Carus Chemical Company packages potassium permanganate only in single-trip metal drums meeting the DOT Section 173.154(2)regulations. THE REUSE OF THESE DRUMS FOR THE TRANSPORT OF POTASSIUM PERMANGANATE IS A VIOLATION OF FEDERAL REGULATIONS.

R.C.R.A. Information

Under RCRA's definition, potassium permanganate, if discarded, meets the criteria of ignitible waste and, as such, is considered hazardous waste.

Section 261.7(b)(1) "A container or inner liner removed from a container that has held any hazardous waste, except a waste that is a compressed gas or that is identified as an acute hazardous waste listed in Section 261.31, 261.32, or 261.33(c) of this chapter is empty if:"

a. "All wastes have been removed that can be removed using the practices commonly employed to remove materials from that type of container, and

b. No more than 2.5 centimeters (one inch) of residue remains on the bottom of the container or inner liner, or

c. No more than 3% by weight of the total capacity of the container remains in the container or inner liner if the container is less than or equal to 110 gallons in size."

CLEANING AND DISPOSAL

Water easily and effectively dissolves potassium permanganate. A triple rinse with water should be adequate. The rinse water could be used for makeup water for a potassium permanganate solution or as dilution water in process.

For further information, refer to the Material Safety Data Sheet, Carus Form #170.

REPORTABLE QUANTITIES

If 100 pounds or more of potassium permanganate are released or spilled to the environment, (air, water and land), report it immediately to the U.S. Coast Guard National Response Center at:

**Toll Free 800/424-8802
or 202/267-2675**

Quantities of less than 100 pounds of CAIROX have to be cleaned up, but there are no reporting requirements. In order to contain spills of any size, clean up immediately. If it is windy or rainy, cover the spilled CAIROX with polyethylene sheeting to minimize fugitive emissions to the air or storm water contamination.

State and local authorities require notification in addition to the federal requirements.

RECOMMENDATION

Before disposing of drums, inspect each drum to ensure it is empty. Then triple rinse with water until the characteristic purple color of potassium permanganate is no longer visible in the rinse water.

 CHEMICAL COMPANY

Division of Carus Corporation
1001 Boyce Memorial Drive
Ottawa, IL 61350

Telephone: 800 435-6856
815 433-9070
Fax: 815 433-9075

Copyright © 1984 Rev. 7/93

FORM #171

MATERIAL SAFETY DATA SHEET
CAIROX® Potassium Permanganate

Section 1 — Chemical Product and Company Identification

PRODUCT NAME: CAIROX® potassium permanganate, $KMnO_4$
SYNONYMS:
Permanganic acid potassium salt
Chameleon mineral
Condy's crystals
Permanganate of potash

MANUFACTURER'S NAME: CARUS CHEMICAL COMPANY

MANUFACTURER'S ADDRESS:
Carus Chemical Company
1500 Eighth Street
P. O. Box 1500
LaSalle, IL 61301

TRADE NAME: CAIROX® potassium permanganate

TELEPHONE NUMBER FOR INFORMATION: 815/223-1500

EMERGENCY TELEPHONE NO: 800/435-6856

AFTER HOURS NO. 815/223-1565
5:00 PM-8:00 AM Central Standard Time
Monday-Friday, Weekends and Holidays

CHEMTREC TELEPHONE NO.:   800/424-9300

Section 2 — Composition/Information on Ingredients

Material or component	CAS No.	%	Hazard Data	
Potassium permanganate	7722-64-7	97% min. $KMnO_4$	PEL-C	5 mg Mn per cubic meter of air
			TLV-TWA	0.2 mg Mn per cubic meter of air

Section 3 — Hazards Identification

1. **Eye Contact**
Potassium permanganate is damaging to eye tissue on contact. It may cause severe burns that result in damage to the eye.

2. **Skin Contact**
Contact of solutions at room temperature may be irritating to the skin, leaving brown stains. Concentrated solutions at elevated temperature and crystals are damaging to the skin.

3. **Inhalation**
Acute inhalation toxicity data are not available. However, airborne concentrations of potassium permanganate in the form of dust or mist may cause damage to the respiratory tract.

4. **Ingestion**
Potassium permanganate, if swallowed, may cause severe burns to mucous membranes of the mouth, throat, esophagus, and stomach.

Responsible Care®
A Public Commitment

CARUS CHEMICAL COMPANY

Section 4   First Aid Measures

1. **Eyes**
   Immediately flush eyes with large amounts of water for at least 15 minutes holding lids apart to ensure flushing of the entire surface. Do not attempt to neutralize chemically. Seek medical attention immediately. Note to physician: Soluble decomposition products are alkaline. Insoluble decomposition product is brown manganese dioxide.

2. **Skin**
   Immediately wash contaminated areas with large amounts of water. Remove contaminated clothing and footwear. Wash clothing and decontaminate footwear before reuse. Seek medical attention immediately if irritation is severe or persistent.

3. **Inhalation**
   Remove person from contaminated area to fresh air. If breathing has stopped, resuscitate and administer oxygen if readily available. Seek medical attention immediately.

4. **Ingestion**
   Never give anything by mouth to an unconscious or convulsing person. If person is conscious, give large quantities of water. Seek medical attention immediately.

Section 5   Fire Fighting Measures

NFPA* HAZARD SIGNAL

Health Hazard	1 =	Materials which under fire conditions would give off irritating combustion products.
(less than 1 hour exposure)		Materials which on the skin could cause irritation.
Flammability Hazard	0 =	Materials that will not burn.
Reactivity Hazard	0 =	Materials which in themselves are normally stable, even under fire exposure conditions, and which are not reactive with water.
Special Hazard	OX =	Oxidizer

*National Fire Protection Association 704

FIRST RESPONDERS:

Wear protective gloves, boots, goggles, and respirator. In case of fire, wear positive pressure breathing apparatus. Approach site of incident with caution. Use Emergency Response Guide NAERG 96 (RSPA P5800.7), Guide No. 140.

FLASHPOINT — None

FLAMMABLE OR EXPLOSIVE LIMITS — Lower: Nonflammable   Upper: Nonflammable

EXTINGUISHING MEDIA — Use large quantities of water. Water will turn pink to purple if in contact with potassium permanganate. Dike to contain. Do not use dry chemicals, CO₂, Halon* or foams.

SPECIAL FIREFIGHTING PROCEDURES — If material is involved in fire, flood with water. Cool all affected containers with large quantities of water. Apply water from as far a distance as possible. Wear self-contained breathing apparatus and full protective clothing.

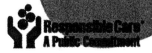

CARUS CHEMICAL COMPANY

Section 6   Accidental Release Measures

STEPS TO BE TAKEN IF MATERIAL IS RELEASED OR SPILLED
Clean up spills immediately by sweeping or shoveling up the material. Do not return spilled material to the original container. Transfer to a clean metal drum. EPA banned the land disposal of D001 ignitable waste oxidizers. These wastes must be deactivated by reduction. To clean floors, flush with abundant quantities of water into sewer, if permitted by Federal, State, and Local regulations. If not permitted, collect water and treat chemically (Section 13).

PERSONAL PRECAUTIONS
Personnel should wear protective clothing suitable for the task. Remove all ignition sources and incompatible materials before attempting clean-up.

Section 7   Handling and Storage

WORK/HYGENIC PRACTICES
Wash hands thoroughly with soap and water after handling potassium permanganate, and before eating or smoking. Wear proper protective equipment. Remove contaminated clothing.

VENTILATION REQUIREMENTS
Provide sufficient area or local exhaust to maintain exposure below the TLV-TWA.

CONDITIONS FOR SAFE STORAGE
Store in accordance with NFPA 430 requirements for Class II oxidizers. Protect containers from physical damage. Store in a cool, dry area in closed containers. Segregate from acids, peroxides, formaldehyde, and all combustible, organic or easily oxidizable materials including anti-freeze and hydraulic fluid.

Section 8   Exposure Controls/Personal Protection

RESPIRATORY PROTECTION
In the case where overexposure may exist, the use of an approved NIOSH-MSHA dust respirator or an air supplied respirator is advised. Engineering or administrative controls should be implemented to control dust.

EYE
Faceshield, goggles, or safety glasses with side shields should be worn. Provide eye wash in working area.

GLOVES
Rubber or plastic gloves should be worn.

OTHER PROTECTIVE EQUIPMENT
Normal work clothing covering arms and legs, and rubber or plastic apron should be worn.

Responsible Care®
A Public Commitment

CARUS CHEMICAL COMPANY

Section 9   Physical and Chemical Properties

APPEARANCE AND ODOR	Dark purple solid with a metallic luster, odorless
BOILING POINT, 760 mm Hg	Not applicable
VAPOR PRESSURE (mm Hg)	Not applicable
SOLUBILITY IN WATER % BY SOLUTION	6% at 20°C (68°F), and 20% at 65°C (149°F)
PERCENT VOLATILE BY VOLUME	Not volatile
EVAPORATION RATE (BUTYL ACETATE=1)	Not applicable
MELTING POINT	Starts to decompose with evolution of oxygen (O_2) at temperatures above 150°C (302°F). Once initiated, the decomposition is exothermic and self-sustaining.
OXIDIZING PROPERTIES	Strong oxidizer
SPECIFIC GRAVITY	2.7 @ 20°C (68°F)
VAPOR DENSITY (AIR=1)	Not applicable

Section 10   Stability and Reactivity

STABILITY Under normal conditions, the material is stable.

CONDITIONS TO AVOID Contact with incompatible materials or heat (>150°C/302°F).

INCOMPATIBLE MATERIALS Acids, peroxides, formaldehyde, anti-freeze, hydraulic fluids, and all combustible organic or readily oxidizable inorganic materials including metal powders. With hydrochloric acid, toxic chlorine gas is liberated.

HAZARDOUS DECOMPOSITION PRODUCTS When involved in a fire, potassium permanganate may liberate corrosive fumes.

CONDITIONS CONTRIBUTING TO HAZARDOUS POLYMERIZATION Material is not known to polymerize.

Section 11   Toxicological Information

Potassium permanganate: Acute oral LD$_{50}$(rat) = 780 mg/kg Male (14 days); 525 mg/kg Female (14 days)
The fatal adult human dose by ingestion is estimated to be 10 grams. (Ref. Handbook of Poisoning: Prevention, Diagnosis & Treatment, Twelfth Edition)

EFFECTS OF OVEREXPOSURE
1. Acute Overexposure
   Irritating to body tissue with which it comes into contact.

2. Chronic Overexposure
   No known cases of chronic poisoning due to potassium permanganate have been reported. Prolonged exposure, usually over many years, to heavy concentrations of manganese oxides in the form of dust and fumes, may lead to chronic manganese poisoning, chiefly involving the central nervous system.

3. Carcinogenicity
   Potassium permanganate has not been classified as a carcinogen by OSHA, NTP, IARC.

4. Medical Conditions Generally Aggravated by Exposure
   Potassium permanganate will cause further irritation of tissue, open wounds, burns or mucous membranes.

   Registry of Toxic Effects of Chemical Substances
   RTECS #SD6476000

CARUS CHEMICAL COMPANY

Section 12   Ecological Information

Entry to the Environment

Potassium Permanganate has a low estimated lifetime in the environment, being readily converted by oxidizable materials to insoluble manganese dioxide (MnO_2).

Bioconcentration Potential

In non-reducing and non-acidic environments manganese dioxide (MnO_2) is insoluble and has a very low bioaccumulative potential.

Aquatic Toxicity

Rainbow trout, 96 hour LC_{50}: 1.8 mg/L.
Bluegill sunfish, 96 hour LC_{50}: 2.3 mg/L.

Section 13   Disposal Consideration

DEACTIVATION OF D001 IGNITABLE WASTE OXIDIZERS BY CHEMICAL REDUCTION

Reduce potassium permanganate in aqueous solutions with sodium thiosulfate (Hypo), or sodium bisulfite or ferrous salt solution. The thiosulfite or ferrous salt may require some dilute sulfuric acid to promote rapid reduction. If acid was used, neutralize with sodium bicarbonate to neutral pH. Decant or filter, and mix the sludge with sodium carbonate and deposit in an approved landfill. Where permitted, the sludge can be drained into sewer with large quantities of water. Use caution when reacting chemicals. Contact Carus Chemical Company for additional recommendations.

Section 14   Transport Information

U.S. DEPARTMENT OF TRANSPORTATION INFORMATION:

Proper Shipping Name:	49 CFR 172.101	Potassium Permanganate
ID Number:	49 CFR 172.101	UN 1490
Hazard Class:	49 CFR 172.101	Oxidizer
Division:	49 CFR 172.101	5.1
Packing Group:	49 CFR 172.101	II

Section 15   Regulatory Information

TSCA	Listed in the TSCA Chemical Substance Inventory
CERCLA	Hazardous Substance
	Reportable Quantity: RQ - 100 lb 40 CFR 116.4; 40 CFR 302.4
RCRA	Oxidizers such as potassium permanganate meet the criteria of ignitable waste. 40 CFR 261.21
SARA TITLE III Information	

Section 302   Extremely hazardous substance: Not listed
Section 311/312   Hazard categories: Fire, acute and chronic toxicity
Section 313   CAIROX® potassium permanganate contains 97% Manganese Compound as part of the chemical structure (manganese compounds CAS Reg. No. N/A) and is subject to the reporting requirements of Section 313 of Title III, Superfund Amendments and Reauthorization Act of 1986 and 40 CFR 372.

Responsible Care®
A Public Commitment

CARUS CHEMICAL COMPANY

Section 15   Regulatory Information (cont.)

STATE LISTS   Michigan Critical Materials Register: Not listed
                   California Proposition 65:   Not listed
                   Massachusetts Substance List: 5 FB
                   Pennsylvania Hazard Substance List: E

FOREIGN LISTS Canadian Domestic Substances List (DSL)   Listed
                   Canadian Ingredient Disclosure List   Listed
                   European Inventory of Existing Chemical Substances (EINECS) 2317603

Section 16   Other Information

NIOSH   National Institute for Occupational Safety and Health
MSHA   Mine Safety and Health Administration
OSHA   Occupational Safety and Health Administration
NTP   National Toxicology Program
IARC   International Agency for Research on Cancer
TSCA   Toxic Substances Control Act
CERCLA   Comprehensive Environmental Response, Compensation and Liability Act of 1980
RCRA   Resource Conservation and Recovery Act
SARA   Superfund Amendments and Reauthorization Act of 1986
PEL-C   OSHA Permissible Exposure Limit-OSHA Ceiling Exposure Limit
TLV-TWA   Threshold Limit Value - Time Weighted Average (American Conference of Governmental Industrial Hygienists)

Kenneth Krogulski
May 2000

CARUS

Responsible Care®
A Public Commitment

The information contained is accurate to the best of our knowledge. However, data, safety standards and government regulations are subject to change; and the conditions of handling, use or misuse of the product are beyond our control. Carus Chemical Company makes no warranty, either express or implied including any warranties of merchantability and fitness for a particular purpose. Carus also disclaims all liability for reliance on the completeness or confirming accuracy of any information included herein. Users should satisfy themselves that they are aware of all current data relevant to their particular uses.

CAIROX® is registered trademark of Carus Corporation.
Responsible Care® is a service mark of the Chemical Manufacturers Association.   Rev. 5/ 00   Form # CX 1028

MATERIAL SAFETY DATA SHEET
LIQUOX™ Sodium Permanganate

Section 1   Product and Company Identification

MANUFACTURER'S NAME: CARUS CORPORATION

MANUFACTURER'S ADDRESS:
Carus Chemical Company
1500 Eighth Street
P. O. Box 1500
LaSalle, IL 61301

TELEPHONE NUMBER FOR INFORMATION: 815/223-1500

EMERGENCY TELEPHONE NO: 800/435-6856

CHEMTREC TELEPHONE NO.: 800/424-9300

PRODUCT NAME: LIQUOX™ sodium permanganate, $NaMnO_4$
TRADE NAME: LIQUOX™ sodium permanganate
SYNONYMS: Permanganic acid sodium salt solution

Section 2   Composition/Information on Ingredients

Material or component	CAS No.	%	Hazard Data	
Sodium Permanganate	10101-50-5	40% - 42%	PEL/C	5 mg Mn per cubic meter of air
			TLV-TWA	0.2 mg Mn per cubic meter of air

Section 3   Hazards Identification

1. **Eye Contact**
   Sodium Permanganate is damaging to eye tissue on contact. It may cause severe burns that result in damage to the eye.

2. **Skin Contact**
   Momentary contact of solution at room temperature may be irritating to the skin, leaving brown stains. Prolonged contact is damaging to the skin.

3. **Inhalation**
   Acute inhalation toxicity data are not available. However, airborne concentrations of sodium permanganate in the form of mist may cause damage to the respiratory tract.

4. **Ingestion**
   Sodium permanganate solution, if swallowed, may cause severe burns to mucous membranes of the mouth, throat, esophagus, and stomach.

Responsible Care®
A Public Commitment

CARUS CHEMICAL COMPANY

Section 4   First Aid Measures

1. Eyes
    Immediately flush eyes with large amounts of water for at least 15 minutes holding lids apart to ensure flushing of the entire surface. Do not attempt to neutralize chemically. Seek medical attention immediately. Note to physician: Decomposition products are alkaline.

2. Skin
    Immediately wash contaminated areas with water. Remove contaminated clothing and footwear (Caution: Solution may ignite certain textiles). Wash clothing and decontaminate footwear before reuse. Seek medical attention immediately, if irritation is severe and persistent.

3. Inhalation
    Remove person from contaminated area to fresh air. If breathing has stopped, resuscitate and administer oxygen if readily available. Seek medical attention immediately.

4. Ingestion
    Never give anything by mouth to an unconscious or convulsing person. If person is conscious, give large quantities of water or milk. Seek medical attention immediately.

Section 5   Fire Fighting Measures

NFPA* HAZARD SIGNAL

Health Hazard (less than 1 hour exposure)	1 =	Materials which under fire conditions would give off irritating combustion products.
Flammability Hazard	0 =	Materials which on the skin could cause irritation. Materials that will not burn.
Reactivity Hazard	0 =	Materials which in themselves are normally stable, even under fire exposure conditions, and which are not reactive with water.
Special Hazard	OX =	Oxidizer

*National Fire Protection Association 704

FIRST RESPONDERS

Wear protective gloves, boots, goggles, and respirator. In case of fire, wear positive pressure breathing apparatus. Approach incident with caution. Use Emergency Response Guide NAERG 96 (RSPA P5800.7), Guide No. 140.

FLASHPOINT   None

FLAMMABLE OR EXPLOSIVE LIMITS   Lower: Nonflammable   Upper: Nonflammable

EXTINGUISHING MEDIA   Use large quantities of water. Water will turn pink to purple, if in contact with sodium permanganate. Dike to contain. Do not use dry chemicals, CO_2, Halon® or foams.

SPECIAL FIRE FIGHTING PROCEDURES   If material is involved in fire, flood with water. Cool all affected containers with large quantities of water. Apply water from as far a distance as possible. Wear self-contained breathing apparatus and full protective clothing.

UNUSUAL FIRE AND EXPLOSION HAZARDS   Powerful oxidizing material. May decompose spontaneously if exposed to intense heat (150°C/270°F). May be explosive in contact with certain other chemicals (Section 10). May react violently with finely divided and readily oxidizable substances. Increases burning rate of combustible material. May ignite wood and cloth.

Responsible Care®
A Public Commitment   CARUS CHEMICAL COMPANY

Section 6   Accidental Release Measures

STEPS TO BE TAKEN IF MATERIAL IS RELEASED OR SPILLED
Contain spill by collecting the liquid in a pit or holding behind a dam (sand or soil). Dilute to approximately 5% with water, then reduce with sodium thiosulfate, a bisulfite or ferrous salt solution. The bisulfite or ferrous salt may require some dilute sulfuric acid (10% w/w) to promote reduction. Neutralize with sodium carbonate to neutral pH, if acid was used. Decant or filter and deposit sludge in an approved landfill. Where permitted, the sludge may be drained into sewer with large quantities of water. To clean contaminated floors, flush with abundant quantities of water into sewer, if permitted by federal, state, and local regulations. If not, collect water and treat as above.

PERSONAL PRECAUTIONS
Personnel should wear protective clothing suitable for the task. Remove all ignition sources and incompatible materials before attempting clean-up.

Section 7   Handling and Storage

WORK/HYGIENIC PRACTICES
Wash hands thoroughly with soap and water after handling permanganate solution, and before eating or smoking. Wear proper protective equipment. Remove contaminated clothing.

VENTILATION REQUIREMENTS
Provide sufficient mechanical and/or local exhaust to maintain exposure below the TLV/TWA.

CONDITIONS FOR SAFE STORAGE
Store in accordance with NFPA 430 requirements for Class II oxidizers. Protect containers from physical damage. Store in a cool, dry area in closed containers. Segregate from acids, peroxides, formaldehyde, and all combustible, organic or easily oxidizable materials including antifreeze and hydraulic fluid.

Section 8   Exposure Controls/Personal Protection

RESPIRATORY PROTECTION
In the case where exposure to mist may occur, the use of an approved NIOSH-MSHA mist respirator or an air supplied respirator is advised. Engineering or administrative controls should be implemented to control mist.

EYE
Faceshield, goggles, or safety glasses with side shields should be worn. Provide eye wash in working area.

GLOVES
Rubber or plastic gloves should be worn.

OTHER PROTECTIVE EQUIPMENT
Normal work clothing covering arms and legs, and rubber, or plastic apron should be worn. Caution: If clothing becomes contaminated, wash off immediately. Spontaneous ignition may occur with cloth or paper.

Responsible Care®
A Public Commitment

Section 9 Physical and Chemical Properties

APPEARANCE AND ODOR	Dark purple solution, odorless
BOILING POINT, 760 mm Hg	105°C
VAPOR PRESSURE (mm Hg)	760 mm at 105°C
SOLUBILITY IN WATER % BY SOLUTION	Miscible in all proportions with water
PERCENT VOLATILE BY VOLUME	60% (as water)
EVAPORATION RATE	Same as water
MELTING POINT	Not applicable
SPECIFIC GRAVITY	1.36
pH	6-7
OXIDIZING PROPERTIES	Strong oxidizer. May ignite wood and cloth.
EXPLOSIVE PROPERTIES	Explosive in contact with sulfuric acid or peroxides, or readily oxidizable substances.

Section 10 Stability and Reactivity

STABILITY Under normal conditions, the material is stable.

CONDITIONS TO AVOID Contact with incompatible materials or heat (135°C/275°F).

INCOMPATIBLE MATERIALS Contact with acids, peroxides, and all combustible organic or readily oxidizable materials including inorganic oxidizable materials and metal powders. With hydrochloric acid, chlorine gas is liberated.

HAZARDOUS DECOMPOSITION PRODUCTS When involved in a fire, sodium permanganate may form corrosive fumes.

CONDITIONS CONTRIBUTING TO HAZARDOUS POLYMERIZATION Material is not known to polymerize.

Section 11 Toxicological Information

SODIUM PERMANGANATE: Acute oral LD_{50} not known.

EFFECTS OF OVEREXPOSURE

1. Acute Overexposure
   Irritating to body tissue with which it comes into contact.

2. Chronic Overexposure
   No known cases of chronic poisoning due to permanganates have been reported. Prolonged exposure (usually over many years) to heavy concentrations of manganese oxides in the form of dust and fumes, may lead to chronic manganese poisoning, chiefly involving the central nervous system.

3. Carcinogenicity
   Sodium permanganate has not been classified as a carcinogen by OSHA, NTP, IARC.

4. Medical Conditions Generally Aggravated by Exposure
   Sodium permanganate solution will cause further irritation of tissue, open wounds, burns, or mucous membranes.

   Registry of Toxic Effects of Chemical Substances
   RTECS # SD6650000

Responsible Care®
A Public Commitment

Section 12   Ecological Information

Entry to the Environment

Permanganate has a low estimated lifetime in the environment, being readily converted by oxidizable materials to insoluble MnO_2.

Bioconcentration Potential

In non-reducing and non-acidic environments, MnO_2 is insoluble and has a very low bioaccumulative potential.

Aquatic Toxicity

No Data.

Section 13   Disposal Consideration

WASTE DISPOSAL

Sodium permanganate is considered a D001 hazardous (ignitable) waste. For disposal of sodium permanganate solutions, follow procedures in Section 6 and deactivate the permanganate to insoluble manganese dioxide. Dispose of it in a permitted landfill. Contact Carus Chemical Company for additional recommendations.

Section 14   Transport Information

U. S. DEPARTMENT OF TRANSPORTATION INFORMATION:

Proper Shipping Name:	49 CFR172.101	Permanganates, inorganic, aqueous solution, n.o.s. (contains sodium permanganate)
ID Number:	49CFR172.101	UN 3214
Hazard Class:	49CFR172.101	Oxidizer
Division:	49CFR172.101	5.1
Packing Group:	49CFR172.101	II

Section 15   Regulatory Information

TSCA	Listed in the TSCA Chemical Substance Inventory.
CERCLA	Not listed.
RCRA	Oxidizers such as sodium permanganate solution meet the criteria of ignitable waste. 40 CFR 261.21.

SARA TITLE III Information

Section 302/303	Extremely hazardous substance: Not listed
Section 311/312	Hazard categories: Fire, acute and chronic toxicity
Section 313	LIQUOX™ Sodium Permanganate contains 40% manganese compounds as part of the chemical infrastructure (manganese compounds CAS Reg. No. N/A) and is subject to the reporting requirements of Section 313 of Title III Superfund Amendments and Reauthorization Act of 1986 and 40 CFR 372.

STATE LISTS		
Michigan Critical Materials Register		Not listed
California Proposition 65		Not listed
Massachusetts Substance List		Not listed
Pennsylvania Hazard Substance List		Not listed

FOREIGN LISTS		
Canadian Ingredient Disclosure List		Not listed
Canadian Non-Domestic Substance List		Listed
EINECS		Listed

Responsible Care®
A Public Commitment

CARUS CHEMICAL COMPANY

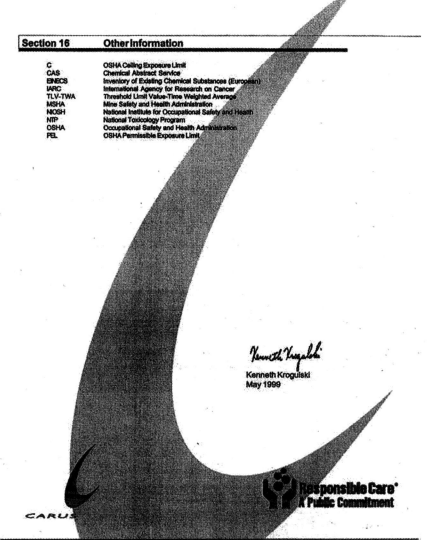

Section 16	Other Information
C	OSHA Ceiling Exposure Limit
CAS	Chemical Abstract Service
EINECS	Inventory of Existing Chemical Substances (European)
IARC	International Agency for Research on Cancer
TLV-TWA	Threshold Limit Value-Time Weighted Average
MSHA	Mine Safety and Health Administration
NIOSH	National Institute for Occupational Safety and Health
NTP	National Toxicology Program
OSHA	Occupational Safety and Health Administration
PEL	OSHA Permissible Exposure Limit

Kenneth Kroguiski
May 1999

Responsible Care®
A Public Commitment

CARUS

The information contained is accurate to the best of our knowledge. However, data, safety standards and government regulations are subject to change; and the conditions of handling, use or misuse of the product are beyond our control. Carus Chemical Company makes no warranty, either express or implied including any warranties of merchantability and fitness for a particular purpose. Carus also disclaims all liability for reliance on the completeness or confirming accuracy of any information included herein. Users should satisfy themselves that they are aware of all current data relevant to their particular use.

LIQUOX™ is trademark of Carus Corporation.
Responsible Care® is a service mark of the Chemical Manufacturers Association   Rev. 5/99   Form # LX 1502

Index

The following conventions are used in this index. For pages containing figures, an *f* follows the page number. For pages containing tables, a *t* follows the page number. The following abbreviations are used: BTEX for benzene, toluene, ethylbenene, and xylene; DNAPL for dense nonaqueous phase liquids; and ISCO for in situ chemical oxidation.

The Authors

Robert L. Siegrist, Ph.D., P.E., Environmental Engineering, University of Wisconsin, is a Professor and Interim Director of the Environmental Science and Engineering Division at the Colorado School of Mines. He was formerly Group Leader for Environmental Engineering at Oak Ridge National Laboratory. Dr. Siegrist has over 20 years of experience in remediation science and engineering in the U.S. and abroad and is a member of national and international committees and advisory panels and a Fellow with the NATO Committee for Challenges to Modern Society. During the past 10 years, Dr. Siegrist has led an integrated program of research and practice concerning in situ chemical oxidation including fundamental studies of oxidation reaction chemistry and kinetics, contaminant mass transfer, and oxidant delivery systems, as well as applied studies involving pilot- and full-scale field applications. *Michael A. Urynowicz*, Ph.D., P.E., Environmental Science and Engineering, Colorado School of Mines, is an Adjunct Assistant Professor of Environmental Science and Engineering at the Colorado School of Mines and President of ENVIROX LLC, an environmental remediation company. Dr. Urynowicz has become a leading authority on the use of permanganate for in situ remediation including the degradation of DNAPLs. *Olivia R. West*, Ph.D., Civil Engineering, MIT, is a Research Staff Member with the Environmental Sciences Division of Oak Ridge National Laboratory. During the past ten years, Dr. West has conducted research of in situ remediation of organics using mass transfer and chemical degradation processes through laboratory and field-scale experimentation and mathematical modeling. *Michelle Crimi* received her M.S. in Environmental Health, Colorado State University, on an Oak Ridge Institute for Science and Engineering fellowship. Having previously worked as a consultant with projects including the development of EHS programs, training, and hazard and risk assessments, she is currently a Ph.D. candidate in Environmental Science and Engineering at the Colorado School of Mines. Ms. Crimi is leading research into particle genesis and metal mobility and fate during in situ chemical oxidation. *Kathryn S. Lowe*, M.S. in Environmental Engineering, University of Colorado, Boulder, is a Senior Research Associate at the Colorado School of Mines and formerly a Staff Scientist with Oak Ridge National Laboratory. Ms. Lowe has over 15 years experience leading field demonstrations to evaluate developing technologies including various permanganate delivery methods for in situ treatment of DNAPL compounds.